Nicolas Jaccard

Bioinformatics

Andrzej Polanski · Marek Kimmel

Bioinformatics

With 119 Figures

 Springer

Authors

Andrzej Polanski
Department of Computer Sciences
Silesian University of Technology
ul. Akademicka 16
44-100 Gliwice, Poland

Polish-Japanese
Institute of Information Technology
ul. Koszykowa 86
02-008 Warszawa, Poland

Andrzej.polanski@polsl.pl
polanski@rice.edu

Marek Kimmel
Department of Statistics
Rice University
6100 Main Street
Houston, TX 77005, USA

System Engineering Group
Department of Automatic Control
Silesian University of Technology
ul. Akademicka 16
44-100 Gliwice, Poland

kimmel@rice.edu

Library of Congress Control Number: 2007924087

ACM Computing Classification (1998): J.3, G.3, F.2.2, G.1.6

ISBN 978-3-540-24166-9 Springer Berlin Heidelberg New York

This work is subject to copyright. All rights are reserved, whether the whole or part of the material is concerned, specifically the rights of translation, reprinting, reuse of illustrations, recitation, broadcasting, reproduction on microfilm or in any other way, and storage in data banks. Duplication of this publication or parts thereof is permitted only under the provisions of the German Copyright Law of September 9, 1965, in its current version, and permission for use must always be obtained from Springer. Violations are liable for prosecution under the German Copyright Law.

Springer is a part of Springer Science+Business Media

springer.com

© Springer-Verlag Berlin Heidelberg 2007

The use of general descriptive names, registered names, trademarks, etc. in this publication does not imply, even in the absence of a specific statement, that such names are exempt from the relevant protective laws and regulations and therefore free for general use.

Typeset by the authors
Production: LE-TeX Jelonek, Schmidt & Vöckler GbR, Leipzig
Cover design: KünkelLopka Werbeagentur, Heidelberg

Printed on acid-free paper   45/3100/YL - 5 4 3 2 1 0

To our families and friends

Preface

This volume is a senior undergraduate-or graduate-level textbook on bioinformatics, which also can be treated as a short monograph. It is focused on the mathematical and computer-science background of bioinformatics. However, each method is developed with attention to the biological problem that inspired it. The material covered includes widely used and well-developed applications in genomics and proteomics. Some important parts of the book are problems to be solved and a thorough bibliography. The book is composed of two parts: I, Mathematical and Computational Methods, and II, Applications.

Part I covers a wide range of mathematical methods and computational algorithms applied in bioinformatics. Most of the mathematical methods and computational algorithms are presented in enough detail to allow the reader to use them practically, for example to develop appropriate computer programs to be used later with some bioinformatic data. Part II covers the applications of bioinformatics. In this part, we start by presenting some biological and biochemical knowledge, before describing how mathematical and computational methods are used to solve problems originating from biology and biochemistry. We aim at enhancing the reader's motivation to pursue research and projects in bioinformatics.

There are already many excellent monographs and textbooks in bioinformatics, [6, 15, 22, 69, 71, 76, 162, 168, 200, 221, 267, 281]. However, our book is situated in an area which, so far, is not fully covered by other texts. We present a wide range of mathematical and computational methods in a self-consistent manner. We also pay attention to the biological and biochemical background of bioinformatics. Therefore, while focusing on mathematical and statistical methodology, we nevertheless present a view on the subject which explains its interdisciplinary character and allows a better understanding of the motivation for pursuing research in bioinformatics. The scope of the book is broad and covers many areas of bioinformatics.

This volume has emerged as a result of several years of teaching courses related to topics in bioinformatics at technical and medical schools, to Ph.D. and undergraduate students, and of taking part in scientific research involving cooperation between engineers, statisticians and mathematicians on one hand and biologists, clinicians, and biochemists on the other. Substantial parts of this texts were written during the first author's visits to the Department of Statistics, Rice University, Houston, USA. We highly appreciate the support obtained from the following institutions: the Department of Statistics, Rice University, Houston, USA, the Department of Computer Sciences and the Department of Automatic Control, Silesian University of Technology, Gliwice, Poland, the Polish-Japanese Institute of Information Technology, Warsaw and Bytom, Poland, Institute of Oncology, Gliwice, Poland; and the Faculty of Pharmacy, Silesian Medical University, Sosnowiec, Poland.

We would like to express our thanks to the people who helped us in writing this text. We extend our warmest thanks to Professor Leonard Bolc, whose constant interest and encouragement was the greatest support in our work. We would also like to thank Professor Piotr Widlak and Dr. Chad Shaw for reading parts of the manuscript and for helpful suggestions. We are grateful to all, of the following people, whose cooperation with us, discussions and criticism improved our understanding of topics in bioinformatics: Keith Baggerly, Sara Barton, Damian Bereska, Adam Bobrowski, Damian Borys, Ranajit Chakraborty, Michal Dabrowski, Kathy Ensor, Krzysztof Fujarewicz, Adam Galuszka, Ryszard Gessing, Rafal Gieleciak, Rudy Guerra, Paul Havlak, Przemyslawa Jarosz - Chobot, Barbara Jarzab, Michal Jarzab, Patrick King, Tadeusz Kulinski, Olivier Lichtarge, Tomasz Lipniacki, Cezary Lossy, Maria Luszczkiewicz, Tomasz Magdziarz, Urszula Mazurek, Peter Olofsson, Marcin Pacholczyk, Pawel Paszek, Rafal Pokrzywa, Joanna Polanska, Rafal Polanski, Krzysztof Psiuk, Krzysztof Puszynski, Alex Renwick, Joanna Rzeszowska, Krzysztof Simek, Jaroslaw Smieja, Heidi Spratt, Zbigniew Starosolski, David Stivers, Andrzej Swierniak, Rafal Tarnawski, Jerzy Tiuryn, Jim Thompson, Maria Widel, and Konrad Wojciechowski. We are also grateful to the publisher's editors for their professional support and patience.

This book was partly supported by the following grants and projects: the NSF/CRCD grant "Bioinformatics: Sequence to Structure"; the Polish Committee for Scientific Research grant 3 T11F01029, "Bioinformatics for Determination of Environmental and Genetic Risk Factors in Endocrine Autoimmune Diseases", and the European Sixth Framework Programme Project, GENEPI-lowRT, "Genetic Pathways for the Prediction of the Effects of Ionising Radiation: Low Dose Radiosensitivity and Risk to Normal Tissue after Radiotherapy".

The Authors contributions were as follows: Chap. 1, M.K.; Chap. 2, A.P. and M.K.; Chap. 3, A.P.; Chap. 4, A.P.; Chap. 5, A.P., Chap. 6, A.P.; Chap. 7, M.K.; Chap. 8, M.K.; Chap. 9, A.P.; Chap. 10, A.P.; Chap, 11, A.P.

Houston, Gliwice *Andrzej Polanski*
September 2006 *Marek Kimmel*

Contents

1   **Introduction** 1
    1.1   The Genesis of Bioinformatics 1
    1.2   Bioinformatics Versus Other Disciplines 2
    1.3   Further Developments: from Linear Information to
        Multidimensional Structure Organization. 4
    1.4   Mathematical and Computational Methods 5
        1.4.1   Why Mathematical Modeling? 6
        1.4.2   Fitting Models to Data 7
        1.4.3   Computer Software 7
    1.5   Applications 8

Part I Mathematical and Computational Methods

2   **Probability and Statistics** 13
    2.1   The Rules of Probability Calculus 13
        2.1.1   Independence, Conditional Probabilities and Bayes'
               Rules 14
    2.2   Random Variables 15
         2.2.1   Vector Random Variables 16
         2.2.2   Marginal Distributions 17
         2.2.3   Operations on Random Variables 17
         2.2.4   Notation 19
         2.2.5   Expectation and Moments of Random Variables 19
         2.2.6   Probability-Generating Functions and Characteristic
               Functions 20
    2.3   A Collection of Discrete and Continuous Distributions 22
         2.3.1   Bernoulli Trials and the Binomial Distribution 22
         2.3.2   The Geometric Distribution 23
         2.3.3   The Negative Binomial Distribution 23
         2.3.4   The Poisson Distribution 24

	2.3.5	The Multinomial Distribution	25
	2.3.6	The Hypergeometric Distribution	25
	2.3.7	The Normal (Gaussian) Distribution	26
	2.3.8	The Exponential Distribution	26
	2.3.9	The Gamma Distribution	27
	2.3.10	The Beta Distribution	27
2.4	Likelihood maximization		28
	2.4.1	Binomial Distribution	29
	2.4.2	Multinomial distribution	29
	2.4.3	Poisson Distribution	29
	2.4.4	Geometric Distribution	30
	2.4.5	Normal Distribution	30
	2.4.6	Exponential Distribution	31
2.5	Other Methods of Estimating Parameters: a Comparison		31
	2.5.1	Example 1. Uniform Distribution	31
	2.5.2	Example 2. Cauchy Distribution	33
	2.5.3	Minimum Variance Parameter Estimation	35
2.6	The Expectation Maximization Method		37
	2.6.1	The Derivations of the Algorithm	38
	2.6.2	Examples of Recursive Estimation of Parameters by Using the EM Algorithm	41
2.7	Statistical Tests		45
	2.7.1	The Idea	45
	2.7.2	Parametric Tests	47
	2.7.3	Nonparametric Tests	48
	2.7.4	Type I and II statistical errors	49
2.8	Markov Chains		49
	2.8.1	Transition Probability Matrix and State Transition Graph	50
	2.8.2	Time Evolution of Probability Distributions of States	51
	2.8.3	Classification of States	52
	2.8.4	Ergodicity	54
	2.8.5	Stationary Distribution	54
	2.8.6	Reversible Markov Chains	55
	2.8.7	Time-Continuous Markov Chains	56
2.9	Markov Chain Monte Carlo (MCMC) Methods		57
	2.9.1	Acceptance–Rejection Rule	59
	2.9.2	Applications of the Metropolis–Hastings Algorithm	59
	2.9.3	Simulated Annealing and MC3	59
2.10	Hidden Markov Models		60
	2.10.1	Probability of Occurrence of a Sequence of Symbols	60
	2.10.2	Backward Algorithm	61
	2.10.3	Forward Algorithm	61
	2.10.4	Viterbi Algorithm	62
	2.10.5	The Baum–Welch algorithm	63

		Contents	XI

2.11 Exercises 63

3 Computer Science Algorithms 67
3.1 Algorithms 67
3.2 Sorting and Quicksort................................... 68
    3.2.1 Simple Sort 69
    3.2.2 Quicksort 69
3.3 String Searches. Fast Search 70
    3.3.1 Easy Search 71
    3.3.2 Fast Search 71
3.4 Index Structures for Strings. Search Tries. Suffix Trees 73
    3.4.1 A Treelike Structure in Computer Memory 74
    3.4.2 Search Tries 75
    3.4.3 Compact Search Tries 76
    3.4.4 Suffix Tries and Suffix Trees........................ 77
    3.4.5 Suffix Arrays 80
    3.4.6 Algorithms for Searching Tries 80
    3.4.7 Building Tries..................................... 83
    3.4.8 Remarks on the Efficiency of the Algorithms 85
3.5 The Burrows–Wheeler Transform 85
    3.5.1 Inverse transform.................................. 86
    3.5.2 BW Transform as a Compression Tool 88
    3.5.3 BW Transform as a Search Tool for Patterns.......... 89
    3.5.4 BW Transform as an Associative, Compressed Memory 90
    3.5.5 Computational Complexity of BW Transform 91
3.6 Hashing 91
    3.6.1 Hashing functions for addressing variables 91
    3.6.2 Collisions... 92
    3.6.3 Statistics of Memory Access Time with Hashing 93
    3.6.4 Inquiring About Repetitive Structure of Sequences, Comparing Sequences and Detecting Sequence Overlap by Hashing 94
3.7 Exercises 95

4 Pattern Analysis... 97
4.1 Feature Extraction 97
4.2 Classification 98
    4.2.1 Linear Classifiers 98
    4.2.2 Linear Classifier Functions and Artificial Neurons 100
    4.2.3 Artificial Neural Networks 100
    4.2.4 Support Vector Machines........................... 102
4.3 Clustering 103
    4.3.1 K-means Clustering................................ 104
    4.3.2 Hierarchical Clustering 105
4.4 Dimensionality Reduction, Principal Component Analysis 107

Contents

- 4.4.1 Singular-Value Decomposition (SVD) 108
- 4.4.2 Geometric Interpretation of SVD 109
- 4.4.3 Partial-Least-Squares (PLS) Method 115
- 4.5 Parametric Transformations 116
  - 4.5.1 Hough Transform 117
  - 4.5.2 Generalized Hough Transforms 118
  - 4.5.3 Geometric Hashing 119
- 4.6 Exercises 119

5 Optimization 123
- 5.1 Static Optimization 124
  - 5.1.1 Convexity and Concavity 126
  - 5.1.2 Constrained Optimization with Equality Constraints ... 128
  - 5.1.3 Constrained Optimization with Inequality Constraints . 131
  - 5.1.4 Sufficiency of Optimality Conditions for Constrained Problems 133
  - 5.1.5 Computing Solutions to Optimization Problems 133
  - 5.1.6 Linear Programming 136
  - 5.1.7 Quadratic Programming 137
  - 5.1.8 Recursive Optimization Algorithms 137
- 5.2 Dynamic Programming 140
  - 5.2.1 Dynamic Programming Algorithm for a Discrete-Time System 141
  - 5.2.2 Tracing a Path in a Plane 143
  - 5.2.3 Shortest Paths in Arrays and Graphs 145
- 5.3 Combinatorial Optimization 147
  - 5.3.1 Examples of Combinatorial Optimization Problems 148
  - 5.3.2 Time Complexity 148
  - 5.3.3 Decision and Optimization Problems 149
  - 5.3.4 Classes of Problems and Algorithms 149
  - 5.3.5 Suboptimal Algorithms 150
  - 5.3.6 Unsolved Problems 150
- 5.4 Exercises 151

Part II Applications

6 Sequence Alignment 155
- 6.1 Number of Possible Alignments 157
- 6.2 Dot Matrices 159
- 6.3 Scoring Correspondences and Mismatches 160
- 6.4 Developing Scoring Functions 162
  - 6.4.1 Estimating Probabilities of Nucleotide Substitution .. 162
  - 6.4.2 Parametric Models of Nucleotide Substitution 163
  - 6.4.3 Computing Transition Probabilities 165

	6.4.4	Fitting Nucleotide Substitution Models to Data 168
	6.4.5	Breaking the Loop of Dependencies 173
	6.4.6	Scaling Substitution Probabilities 173
	6.4.7	Amino Acid Substitution Matrices 173
	6.4.8	Gaps 177
6.5	Sequence Alignment by Dynamic Programming 178
	6.5.1	The Needleman–Wunsch Alignment Algorithm 178
	6.5.2	The Smith–Waterman Algorithm 181
6.6	Aligning Sequences Against Databases 182
6.7	Methods of Multiple Alignment 183
6.8	Exercises 184

7 Molecular Phylogenetics 187
- 7.1 Trees: Vocabulary and Methods 187
  - 7.1.1 The Vocabulary of Trees 188
- 7.2 Overview of Tree-Building Methodologies 189
- 7.3 Distance-Based Trees 190
  - 7.3.1 Tree-Derived Distance 191
  - 7.3.2 Ultrametric Distances and Molecular-Clock Trees 191
  - 7.3.3 Unweighted Pair Group Method with Arithmetic Mean (UPGMA) Algorithm 193
  - 7.3.4 Neighbor-Joining Trees 193
- 7.4 Maximum Likelihood (Felsenstein) Trees 194
  - 7.4.1 Hypotheses and Steps: 196
  - 7.4.2 The Pulley Principle 197
  - 7.4.3 Estimating Branch Lengths 197
  - 7.4.4 Estimating the Tree Topology 198
- 7.5 Maximum-Parsimony Trees 198
  - 7.5.1 Minimal Number of Evolutionary Events for a Given Tree 199
  - 7.5.2 Searching for the Optimal Tree Topology 199
- 7.6 Miscellaneous Topics in Phylogenetic Tree Models 200
  - 7.6.1 The Nonparametric Bootstrap Method 200
  - 7.6.2 Variable Substitution Rates, the Felsenstein-Churchill Algorithm and Related Methods 201
  - 7.6.3 The Evolutionary Trace Method and Functional Sites in Proteins 201
- 7.7 Coalescence Theory 202
  - 7.7.1 Neutral Evolution: Interaction of Genetic Drift and Mutation 202
  - 7.7.2 Modeling Genetic Drift 203
  - 7.7.3 Modeling Mutation 204
  - 7.7.4 Coalescence Under Different Demographic Scenarios ... 204
  - 7.7.5 Statistical Inference on Demographic Hypotheses and Parameters 207

		7.7.6	Markov Chain Monte Carlo (MCMC) Methods 207
		7.7.7	Approximate Approaches 208
	7.8	Exercises 212

8 Genomics 213
8.1 The DNA Molecule and the Central Dogma of Molecular Biology 214
8.2 Genome Structure 220
8.3 Genome Sequencing 223
    8.3.1 Restriction Enzymes 224
    8.3.2 Electrophoresis 224
    8.3.3 Southern Blot 224
    8.3.4 The Polymerase Chain Reaction 225
    8.3.5 DNA Cloning 226
    8.3.6 Chain Termination DNA Sequencing 226
    8.3.7 Genome Shotgun Sequencing 228
8.4 Genome Assembly Algorithms 230
    8.4.1 Growing Contigs from Fragments 230
    8.4.2 Detection of Overlaps Between Reads 230
    8.4.3 Repetitive Structure of DNA 232
    8.4.4 The Shortest Superstring Problem 233
    8.4.5 Overlap Graphs and the Hamiltonian Path Problem ... 234
    8.4.6 Sequencing by Hybridization 235
    8.4.7 De Bruijn Graphs 238
    8.4.8 All l-mers in the Reads 238
    8.4.9 The Euler Superpath Problem 239
    8.4.10 Further Aspects of DNA Assembly Algorithms 240
8.5 Statistics of the Genome Coverage 243
    8.5.1 Contigs, Gaps and Anchored Contigs 244
    8.5.2 Statistics with Minimum Overlaps Between Fragments, Anchored Contigs 246
    8.5.3 Genome Length and Structure Estimation by Sampling l-mers 247
    8.5.4 Polymorphisms 252
8.6 Genome Annotation 252
    8.6.1 Research Tools for Genome Annotation 254
    8.6.2 Gene Identification 254
    8.6.3 DNA Motifs 257
    8.6.4 Annotation by Words and Comparisons of Genome Assemblies 258
    8.6.5 Human Chromosome 14 258
8.7 Exercises 259

9 Proteomics ... 261
9.1 Protein Structure ... 262
- 9.1.1 Amino Acids ... 262
- 9.1.2 Peptide Bonds ... 265
- 9.1.3 Primary Structure ... 266
- 9.1.4 Secondary Structure ... 266
- 9.1.5 Tertiary Structure ... 268
- 9.1.6 Quaternary Structure ... 271

9.2 Experimental Determination of Amino Acid Sequences and Protein Structures ... 271
- 9.2.1 Electrophoresis ... 272
- 9.2.2 Protein 2D Gels ... 272
- 9.2.3 Protein Western Blots ... 273
- 9.2.4 Mass Spectrometry ... 273
- 9.2.5 Chemical Identification of Amino Acids in Peptides ... 274
- 9.2.6 Analysis of Protein 3D Structure by X Ray Diffraction and NMR ... 275
- 9.2.7 Other Assays for Protein Compositions and Interactions 275

9.3 Computational Methods for Modeling Molecular Structures ... 275
- 9.3.1 Molecular-Force-Field Model ... 276
- 9.3.2 Molecular Dynamics ... 281
- 9.3.3 Hydrogen Bonds ... 281
- 9.3.4 Computation and Minimization of RMSD ... 282
- 9.3.5 Solutions to the Problem of Minimization of RMSD over Rotations ... 284
- 9.3.6 Solutions to the Problem of Minimization of RMSD over Rotations and Translations ... 290
- 9.3.7 Solvent-Accessible Surface of a Protein ... 290

9.4 Computational Prediction of Protein Structure and Function ... 290
- 9.4.1 Inferring Structures of Proteins ... 291
- 9.4.2 Protein Annotation ... 292
- 9.4.3 De Novo Methods ... 292
- 9.4.4 Comparative Modeling ... 293
- 9.4.5 Protein–Ligand Binding Analysis ... 295
- 9.4.6 Classification Based on Proteomic Assays ... 295

9.5 Exercises ... 296

10 RNA ... 299
10.1 The RNA World Hypothesis ... 300
10.2 The Functions of RNA ... 300
10.3 Reverse Transcription, Sequencing RNA Chains ... 301
10.4 The Northern Blot ... 302
10.5 RNA Primary Structure ... 302
10.6 RNA Secondary Structure ... 302
10.7 RNA Tertiary Structure ... 302

XVI   Contents

      10.8 Computational Prediction of RNA Secondary Structure 303
           10.8.1 Nested Structure 304
           10.8.2 Maximizing the Number of Pairings Between Bases 304
           10.8.3 Minimizing the Energy of RNA Secondary Structure ... 306
           10.8.4 Pseudoknots 310
      10.9 Prediction of RNA Structure by Comparative Sequence
          Analysis.. 311
      10.10 Exercises 311

11 DNA Microarrays 313
      11.1 Design of DNA Microarrays................................ 315
      11.2 Kinetics of the Binding Process 318
      11.3 Data Preprocessing and Normalization 320
           11.3.1 Normalization Procedures for Single Microarrays 321
           11.3.2 Normalization Based on Spiked-in Control RNA 323
           11.3.3 RMA Normalization Procedure 326
           11.3.4 Correction of Ratio–Intensity Plots for cDNA 328
      11.4 Statistics of Gene Expression Profiles 328
           11.4.1 Modeling Probability Distributions of Gene Expressions 331
      11.5 Class Prediction and Class Discovery 336
      11.6 Dimensionality Reduction 337
           11.6.1 Example of Application of PCA to Microarray Data ... 338
      11.7 Class Discovery 338
           11.7.1 Hierarchical Clustering 339
      11.8 Class Prediction. Differentially Expressed Genes 340
      11.9 Multiple Testing, and Analysis of False Discovery Rate (FDR) 341
           11.9.1 FDR analysis in ALL versus AML gene expression data 344
      11.10 The Gene Ontology Database 344
           11.10.1 Structure of GO 345
           11.10.2 Other Vocabularies of Terms 346
           11.10.3 Supporting Results of DNA Microarray Analyses
                with GO and other Vocabulary Terms 347
      11.11 Exercises 347

**12 Bioinformatic Databases and Bioinformatic Internet
Resources** 349
      12.1 Genomic Databases....................................... 350
      12.2 Proteomic Databases..................................... 350
      12.3 RNA Databases 350
      12.4 Gene Expression Databases 351
      12.5 Ontology Databases 351
      12.6 Databases of Genetic and Proteomic Pathways 351
      12.7 Programs and Services 352
      12.8 Clinical Databases 352

References 355

Index 371

1
Introduction

1.1 The Genesis of Bioinformatics

Bioinformatics is a discipline which originally arose for the utilitarian purpose of introducing order into the massive data sets produced by the new technologies of molecular biology. These techniques originated with large-scale DNA sequencing and the need for tools for sequence assembly and for sequence annotation, i.e., determination of locations of protein-coding regions in DNA. A parallel development was the construction of sequence repositories. The crowning achievement has been the sequencing of the human genome and, subsequently of many other genomes.

Another new technology, which has started to provide wealth of new data, is the measurement of multiple gene expression. It employs various physical media, including glass slides, nylon membranes, and other media. The idea is to expose a probe (a DNA chip) including thousands of DNA nucleotide sequences, each uniquely identifying a gene, to a sample of coding DNA extracted from a specimen of interest. Multiple-gene-expression techniques are usually employed to identify subsets of genes discriminating between two or more biological conditions (supervised classification), or to identify clusters in the gene sample space, which leads to a classification of both samples and genes (unsupervised classification). Analysis of gene expression data has led to new developments in computational algorithms: existing computational techniques, with their origin in computer science, such as self-organizing maps and support vector machines, and of statistical origin such as principal-component analysis and analysis of variance, have been adapted, and new techniques have been developed.

The next step in the development of the technology includes proteomic techniques, which allow measurements of the abundance and activity of thousands of protein species at once. These are usually multistep procedures. The initial phase involves physical separation of proteins from the sample according to one or more (typically two) variables, for example molecular weight and isoelectric point. This is physically accomplished using two-dimensional gels, on which different proteins can be spotted as individual clusters. The next step involves identification of proteins sampled from different spots on the gel. This involves cleavage of amino acid chains and producing mass spectra using extremely precise mass spectrometry machines. Finally, on the basis of the distribution of molecular weights of the fragmented chains, it is possible to identify known proteins or even to sequence unknown ones. Various more refined versions of the technology exist, which allow the labeling of activated proteins, various protein subsets, and so forth.

The interpretation of proteomic data has led to the development of warping and deconvolution techniques. Two-dimensional protein gels are distorted with respect to the perfect Cartesian coordinates of the two variables describing each protein. To allow comparison with standards and with results obtained under other experimental conditions, it is necessary to transform the gel coordinates into Cartesian ones, a procedure known as warping. As mentioned above, after this is accomplished, we may analyze a gel spot representing a protein, using mass spectrometry. Deciphering the sequence of the polypeptide chain using mass spectrometry of fragments 5 – 10 amino acids long is accomplished using deconvolution.

One of the more notable consequences of the developments in genomics and proteomics has been an explosion in the methodology of genetic and metabolic networks. As is known, the expression of genes is regulated by proteins, which are activated by cascades of reactions involving interactions with other proteins, as well as the promotion or inhibition of the expression of other genes. The resulting feedback loops are largely unknown. They can be identified by perturbing the system in various ways and synthesizing a network on the basis of genomic and proteomic measurements in the presence of perturbations. A variety of network types can be used, varying from Boolean networks (discrete automata) and probabilistic versions of them, to Bayesian networks and others. Although these techniques are still unsatisfactory in practice, in many cases they have allowed us to gain insight into the structure of the feedback loops, which than can be analyzed using more conventional tools, including, for example, systems of nonlinear differential equations.

1.2 Bioinformatics Versus Other Disciplines

Bioinformatics has been developed in the space, which was already occupied by a number of related disciplines. These include quantitative sciences such as

- mathematical and computational biology,
- biometry and biostatistics,
- computer science,
- cybernetics,

    as well as biological sciences such as

- molecular evolution,
- genomics and proteomics,
- genetics, and
- molecular and cell biology.

It might be argued that bioinformatics is a direct extension of mathematical and computational biology into the realm of new, massive data sets. However, the sheer size of this mass of data creates qualitatively new situations. For example, any serious query of the data requires writing computer code and/or placing the data within a suitable database. The complexity of the databases varies enormously, reaching the highest proportions in databases designed to handle information about metabolic pathways. Even determining what should be the subject of a query involves computer-intensive methods.

As an example, let us consider the problem of finding enough homologous DNA sequences to carry out an evolutionary analysis of homologous proteins coded by these sequences in different organisms. To accomplish this, one has to use a set of computerized tools, known as BLAST, which has the ability to search for sequences above a certain level of similarity and to assign statistical similarity scores to potential homologs. The probabilistic theory of BLAST involves considerations of how unlikely it is for two sequences of given length to display a given level of similarity.

Another interesting example concerns carrying out statistical comparisons between gene expression levels obtained using DNA microarrays. Here, we have to deal with comparisons of a limited number of microarrays, each yielding a data vector of high dimension. This is a situation which is exactly opposite to the usual statistical paradigm, according to which a large sample of low-dimensional data is considered most useful. Even worse, comparisons are frequently carried out gene-by-gene, leading to potential repeated-testing problems. This problem becomes even more serious when we realize that large subsets of genes may have correlated expressions. Under such circumstances, the only statistical tools which make it possible to determine whether differences found are significant, are permutation tests. These latter are often computationally intensive,

A major issue in bioinformatics is the combinatorial complexity of algorithms, which can be insurmountable. An example stemming from the field of molecular evolution is the construction of phylogenetic trees of sequences using the maximum-likelihood method. The space of trees with more than 10 nodes is so enormous that there is no way an exhaustive search might be carried out. Instead, various heuristics and suboptimal searches are used. This is an important point, since, as noted later, evolutionary changes of biological sequences can be treated as a result of an experiment not requiring a new laboratory. This is discussed later in the context of identification of active sites of proteins.

Another example of a typically bioinformatic problem is provided by polymorphisms in the human genome. As is known, any two human DNA sequences differ at random points located, on average several hundred nucleotides apart. These are the single-nucleotide polymorphisms (SNPs). Therefore, there exists an enormous multitude of sequence variants. At the same time, the human genome sequence is based on only a few individuals. This illustrates the difference with respect to classical human genetics, which attempts to elucidate the role of genetic variability at a limited number of loci at a time. With the onset of mass sequencing of either entire genomes or major portions of genomes, analysis of their genetic and evolutionary relationships will require increased computational power and new data structures.

1.3 Further Developments: from Linear Information to Multidimensional Structure Organization.

Many widely used methods of bioinformatics hinge upon the linear structure of genomic information. This includes sequencing and annotation, but also sequence retrieval and comparison. A natural toolbox for problems of this nature is provided by hidden Markov models (HMMs) and the Viterbi algorithm based on dynamic programming. The idea of the Viterbi algorithm is to find the most likely estimate of the Markov process underlying a given biological process, based on the so-called emissions, i.e., the limited available observations of the process. The solution is obtained recursively, following the dynamic programming paradigm. A typical application of the Viterbi algorithm arises when the Markov process describes some feature of the genetic/genomic information distributed along the DNA sequence (this can be some functionality of the DNA region) and the emissions are constituted by the sequence of DNA nucleotides. An example is the identification of promoter regions of genes. However, the Viterbi algorithm can be defined for Markov processes evolving on very general spaces. For example, consider the space of nested quasi-palindromic motifs, which is equivalent to all possible secondary structures of RNA molecules, endowed with a Markov process defined as a stochastic algebra of admissible rules by which the motifs can be created. This framework makes possible to define a Viterbi algorithm for identification of the structure, based on the sequence. Other interesting applications of the Viterbi algorithm arise when we attempt to build phylogenetic trees of sequences involving a variable substitution rate along the sequence. This extension to branching structures is the foundation of the Felsentein–Churchill algorithm for maximum likelihood trees, discussed later.

Biological information is translated into the structure and function of biomolecules, which in turn form higher-level structures. The simplest example is protein folding. Proteins are active because of their spatial conformation and the occurrence of active centers, which interact with other molecules. Quantitative studies of these features can be accomplished in various ways. A direct approach involves computations of protein folding based on energy functions. Again, dynamic programming can be used to reduce the computational burden. If this is accomplished, or if the structure is known from X-ray crystallography, it is possible to consider computations of active centers of proteins based on the geometry of their surfaces. The interaction of proteins may be approached computationally by solving the docking problem, employing methods of computational geometry similar to those used in robotics. These and related computations are involved and time-and memory-consuming.

An alternative approach is based on the notion of evolution as a laboratory. By following the evolution of biomolecules, it is possible to infer their function and the relationships between them. Example of this approach is the evolutionary trace method of Lichtarge. In this method, homologous amino acid sequences from a number of species are used to infer a phylogeny. Subsequently, this phylogeny forms a basis for classification of the amino acids in the sequence, based on their conservation in branches of the tree of different order. The amino acids which are conserved best are likely to belong to the active center. This method has led to confirmed predictions of the active sites. Similarly, the Felsentein–Churchill algorithm mentioned above allows identification of amino acids, which have evolved slowly. These will be candidates for belonging to the active center.

The new branches of bioinformatics will require the creation of new databases and continued work on purely informatic structures such as ontologies, which allow retrieval of information with a very rich structure.

1.4 Mathematical and Computational Methods

At present, virtually all branches of science use mathematical methods as parts of their research tools. Science has entered a phase of mathematics invading other disciplines. This is because the concepts in all areas of science are becoming more and more mature and precise, and mathematical tools are flexible and generalizable.

Without exaggeration, we can say that the majority of the methods of applied mathematics are used as tools in bioinformatics. So, is there anything peculiar about using mathematical modeling in bioinformatics? Among the tools of applied mathematics some are of special importance, namely probability theory and statistics and algorithms in computer science. A large amount of research in bioinformatics uses and combines methods from these two areas. Computer-science algorithms form the technical background for bioinformatics, in the sense that the operation and maintenance of bioinformatic databases require the most up-to-date algorithmic tools. Probability and statistics, besides being a tool for research, also provides a language for formulating results in bioinformatics.

Other mathematical and computational tools, such as optimization techniques with dynamic programming, discrete-mathematics algorithms, and pattern analysis methods, are also of basic importance in ordering bioinformatic data and in modeling biological mechanisms at various levels.

The first part of the book, on mathematical and computational methods is intended to cover the tools used in the book. The presentations of methods in this part are oriented towards their applications in bioinformatics. In the second part of this book, practical uses of these methods are illustrated on the basis of the rather large number of research papers devoted to the analysis of bioinformatic data. Sometimes some further developments of methods are presented, together with the problem they apply to, or some references are given to the derivation of the algorithm. Description of applied mathematical methods is organized into several sections corresponding to logical grouping of methods.

Our presentation of the mathematical approaches is rather descriptive. When discussing mathematical methods we appeal to comprehension and intuitive understanding, to their relations to bioinformatic problems and to cross-applications between items we discuss. This approach allows us to go through a variety of methods and, hopefully, to sketch a picture of bioinformatics. Despite avoiding much of the mathematical formalism we have tried to keep the presentation sufficiently clear and precise. All chapters are accompanied by exercises and problems, which are intended to support understanding of the material and often show further developments. Their levels of difficulty varies, but generally they are rather non trivial.

1.4.1 Why Mathematical Modeling?

What is mathematical modeling? By mathematical modeling, we understand describing and reflecting reality by using formalized tools. Models can be of very different types: stochastic or deterministic, descriptive or mechanistic, dynamic or static. Mathematical models can pertain to phenomena in many different areas, for instance physics, chemistry, biology, engineering, or economics.

How do we develop models? Models are developed by combining, comparing, or verifying hypotheses versus empirical observations. We develop models by using the laws of nature, physics, chemistry, and biology. We apply principles of conservation and/or variational extremum principles, which lead to balances and to differential or difference equations for the evolution of the state of a system. Models can include discrete events and random phenomena.

What is the benefit of using mathematical models? Using mathematical models allows us to achieve a better understanding and to organize better our knowledge about the underlying mechanisms and phenomena. Sometimes models can change qualitative understanding to quantitative knowledge. Models can allow us to predict future events from present observations. Models can be helpful in programming and planning our control and design actions.

What is specific in modeling in biology and molecular biology? Compared with models in physics and classical chemistry, models in (molecular) biology pertain to more complex phenomena. Following from this there is usually a greater extent of simplification that needs to be applied when building the model. The large individual variation leads to a substantial element of randomness, which needs to be incorporated into the model.

1.4.2 Fitting Models to Data

An element which is present in all models is simplifying hypotheses. The benefit in using a mathematical model is often related to solving the compromise between the extent of simplification in the model and the precision in predicting data. Complicated models are usually less reliable and less comprehensive. Oversimplified models can ignore important phenomena.

The research work that forms part of modeling involves model building or model learning, applying the model to the data and model modification. After enough experience has been gained by repeated application of these elements of modeling research, models often start bringing benefits.

One crucial element is verifying a model versus the data, which very often starts from fitting free parameters of the model. This involves tasks such as identification and parameter estimation, solved by various methods of static, dynamic or stochastic optimization. Among optimization methods the least squares method deserves special attention owing to its reliability and very vast range of application.

If one assumes model with many free parameters, one has substantial flexibility in fitting the model to the data. The extreme case is called "black box modeling", which means fitting the parameters of standardized models to the measurements without inquiring about the nature of the underlying processes and phenomena.

1.4.3 Computer Software

Both fitting to data and analyzing the predictions of mathematical models is done by using computers with appropriate software. There is a variety of computer software environments for all platforms, and choosing the appropriate program for the computational aspects of the research being done is an important issue. Some very useful programming environments are the high-level programming languages for supporting engineering and scientific computations Matlab, Mathematica, Maple, R. Several computational examples in this book were programmed using Matlab. Matlab can be equipped with toolboxes, which include many of the algorithms described in this book. For some specialized tasks one may need programming languages of lower level, such as C, C++, Delphi, Java.

We should also mention the numerous Internet servers offering specialized computations in the field of bioinformatics, such as aligning sequences against databases, predicting 2D and 3D structures of proteins and RNA and so forth. Some of them are mentioned or discussed later in this book.

1.5 Applications

Facts in biology and biochemistry become established when they are seen in a biological or a biochemical experiment or, better, in several independent experiments. Knowledge develops in biology and biochemistry in this way. There are two aspects of the development of biology and biochemistry concerning its relation to bioinformatics. First, with the development of experimental techniques, the number of findings and discoveries in biology and biochemistry has become so large, that efficient access to the information requires the use of the most advanced informatic tools. Second, browsing and analyzing data in bioinformatic databases allows or helps us to predict facts in biology and biochemistry or to propose new hypotheses. These hypotheses can be used for designing new experiments. There are several well-established paths in which bioinformatics can be applied in this second way. After the genome of a new organism has been sequenced, then by using knowledge about the structure and organization of genomes and the contents of genomic databases, researchers can find the genes and compare them with their homologs in other organisms. Inside genes, coding sequences can be identified, leading to amino acid sequences of proteins. These approaches can be used in a variety of types of research. Information obtained from comparing genomes can be used for inferring the ancestry of organisms and also for predicting the functions of genes and proteins. Comparing sequences of amino acids in proteins in different organisms allows one to infer their functionally important sites and active sites. By combining computational methods with browsing protein databases, one can improve the methods for drug design. For example, when the sequence of a virus causing a disease has been found then it is often searched for regions coding for proteins. Next, using the hypothesis that these proteins are important in the activity of the virus in the human organism, design of the appropriate treatment can focus on drugs blocking their activity.

Bioinformatic databases contain massive amounts of experimental data. Browsing and analyzing these data is fascinating and will surely lead to many interesting discoveries. The developing projects concerned with searching for interesting information in bioinformatic databases belong to the most vital area in scientific research.

It is important to stress here the interdisciplinary aspects of the research in bioinformatics. A search through bioinformatic databases is often initiated by posing a question related to some biological problem. The bioinformatic project then involves designing the computational and algorithmic aspects of the search or browsing. The results are most valuable when they lead to answering the question, to improved understanding or to interesting biological interpretations.

In the second part of this book, we have organized the material such that the biological and biochemical aspects are treated with enough care to explain the motivation for pursuing research in bioinformatics. The second part of the book includes seven chapters, each devoted to a specific area. We start with two chapters, on sequence alignment and molecular phylogenetics, devoted to specific methodologies applicable in many contexts, which are discussed later. In the chapter on sequence alignment, we present the methodologies and their relation to optimization and to computer-science algorithms. In the chapter on molecular phylogenetics we discussed basic approaches of reconstructing phylogenetic trees, using appropriate tools of optimization and statistics. We also included a section on coalescence, which (i) allows us to understand the processes behind the formation of phylogenetic trees, and (ii) illustrates some new applications of phylogenetics, such as inferring demographic scenarios from molecular data. The next three chapters are devoted to biological items, namely genomics, proteomics and RNA. These chapters include, in their introductory parts, the basics of the underlying biological and biochemical facts. Next, mathematical modeling methods and their relations to experimental approaches are presented. The chapter on DNA microarrays is focused on the biological process of gene expression and the associated technology of biological assays, as well as related mathematical and computational approaches. Owing to its importance and the large number of research papers and monographs in the field, it deserves special attention. We have provided a description of DNA microarray technology in the introductory part. Then we discuss mathematical modeling in the context of analyzing gene expression profiles. Finally, the last chapter is devoted to bioinformatic databases and other bioinformatic Web sites and services. In this short chapter, we have aimed to give an overview of some of the internet resources related to bioinformatics.

Most of the chapters have a set of exercises at the end. Some exercises are problems aimed at supporting understanding of presented ideas and often completing or adding some elements of derivations of methods. Other exercises are projects, which often involve issues such as developing computer programs and studying their application to solving problems. Many of the projects suggest downloading publicly available software and/or using some of internet bioinformatic depositories on the Internet. In these projects we have suggested many possibilities, which we are fairly sure will help to develop our understanding of some problems and may lead to interesting results.

Part I

Mathematical and Computational Methods

2

Probability and Statistics

Probability calculus involves computing the probabilities of random or stochastic events, defining probability distributions, and performing manipulations using random variables. Statistics is concerned with presentation, analysis, and inference based on data obtained from experiments with an element of randomness. In a random setting events may happen or not, and defining the conditions of experiments involves assigning probabilities to events, for example $P(A) = 0.3$. The probability of an event is a measure of the frequency of its occurrence under repeated, controlled conditions. Large collections of probability distributions associated with typical physical or biological experiments with a random component, as well as the rules of probability calculus are presented in large number of references, for example [37, 78, 87, 134, 144, 212, 275, 297]. In this chapter we discuss several topics in probability calculus and statistics, to be used later in the processing of bioinformatic data.

When fitting probabilistic models to data, it is commonly necessary to estimate parameters of distributions of random variables. We cover methods of parameter estimation, with emphasis on maximum likelihood. We also develop some more practical computational aspects including the EM algorithm with examples of its applications. We include some topics on testing statistical hypotheses. We also present material regarding Markov chains and computational techniques which have Markov chains as theoretical background, namely Markov chain Monte Carlo Methods and hidden Markov models.

2.1 The Rules of Probability Calculus

Probability is a set function, assigning a number from the interval $[0, 1]$ to a set A. We say that $P(A)$ is the probability of a set, or event, A. Usually it is assumed that the events are selected from a family $\mathcal{A}$ of subsets of a sample space Ω. Such family, if it is closed with respect to set complementation (i.e., if $A \in \mathcal{A} \Rightarrow A^c \in \mathcal{A}$) and with respect to denumerable summation (i.e., if

$A_i \in \mathcal{A}$, $i = 1, 2, \ldots \Rightarrow (\bigcup_{i=1}^{\infty} A_i) \in \mathcal{A}$), and if it contains the empty set $\emptyset$ (and so also $\Omega = \emptyset^c$) is called a σ-field or a σ-algebra of subsets of Ω.

For the definition of the probability to be intuitively consistent, we require the following properties called the axioms of probability

(i) $P(A) \in [0, 1]$, $A \in \mathcal{A}$
(ii) $P(\Omega) = 1$
(iii) $P(\sum_{i=1}^{\infty} A_i) = \sum_{i=1}^{\infty} P(A_i)$, $A_i \in \mathcal{A}$, $A_i \cap A_j = \emptyset$, $i, j = 1, 2, \ldots$

Axiom (ii) is the probability-norming property and axiom (iii) is the denumerable additivity. Finite additivity is a consequence, for example

$$P(A \cup B) = P(A) + P(B), \quad A, B \in \mathcal{A}, \quad A \cap B = \emptyset.$$

Also,

$$P(\Omega) = P(A) + P(A^c) = 1,$$

and hence

$$P(A^c) = 1 - P(A).$$

For the general case, where A and B need not be mutually exclusive, we have

$$P(A \cup B) = P(A) + P(B) - P(A \cap B),$$

where $A \cap B$ is the intersection of the sets A and B also denoted as AB. Frequently we write $P(A, B)$ instead of $P(A \cap B)$ or $P(AB)$.

2.1.1 Independence, Conditional Probabilities and Bayes' Rules

An important concept in probability and statistics, which originates from intuition and from empirical observations is the independence of the probability of an event A from the occurrence of another event B. Mathematically, the events A and B are independent if and only if

$$P(A \cap B) = P(A)P(B).$$

For two events A and B, probability of A, given B, or in other words, the probability conditional on B is frequently defined as

$$P(A \mid B) = \frac{P(A \cap B)}{P(B)}. \tag{2.1}$$

If A and B are independent, then $P(A \mid B) = P(A)$. From (2.1), we have Bayes theorem,

$$P(A \mid B)P(B) = P(B \mid A)P(A). \tag{2.2}$$

If the events $B_1, B_2, \ldots$ are all mutually exclusive ($B_i \cap B_j = \emptyset$) and are collectively exhaustive ($\bigcup_{k=1}^{\infty} B_k = \Omega$), then the expression for the probability $P(A) = P(A \cap \Omega)$ can be decomposed as

$$P(A) = \sum_{k=1}^{\infty} P(A \cap B_k) = \sum_{k=1}^{\infty} P(A \mid B_k) P(B_k). \qquad (2.3)$$

The above expression is called the law of total probability. We should also note that instead of requiring collective exhaustiveness of $B_1, B_2, \ldots$, it is enough that they are mutually exclusive and $A \subset \bigcup_{k=1}^{\infty} B_k$ for validity of (2.3).

Using (2.3), we can also compute the conditional probability $P(B_k \mid A)$:

$$P(B_k \mid A) = \frac{P(A \cap B_k)}{P(A)} = \frac{P(A \mid B_k) P(B_k)}{\sum_{k=1}^{\infty} P(A \mid B_k) P(B_k)}. \qquad (2.4)$$

$P(B_k \mid A)$ is called the posterior probability of B_k, and (2.4) is called Bayes' second formula.

2.2 Random Variables

In deterministic computations, one assigns a fixed value to a variable. In the probabilistic setting we use random variables, which can assume different, random values. Random variables are functions or mappings from the sample space Ω into the space R of real numbers.

We shall not develop a consistent mathematical theory of random variables in this section; instead we shall focus on rules that can be applied in practice. We shall begin with a discrete random variable, which assumes values from an enumerable subset of R. A discrete random variable X assumes values

$$x_0, x_1, x_2, \ldots, x_k, \ldots \qquad (2.5)$$

with corresponding probabilities

$$p_0, p_1, p_2, \ldots, p_k, \ldots, \qquad (2.6)$$

which satisfy the norming condition

$$\sum_{k=0}^{\infty} p_k = 1.$$

The infinite or finite sequence $\{p_0, p_1, \ldots\}$ is called the distribution of X. Frequently, discrete random variables assume values from the set of integers or nonnegative integers. They are then called integer or nonnegative integer random variables.

In contrast, a continuous random variable X assumes values from subintervals of the real axis R. For continuous random variables the role of the distribution is played by the cumulative probability distribution function $F_X(x)$,

$$F_X(x) = P[X \leq x],$$

which, for given x, is equal to the probability of the event $[X \leq x]$. The basic properties of $F_X(.)$ are (i) $F_X(.)$ is nondecreasing, (ii) $F_X(-\infty) = 0$, and (iii) $F_X(+\infty) = 1$. Intervals of constancy of $F_X(x)$ coincide with intervals "prohibited" for X while jumps in $F_X(x)$ coincide with discrete atoms of probability distribution of X.

If $F_X(x)$ is differentiable, then its derivative is called the probability density function (pdf) $f_X(x)$, where

$$f_X(x) = \lim_{\Delta x \to 0} \frac{F(x < X \leq x + \Delta x)}{\Delta x} = \frac{dF_X(x)}{dx}. \qquad (2.7)$$

We also have

$$\int_{-\infty}^{x} f_X(\xi) d\xi = F_X(x),$$

and consequently, since $F_X(+\infty) = 1$, we obtain the normalization condition for the distribution of the continuous random variable X,

$$\int_{-\infty}^{+\infty} f_X(x) dx = \lim_{x \to +\infty} F_X(x) = 1.$$

2.2.1 Vector Random Variables

It is often necessary to analyze distributions of two or more random variables jointly, which leads to vector random variables. For discrete random variables X assuming values $x_0, x_1, x_2, \ldots, x_k, \ldots$, and Y assuming values $y_0, y_1, y_2, \ldots, y_k, \ldots$, the joint probability distribution is given by the array of probabilities

$$p_{ij} = P[X = x_i, Y = y_j], \qquad (2.8)$$

with the norming condition

$$\sum_{i=0}^{\infty} \sum_{j=0}^{\infty} p_{ij} = 1.$$

For the continuous case, the joint cumulative distribution function of the random variables X and Y is

$$F_{X,Y}(x, y) = P[X \leq x, Y \leq y],$$

and the joint probability density function, corresponding to an absolutely continuous $F_{X,Y}(x, y)$, is

$$f_{X,Y}(x,y) = \lim_{\Delta x \to 0, \Delta y \to 0} \frac{P(x < X \le x + \Delta x, y < Y \le y + \Delta y)}{\Delta x \Delta y}$$
$$= \frac{\partial^2 F_{X,Y}(x,y)}{\partial x \partial y} \qquad (2.9)$$

with the condition

$$\int_{-\infty}^{+\infty} \int_{-\infty}^{+\infty} f_{X,Y}(x,y) dx dy = \lim_{x \to +\infty, y \to +\infty} F_{X,Y}(x,y) = 1.$$

2.2.2 Marginal Distributions

Two-dimensional (and multidimensional) distributions can be reduced to one-dimensional distributions by computing marginals. For a discrete random variable X, jointly distributed with Y according to (2.8), we have

$$p_i = P(X = x_i) = \sum_{j=0}^{\infty} p_{ij},$$

whereas for a continuous random variable X, jointly distributed with Y according to (2.9) the marginal distribution is given by

$$F_X(x) = F_{X,Y}(x, \infty),$$

and the probability density function by

$$f_X(x) = \int_{-\infty}^{+\infty} f_{X,Y}(x,y) dx.$$

The above formulas for two-dimensional random variables generalize in an obvious manner for dimensions greater than two.

2.2.3 Operations on Random Variables

The use of probabilistic and statistical tools includes performing manipulations involving random variables. Some of these are discussed below.

Independence and Conditional Distributions

For independent random variables X and Y, their joint probability distribution satisfies

$$p_{ij} = p_i p_j$$

in the discrete case (2.8) and

$$F_{X,Y}(x,y) = F_X(x) F_Y(y)$$

or
$$f_{X,Y}(x,y) = f_X(x)f_Y(y)$$
in the continuous case.

The conditional distribution of X given $[Y = y]$, is given by a formula analogous to (2.1),
$$f_{X|Y}(x \mid y) = \frac{f_{X,Y}(x,y)}{f_Y(y)}.$$
When computing conditional distributions, the following rule is often helpful:
$$f_{X,Y|Z}(x,y \mid z) = f_{X|Y,Z}(x \mid y, z) \, f_{Y|Z}(y \mid z). \tag{2.10}$$
This rule follows from
$$\frac{f_{X,Y,Z}(x,y,z)}{f_Z(z)} = \frac{f_{X,Y,Z}(x,y,z)}{f_{Y,Z}(y,z)} \frac{f_{Y,Z}(y,z)}{f_Z(z)}.$$

Algebraic Operations

If X and Y are random variables with a joint distribution density $f_{X,Y}(x,y)$, and we define a new random variable
$$Z = X + Y,$$
then the distribution of Z can be obtained by integrating over the density $f_{X,Y}(x,y)$, namely
$$f_Z(z) = \iint_{x+y=z} f_{X,Y}(x,y) = \int_{-\infty}^{+\infty} f_{X,Y}(x, z-x)dx. \tag{2.11}$$
A similar calculus applies to products, ratios, etc. When X and Y are independent, (2.11) transforms to a convolution integral:
$$f_Z(z) = \int_{-\infty}^{+\infty} f_X(x) f_Y(z-x)dx. \tag{2.12}$$

Assume that the random variable X has a probability density function $f_X(x)$, and define
$$Y = g(X).$$
What is the probability distribution of Y? We assume that $g(.)$ is strictly monotonic. This assumption can be relaxed, and computations for the general case follow from applications of the main idea separately in each interval in which the function is monotonous. From strict monotonicity, there follows the invertibility of the function $g(.)$:
$$y = g(x) \Rightarrow x = g^{-1}(y).$$

Using the inverse function, one can represent the cumulative distribution of Y, $F_Y(y)$, in terms of the realizations of random variable X, and consequently in terms of the cumulative distribution of X, $F_X(x)$

$$F_Y(y) = P(Y \le y) = P[g(X) \le y]$$
$$= \begin{cases} P[X \le g^{-1}(y)] = F_X[g^{-1}(y)] \text{ for } g(x) \text{ increasing,} \\ P[X \ge g^{-1}(y)] = 1 - F_X[g^{-1}(y)] \text{ for } g(x) \text{ decreasing,} \end{cases} \quad (2.13)$$

or, in terms of densities (if they exist)

$$f_Y(y) = \left| \frac{d}{dy} g^{-1}(y) \right| f_X[g^{-1}(y)].$$

2.2.4 Notation

Here we recall some of the conventions that we are already using, and also announce a convention to be used in forthcoming sections. If X denotes a random variable, then we represent possible values corresponding to X (realizations, observations, or measurements) by the lower case letter x. Probability distribution functions for the random variable X are indexed correspondingly, for example $f_X(x)$. However, for notational ease, the index representing the random variable is often dropped. So, we may write $f(x)$ instead of $f_X(x)$ in instances where it does not lead to confusion.

2.2.5 Expectation and Moments of Random Variables

The expectation of a function $g(x)$ with respect to the distribution of a discrete random variable X, defined by the values (2.5) and probabilities (2.6), is

$$E[g(X)] = \sum_{k=1}^{\infty} p_k g(x_k). \quad (2.14)$$

The expectation of a function $g(x)$ with respect to a continuous random variable X or, equivalently, with respect to its distribution $f_X(x)$ is the following integral:

$$E[g(X)] = \int_{-\infty}^{+\infty} g(x) f_X(x) dx. \quad (2.15)$$

When $g(x) = x$, (2.15) becomes the expectation of X, or the first moment of the random variable X:

$$E(X) = \sum_{k=0}^{\infty} p_k x_k \quad (2.16)$$

for the discrete case, and

$$E(X) = \int_{-\infty}^{+\infty} x f_X(x) dx \quad (2.17)$$

for the continuous case.

Higher moments of the random variable X are defined analogously to (2.14)–(2.17), with $g(X) = X^n$ for the nth moment and $g(X) = [X - E(X)]^n$ for the nth central moment of the random variable X. Among these moments, the second central moment called the variance of the random variable, is of special importance:

$$\text{Var}(X) = \sum_{k=0}^{\infty} p_k [x_k - E(X)]^2 \tag{2.18}$$

(discrete case), and

$$\text{Var}(X) = \int_{-\infty}^{+\infty} [x - E(X)]^2 f_X(x) dx \tag{2.19}$$

(continuous case). This serves as an indicator of the dispersion of the random variable around its expected value. The square root of the variance, called the standard deviation and denoted by

$$\sigma(X) = \sqrt{\text{Var}(X)}$$

is the scale parameter of the distribution $X - E(X)$.

For expectations of functions or moments of random variables to exist, the corresponding series or integrals must be convergent. If the function $g(x)$ in (2.14) and (2.15) increases too fast with x, the series or integrals may not converge. Also, if the distribution of a random variable has tails that are too heavy, certain moments of the random variables may not exist; well-known examples are Cauchy or Student t distributions.

Some important properties of the expectation and variance concern sums of random variables. The expectation of the sum of two random variables is the sum of their expectations:

$$E(X + Y) = E(X) + E(Y).$$

The variance of the sum of two independent random variables is the sum of their variances:

$$\text{Var}(X + Y) = \text{Var}(X) + \text{Var}(Y), \quad X, Y \text{ independent}.$$

2.2.6 Probability-Generating Functions and Characteristic Functions

The transformational approach is very useful in many situations in many areas of scientific research [66, 289]. In probability and statistics, the transformational approach is used in the analysis of probability distributions, for example for performing efficient computations of probabilities, moments, and distributions of random variables, and for discovering and proving theorems concerning limit properties of distributions. Below we provide definitions and some facts. We also show an example of the use of a probability-generating function for computing a distribution of a random variable.

With a discrete random variable X assuming values (2.5) with probabilities (2.6), we associate a function $P_X(z)$ of a complex argument z in the following way:

$$P_X(z) = \sum_{k=0}^{\infty} z^k p_k. \tag{2.20}$$

The above function $P(z)$ is called the probability-generating function of the discrete random variable X. By the normalization property of discrete probability distributions, the probability generating function $P(z)$ is well defined for all values of z in the closed unit disk. From (2.20), we have $P(1) = 1$ and

$$\left[\frac{d}{dz} P_X(z)\right]\bigg|_{z=1} = \sum_{k=0}^{\infty} k p_k = E(X), \tag{2.21}$$

and so by differentiating $P_X(z)$, we can obtain the expectation of X. Similarly, higher derivatives can help in computing higher moments. If two random variables X and Y are independent, then the probability-generating function of their sum is the product of their probability-generating functions:

$$P_{X+Y}(z) = P_X(z) P_Y(z). \tag{2.22}$$

For a continuous random variable X with a probability density function $f(x)$ as in (2.7), we define the associated characteristic function $F(j\omega)$ by

$$F_X(\omega) = \int_{-\infty}^{+\infty} f(x) \exp(-j\omega x) dx, \tag{2.23}$$

where j is the imaginary unit $\sqrt{-1}$ and ω is a real number. The characteristic function of a random variable X is the Fourier transform of its probability density function and has properties analogous to those demonstrated above for the probability-generating function: $F_X(j0) = 1$,

$$\left[\frac{d}{d\omega} F_X(\omega)\right]\bigg|_{\omega=0} = \int_{-\infty}^{+\infty} jx f(x) = j E(X),$$

and

$$F_{X+Y}(\omega) = F_X(\omega) F_Y(\omega)$$

for independent random variables X and Y.

The above concepts and their properties can be used for solving numerous problems concerning manipulations of random variables.

Example. Consider two discrete, independent random variables X, defined by a geometric distribution with parameter $p = 0.5$, and Y, defined by a geometric distribution with parameter $p = 0.2$. The problem to be solved is to compute the probability distribution of $X + Y$. By using (2.27) and (2.22), we have

$$P_X(z) = \frac{0.5}{1 - 0.5z}, \quad P_Y(z) = \frac{0.2}{1 - 0.8z}$$

and

$$P_{X+Y}(z) = \frac{0.1}{(1 - 0.5z)(1 - 0.8z)}.$$

Expanding $P_{X+Y}(z)$ into partial fractions,

$$\frac{0.1}{(1 - 0.5z)(1 - 0.8z)} = \frac{A}{1 - 0.5z} + \frac{B}{1 - 0.8z},$$

where $A = -1/6$ and $B = 4/15$, we obtain

$$P[(X + Y) = k] = \frac{4}{15} 0.8^k - \frac{1}{6} 0.2^k, \quad k = 0, 1, 2, \ldots.$$

2.3 A Collection of Discrete and Continuous Distributions

In this section we present a collection of discrete and continuous distributions, which will often be referred to both in this and in later parts of this book.

2.3.1 Bernoulli Trials and the Binomial Distribution

Bernoulli trials are among the most important sampling schemes in probability theory. A Bernoulli trial is an experiment which has two possible random outcomes, called success and failure. The binomial distribution describes the probabilities p_k of obtaining k successes in K independent Bernoulli trials with no regard to order,

$$p_k = \binom{K}{k} p^k (1 - p)^{K-k}, \tag{2.24}$$

where p denotes the probability of success in one trial. In the above expression, $\binom{K}{k}$ stands for the binomial symbol

$$\binom{K}{k} = \frac{K!}{k!(K - k)!}.$$

The binomial random variable X can be represented by

$$X = \sum_{k=1}^{K} X_k, \tag{2.25}$$

where the X_k are Bernoulli random variables, $P[X_k = 1] = p$, and $P[X_k = 0] = q = 1 - p$. The numbers of successes in repeated experiments, such as coin tossing and dice rolling are well explained by discrete random variables with a binomial distribution. Binomial distribution is also an underlying element or starting point for other distributions and probabilistic models, some of them discussed below.

The moments of a random variable X distributed binomially are

$$E(X) = Kp, \ \text{Var}(X) = Kp(1-p),$$

and its probability-generating function is

$$P(z) = (q + pz)^K,$$

where $q = 1 - p$.

2.3.2 The Geometric Distribution

A discrete random variable X has a geometric distribution if it assumes values $0, 1, \ldots, k, \ldots$, with probabilities

$$p_k = (1-p)^k p. \tag{2.26}$$

The geometric distribution corresponds to a situation where Bernoulli trials are repeated until the first success. The event $[X = k]$, whose probability p_k is given in (2.26), can be identified with k failures followed by a success.

The moments of a random variable X distributed geometrically are

$$E(X) = \frac{1-p}{p}, \ \text{Var}(X) = \frac{1-p}{p^2},$$

and its probability-generating function is

$$P(z) = \frac{p}{1 - (1-p)z}. \tag{2.27}$$

2.3.3 The Negative Binomial Distribution

The negative-binomial discrete random variable X is also related to Bernoulli trials. Here X is equal to the number of trials needed for r successes to occur. The probability of the event $[X = k]$ (the rth success in the kth trial) is equal to

$$p_k = \binom{k-1}{r-1} p^r (1-p)^{k-r}, \tag{2.28}$$

which follows from the fact that the event (the rth success in the kth trial) is an intersection (product) of two independent events, (a) $r-1$ successes in $k-1$ trials, for which the expression describing its probability, $\binom{k-1}{r-1} p^{r-1}(1-p)^{k-r}$, follows from the binomial-distribution formula, and (b) a success in the last (kth) trial.

The moments of random variable X described by negative binomial distribution are
$$E(X) = r\frac{1-p}{p}, \ \text{Var}(X) = r\frac{1-p}{p^2},$$
and its probability-generating function is
$$P(z) = \left(\frac{pz}{1-(1-p)z}\right)^r.$$

2.3.4 The Poisson Distribution

Poisson random variables are often used for modeling experiments involving observing the numbers of occurrences of events that happen at random moments, over a fixed interval of time, for example the number of clicks from a Geiger counter, the number of emergency calls, or the number of car accidents. Poisson random variables can be rigorously derived using a stochastic mechanism called the Poisson point process [151], in which discrete epochs (points) occur on a finite interval I such that (i) in two disjoint subintervals of I the numbers of points are independent, and (ii) the probability that an event occurs in a short interval $(t, t + \Delta t)$ is equal to $\lambda \Delta t + o(\Delta t)$, where λ is an intensity parameter and $o(\Delta t)$ is small compared with $o(\Delta t)$, i.e., $\lim_{\Delta t \to 0} o(\Delta t)/\Delta t = 0$.

A Poisson random variable X assumes integer values $0, 1, \ldots, k, \ldots$ with probabilities
$$p_k = P[X = k] = \exp(-\lambda)\frac{\lambda^k}{k!}, \tag{2.29}$$
where λ is a parameter. There is a relation between Poisson and binomial random variables. If (i) we have an infinite sequence of binomial random variables
$$X_1, X_2, \ldots, X_n, \ldots, \tag{2.30}$$
with parameters p_n and K_n for probability of success and number of trials in the distribution of X_n, and (ii) the sequence of parameters has the property $\lim_{n\to\infty} p_n = 0$, $\lim_{n\to\infty} K_n = \infty$, and $\lim_{n\to\infty} p_n K_n = \lambda$, then the "limit" of (2.30) is a Poisson random variable X distributed according to (2.29).

The moments of a random variable X described by the Poisson distribution are
$$E(X) = \lambda, \ \text{Var}(X) = \lambda,$$
and its probability-generating function is
$$P(z) = \exp[\lambda(z-1)].$$

2.3.5 The Multinomial Distribution

This concerns a generalization of Bernoulli trials in which each independent experiment has M possible outcomes, with corresponding probabilities

$$p_m = P(\text{result of experiment} = m), \ m = 1, 2, \ldots, M, \quad (2.31)$$

where

$$\sum_{m=1}^{M} p_m = 1. \quad (2.32)$$

A vector random variable $X = [X_1, \ldots, X_M]$ is called multinomial with parameters $p_1, \ldots, p_M$ and number of repeats K if the random variables X_m count the number of outcomes m in K trials. The multinomial distribution has the form

$$P(k_1, k_2, \ldots, k_M) = \frac{K!}{k_1! k_2! \ldots k_M!} p_1^{k_1} p_2^{k_2} \ldots p_M^{k_M}, \quad (2.33)$$

where $k_1, k_2, \ldots, k_M$ are counts of outcomes and $\sum_{m=1}^{M} k_m = K$.

2.3.6 The Hypergeometric Distribution

The Hypergeometric distribution describes the number of successes in random sampling, without replacement, from a finite population with two types of individuals, 1 and 0. For a hypergeometrically distributed random variable X, with parameters N, M, n, the event $[X = k]$ is interpreted as k characters of type 1 in a sample of size n, drawn randomly from a finite population of N individuals, of which M are of type 1 and $N-M$ of type 0. The hypergeometric distribution has the form

$$p_k = P[X = k] = \frac{\binom{M}{k}\binom{N-M}{n-k}}{\binom{N}{n}}. \quad (2.34)$$

Equation (2.34) follows from the fact that among all possible samples (their number is given by the number of combinations $\binom{N}{n}$), those with k successes are obtained by combining any k individuals of type 1 drawn from a set of M individuals with $n-k$ individuals of type 0 drawn from a set of $N-M$ individuals. The normalization condition for the hypergeometric distribution becomes

$$\sum_{k=0}^{\min(n,M)} \frac{\binom{M}{k}\binom{N-M}{n-k}}{\binom{N}{n}} = 1.$$

The moments of a random variable X described by the hypergeometric distribution are

$$E(X) = n\frac{M}{N}, \ \text{Var}(X) = n\frac{M(N-M)(N-n)}{N^2(N-1)}$$

Its probability-generating function can be computed by using a hypergeometric series. We shall not provide the exact formula.

2.3.7 The Normal (Gaussian) Distribution

The normal distribution is the most important of all continuous distributions. Its role stems from a fact expressed mathematically as central limit theorem, which states that the sum of many independent random components with finite variances will have, approximately, a normal distribution. Therefore variables describing measurement errors, as well as many parameters describing individuals in populations, such as lengths, weights, and areas, are modeled by use of the normal distribution. From (2.25), we can see that the binomial distribution, when K is large, converges to the normal distribution. Sums of independent normal variables are again normal.

The normal distribution is supported on the whole space of reals, R, and the probability density function of a random variable X distributed normally is

$$f(x) = \frac{1}{\sigma\sqrt{2\pi}} \exp\left[-\frac{1}{2}\left(\frac{x-\mu}{\sigma}\right)^2\right], \tag{2.35}$$

where μ and σ are parameters equal to expectation and standard deviation, respectively.

The moments of a normal random variable X are

$$E(X) = \mu, \; \text{Var}(X) = \sigma^2,$$

and its characteristic function is

$$F(\omega) = \exp(j\mu\omega - \frac{\omega^2\sigma^2}{2}).$$

2.3.8 The Exponential Distribution

The exponential distribution is a continuous counterpart of the geometric distribution described earlier. It is often used for modeling random times, for example waiting times, times between failures, and survival times. The time between the occurrence of two successive events in the Poisson point process mentioned earlier in this chapter is also distributed exponentially. Exponential distribution is supported on the interval $[0, \infty)$. A random variable $T \geq 0$ distributed exponentially has a probability density function

$$f(t) = a \exp(-at). \tag{2.36}$$

The parameter $a > 0$ is called the rate parameter.

The moments of an exponential random variable T are

$$E(T) = \frac{1}{a}, \; \text{Var}(T) = \frac{1}{a^2},$$

and its characteristic function is

$$F(\omega) = \frac{a}{a - j\omega}.$$

2.3.9 The Gamma Distribution

The gamma distribution is a continuous counterpart of the negative binomial distribution described earlier. It is supported on the interval $[0, \infty)$. It can be interpreted as a random time with a composite structure, for example a sum of K identical, independent exponential random variables is a random variable with a gamma distribution. The probability density function of a random variable X which has the gamma distribution is

$$f(x) = x^{k-1} \frac{\exp(-x/\theta)}{\theta^k \Gamma(k)}. \tag{2.37}$$

In the above, $\Gamma(k)$ denotes Euler's gamma function

$$\Gamma(z) = \int_0^\infty t^{z-1} \exp(-t) dt, \tag{2.38}$$

and $k > 0$ and $\theta > 0$ are the parameters of the gamma distribution, respectively called the shape and the scale parameter. When $k = 1$, (2.37) represents an exponential probability density function. If we assume (2.37) $k = n/2$ and $\theta = 2$ in (2.37) we obtain the probability density function of a χ-square distribution with n degrees of freedom.

The moments of a random variable X described by the gamma distribution are

$$E(X) = k\theta, \; \text{Var}(X) = k\theta^2,$$

and its characteristic function is

$$F(\omega) = \frac{1}{(1 - j\theta\omega)^k}.$$

2.3.10 The Beta Distribution

The Beta Distribution is supported on the interval $[0, 1]$. The corresponding probability density function is

$$f(x) = \frac{\Gamma(a+b)}{\Gamma(a)\Gamma(b)} x^{a-1} (1-x)^{b-1} \tag{2.39}$$

where $x \in (0, 1)$ and $a > 0$, $b > 0$ are parameters and $\Gamma(.)$ is again the gamma function defined in (2.38). By changing a and b we obtain different shapes of the graph of the probability density function (2.39). When $a > 1$, $b > 1$ the graph is bell shaped; when $a < 1$, $b < 1$ the graph is U-shaped. When $a = 1$ and $b = 1$, the probability density function in expression (2.39) describes uniform distribution over the interval $(0, 1)$.

The moments of a random variable X described by the beta distribution are

$$E(X) = \frac{a}{a+b}, \; \text{Var}(X) = \frac{ab}{(a+b)^2(a+b+1)}.$$

Its characteristic function is given by a sum of a hypergeometric series; we shall not give its exact form here.

2.4 Likelihood maximization

It is a frequent situation that we try to determine from what distribution the data at our disposal were sampled. This is known as the estimation problem. Estimation theory is a major part of statistics, with a wide range of methods available. In practical applications, the method of maximum likelihood (ML) is the most frequently used. We shall outline the principle of ML and provide some examples of estimation. We shall use the parametric form of ML, in which it is assumed that the observations were sampled from a distribution belonging to a known parametric family. In other words, the observations $x_1, x_2 \ldots x_N$ are independent, identically distributed (i.i.d.) realizations of a random variable X with a distribution $f(x, p)$, where $f(.,.)$ may denote either a discrete distribution, a distribution density, or a cumulative distribution. The function $f(x_n, p)$ treated as a function of the parameter p with a fixed x_n, is called the likelihood of observation x_n. The functional form of $f(.,.)$ is known but not the value of parameter(s) p.

To estimate parameter(s) p of probability distributions on the basis of observed realizations of a random variable or vector X we use the ML principle, which states that since events with high probability happen more often than those with low probability then it is natural to assume that *what happened was the most likely*. Therefore the best estimate of p is the value $\hat{p}$ that maximizes the likelihood of the sample, i.e.,

$$L(p, x) = L(p) = f(x_1, x_2, \ldots, x_N, p) = \prod_{n=1}^{N} f(x_n, p),$$

where $x = x_1, x_2, \ldots, x_N$ and the product form follows from the independence of the observations. Mathematically,

$$\hat{p} = \arg\max \prod_{n=1}^{N} f(x_n, p).$$

It is common to use a log-likelihood function $l(x_1, x_2, \ldots, x_N)$,

$$l(x_1, x_2, \ldots, x_N, p) = \ln[L(x_1, x_2, \ldots, x_N, p)] = \sum_{n=1}^{N} \ln[f(x_n, p)],$$

which changes the product to a sum and, owing to the monotonicity of the logarithm function, leads to the same $\hat{p}$ as $L(x_1, x_2, \ldots, x_N, p)$ does. The idea applies both to continuous and to discrete distributions.

In what follows, we provide several examples. In all the examples, maximization of likelihoods is readily accomplished by differentiating with respect to the parameters and equating the derivatives obtained to zero. As a remark, in general, it is necessary to verify both of the conditions for the maximum $dl/dp = 0$ and $d^2l/dl^2 < 0$ (see Chap. 5).

2.4.1 Binomial Distribution

For a random variable X distributed binomially as in (2.24), assuming that the observed realization included k successes in K trials, by maximizing the likelihood (2.24) with respect to p, we obtain the maximum likelihood estimate

$$\hat{p} = \frac{k}{K}.$$

In a more general situation an experiment with K Bernoulli trials is repeated N times and the numbers of successes $k_1, k_2, \ldots, k_N$ are recorded, which leads to the log-likelihood function

$$l(k_1, k_2, \ldots, k_N, p) = \sum_{n=1}^{N} \left[k_n \ln p + (K - k_n) \ln(1 - p) + \ln \binom{K}{k_n} \right]. \quad (2.40)$$

By maximizing expression (2.40) with respect to p, one obtains the estimate

$$\hat{p} = \frac{\sum_{n=1}^{N} k_n}{NK}.$$

2.4.2 Multinomial distribution

The log-likelihood corresponding to the multinomial distribution (2.33) is equal to

$$l(k_1, k_2, \ldots, k_M, p_1, p_2, \ldots, p_M) = \ln \frac{K!}{k_1! k_2! \ldots k_M!} + \sum_{m=1}^{M} k_m \ln p_m. \quad (2.41)$$

When maximizing (2.41) with respect to the parameters we must take the constraint (2.32) into account, which leads to the construction of a Lagrange function (see Sect. 5.1.2)

$$L = (k_1, k_2, \ldots, k_M, p_1, p_2, \ldots, p_M, \lambda) \quad (2.42)$$

$$= \ln \frac{K!}{k_1! k_2! \ldots k_M!} + \sum_{m=1}^{M} k_m \ln p_m - \lambda \left(\sum_{m=1}^{M} p_m - 1 \right)$$

where λ stands for a Lagrange multiplier, and the resulting ML estimates are

$$\hat{p}_m = \frac{k_m}{K}. \quad (2.43)$$

2.4.3 Poisson Distribution

Assume that X is a Poisson random variable with the distribution (2.29). For N independent realizations $k_1, k_2, \ldots, k_N$ of X we have the log-likelihood function

$$l(k_1, k_2..., k_N, \lambda) = \sum_{i=1}^{N}[-\lambda + k_i \ln(\lambda) - \ln(k_i!)]. \qquad (2.44)$$

Log-likelihood function (2.44) has its maximum at

$$\hat{\lambda} = \frac{\sum_{i=1}^{N} k_i}{N}$$

which is the maximum likelihood estimate of the Poisson parameter λ.

2.4.4 Geometric Distribution

Here, the random variable X is distributed as in (2.26). The parameter to be estimated is $p \in [0, 1]$. Assume that N independent realizations, $k_1, k_2, \ldots, k_N$, of X have been observed. The log-likelihood function becomes

$$l(k_1, k_2, \ldots, k_N, p) = \sum_{n=1}^{N}[(k_n - 1)\ln(1-p) + \ln p]$$

and the maximum likelihood estimate of probability $\hat{p}$ is

$$\hat{p} = \begin{cases} \dfrac{N}{\sum_{n=1}^{N} k_n} & \text{if } \sum_{n=1}^{N} k_n \geq 1 \\ 1 & \text{if } \sum_{n=1}^{N} k_n = 0. \end{cases}$$

2.4.5 Normal Distribution

The probability density function of continuous random variable X distributed normally is (2.35). The log-likelihood function resulting from N independent observations $x_1, x_2, \ldots, x_N$ of X is

$$l(x_1, x_2, \ldots, x_N, \mu, \sigma) = \sum_{n=1}^{N}\left[-\frac{1}{2}\ln(2\pi) - \ln(\sigma) - \frac{(x_n - \mu)^2}{2\sigma^2}\right], \qquad (2.45)$$

and the maximum value of $l(x_1, x_2, \ldots, x_N, \mu, \sigma)$ in (2.45) is attained at the values of $\hat{\mu}$ and $\hat{\sigma}$ given by the sample mean and variance,

$$\hat{\mu} = \frac{1}{N}\sum_{n=1}^{N} x_i \qquad (2.46)$$

and

$$\hat{\sigma}^2 = \frac{1}{N}\sum_{n=1}^{N}(x_i - \hat{\mu})^2.$$

2.4.6 Exponential Distribution

Assume that N independent realizations $t_1, t_2, \ldots, t_N$ of a random variable T distributed exponentially as in (2.36), have been recorded. The log-likelihood function for the experiment is

$$l(t_1, t_2, \ldots, t_N, a) = \sum_{n=1}^{N}[-at_n + \ln a],$$

and by maximizing it with respect to a we obtain

$$\hat{a} = \frac{\sum_{n=1}^{N} t_n}{N}.$$

2.5 Other Methods of Estimating Parameters: a Comparison

The ML method is often considered a gold standard in parameter estimation. Yet there may exist arguments for applying methods other than ML. One instance arises when employing the principle of maximum likelihood leads to problems of high complexity, high computational cost, or multiple local maxima. Another such argument is related to the fact that ML estimates are generally only asymptotically unbiased. In this section, we present other methods of parameter estimation and a comparison of these approaches.

A method often applied is the *method of moments* which is based on the law of large numbers. Consider a random variable X, with a probability density function $f_X(x, p)$ depending on a parameter p. The expectation of X, $E(X, p) = \int x f_X(x, p) dx$, depends on the value of p and can be estimated by a sample mean. The law of large numbers guarantees, under some regularity conditions, that for large sample sizes, the sample mean will be close to the expectation of X. Consequently, the moment estimator $\hat{p}$ can be obtained by solving the following equation with respect to p:

$$\frac{1}{N} \sum_{n=1}^{N} x_n = \int_{-\infty}^{+\infty} x f_X(x, p) dx.$$

It turns out that in all the examples above, the moment estimators of the parameters coincide with ML estimators. However, this is not always the case. The analysis of examples of parameter estimation problems, shown below will help in understanding the differences between the two types of estimators.

2.5.1 Example 1. Uniform Distribution

Let us consider the uniform distribution, with the probability density function shown in Fig. 2.1. The left boundary of the interval supporting the distribution

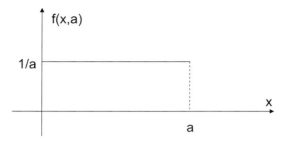

Fig. 2.1. Probability density function of a random variable X, uniformly distributed on the interval $[0, a]$

is fixed at zero and the right boundary is the parameter, a, to be estimated. Let us first compute the ML estimator of a. In Fig. 2.2 the values of $x_1, x_2, \ldots, x_N$, $N = 6$, are marked by short lines perpendicular to the x axis, and three hypothetical values a_1, a_2, and a_3 of the parameter a are assumed, where $a_1 < \max_{1 \leq n \leq N} x_n$, $a_2 = \max_{1 \leq n \leq N} x_n$ and $a_3 > \max_{1 \leq n \leq N} x_n$. Observe that corresponding log-likelihoods are equal to

$$l(x_1, x_2, \ldots, x_N, a_1) = -\infty$$

since two of the observations are impossible given $\hat{a} = a_1$, and

$$l(x_1, x_2, \ldots, x_N, a_i) = -N \ln a_i, \ i = 2, 3.$$

The above leads to the conclusion that the ML estimate of a equals to a_2, or in other words

$$\hat{a}_{ML} = \max_{1 \leq n \leq N} x_n.$$

Using the expression for the expectation of the uniformly distributed random variable X shown in Fig. 2.2, $E(X) = a/2$, we obtain the moment estimator

$$\hat{a}_{mom} = \frac{2}{N} \sum_{n=1}^{N} x_n. \tag{2.47}$$

In order to compare $\hat{a}_{ML}$ and $\hat{a}_{mom}$, let us compute some expectations and variances. We have the expressions

$$E(\hat{a}_{ML}) = E\left(\max_{1 \leq n \leq N} X_n\right) = a \frac{N}{N+1}$$

and

$$E(\hat{a}_{mom}) = E\left(\frac{2}{N} \sum_{n=1}^{N} X_n\right) = a$$

for the expected values and

2.5 Other Methods of Estimating Parameters: a Comparison

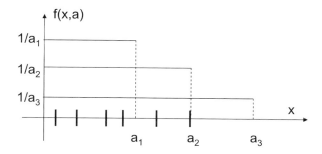

Fig. 2.2. $N = 6$. The recorded values of x_n, $n = 1, 2, \ldots, 6$ are marked by short lines perpendicular to x axis. Three hypothetical values of a, i.e., a_1, a_2, and a_3 are considered, where $a_1 < \max_{1 \leq n \leq N} x_n$, $a_2 = \max_{1 \leq n \leq N} x_n$, and $a_3 > \max_{1 \leq n \leq N} x_n$

$$\mathrm{Var}(\hat{a}_{ML}) = \mathrm{Var}\left(\max_{1 \leq n \leq N} X_n\right) = \frac{Na^2}{(N+1)^2(N+2)}$$

and

$$\mathrm{Var}(\hat{a}_{mom}) = \mathrm{Var}\left(\frac{2}{N}\sum_{n=1}^{N} X_n\right) = \frac{a^2}{3N^2}$$

for the variances. The results of computations show that the variance of the ML estimator is smaller than variance of moment estimator. Their ratio is, approximately, proportional to the sample size N. However, contrary to the moment-based estimator, the ML estimator is biased, since its expectation is not equal to a.

Observe that we can base the estimator of a on moments higher than 1. For kth moment of a random variable X distributed uniformly, as in Fig. 2.1, we have

$$E(X^k) = \frac{a^{k+1}}{k+1},$$

which leads to the following kth moment estimator of a:

$$\hat{a}_{mom,k} = \left[\frac{k+1}{N}\sum_{n=1}^{N} x_n^k\right]^{\frac{1}{k+1}}. \tag{2.48}$$

One can verify that the above statistics converge to the ML estimator of a (Exercise 3) when k tends to infinity.

2.5.2 Example 2. Cauchy Distribution

We consider the probability density function of a random variable X with the Cauchy distribution

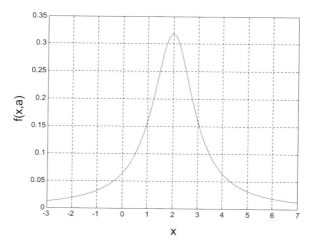

Fig. 2.3. Plot of the probability density function of Cauchy distribution with location parameter $a = 2$

$$f(x,a) = \frac{1}{\pi[1 + (x-a)^2]}, \qquad (2.49)$$

with an unknown location parameter a to be estimated. The Cauchy distribution is different from those analyzed earlier in that it has no finite moments, which follows from the fact that for all $k \geq 1$, $E(|X^k|)$ is expressed by an improper integral, which is not convergent:

$$\int_{-\infty}^{+\infty} \frac{|x|}{\pi[1 + (x-a)^2]} dx = \infty. \qquad (2.50)$$

Therefore moment estimators do not make sense.

A plot of the Cauchy pdf (2.49) is given in Fig. 2.3. Looking at Fig. 2.3, it may seem that if N independent realizations $x_1, x_2, \ldots, x_N$ of X were observed, then the sample mean $(1/N)\sum_{n=1}^{N} x_i$ would be a reasonable estimate of a. This is, however, not true, since owing to (2.50) variance of the statistics $(1/N)\sum_{n=1}^{N} x_i$ is infinite for all N.

Differentiating with respect to a and equating to zero the log-likelihood function

$$l(x_1, x_2, \ldots, x_N, a) = \sum_{n=1}^{N} -\ln \pi - \ln[1 + (x_n - a)^2]$$

leads to the following equation for the ML estimate $\hat{a}$:

$$\sum_{n=1}^{N} \frac{x_n - \hat{a}}{1 + (x_n - \hat{a})^2} = 0. \qquad (2.51)$$

Except for $N = 1$ and $N = 2$, this equation must be solved numerically to obtain the ML estimate $\hat{a}$. However, it can be proven [112] that, starting from

$N = 3$, the estimator resulting from solving (2.51) is unbiased and has a finite variance. Another estimate of a, simpler than the ML estimate, is the sample median [112], which also is unbiased and of finite variance.

2.5.3 Minimum Variance Parameter Estimation

When comparing and scoring estimators above, we were using their means and variances. A question arises, "Do estimators with a lower variance than ML estimators exist?" In many instances, such as those in Sects. 2.3.1–2.4.6, the ML estimator is also the minimum-variance estimator. Maximum likelihood estimators achieve minimum variance in the limit as the sample size tends to infinity. However, for finite sample sizes, there may exist statistics with lower variances than the ML estimate. One example is that for the Cauchy distribution analyzed in the previous subsection. By numerical computations [112], a minimum-variance estimator of a can be obtained, whose variance is lower than the variance of the ML estimator for finite N.

There are well-known results concerning the variances of parameter estimators, which we present below.

The Fisher Information

We first introduce the Fisher information $I(p)$, where p is a parameter of the probability distribution $f(x, p)$

$$I(p) = E\left([\frac{\partial}{\partial p} \log f(x,p)]^2\right) = -E\left[\frac{\partial^2}{\partial p^2} \log f(x,p)\right]. \quad (2.52)$$

From its definition, it follows that the Fisher information is additive with respect to repeated independent measurements, i.e.,

$$I_{X_1, X_2}(p) = I_{X_1}(p) + I_{X_2}(p) = 2I_{X_1}(p), \quad (2.53)$$

where indices have been were added to denote different measurements, and the second equality in (2.53) is valid for two identically distributed independent measurements.

Cramer–Rao Theorem

The Cramer–Rao theorem (Cramer–Rao bound) states that every unbiased estimator $\hat{a}$ of a parameter a must satisfy

$$\text{Var}(\hat{p}) \geq \frac{1}{I(p)}. \quad (2.54)$$

Using (2.54), we can compute a lower bound for the variance of any unbiased estimator.

For the location parameter of the normal distribution, μ, we obtain from (2.52) the Fisher information given by

$$I_{X_1,X_2,...,X_N}(\mu) = \frac{N}{\sigma^2}.$$

For the estimator $\hat{\mu}$ of this parameter, given by (2.46), we can compute $\text{Var}(\hat{\mu}) = \sigma^2/N$, which, on the basis on Cramer–Rao bound, proves that $\hat{\mu}$ has the minimum variance and no better estimate of μ can be obtained.

Let us analyze again the Cauchy distribution and its location parameter a. On the basis of (2.52) we compute the Fisher information corresponding to measurements $X_1, X_2, \ldots, X_N$,

$$I_{X_1,X_2,...,X_N}(a) = \frac{N}{2}.$$

This leads to a lower bound on the variance of any unbiased estimator $\hat{a}$,

$$\text{Var}(\hat{a}) \geq \frac{2}{N}.$$

By numerical computations one can demonstrate that estimates of the parameter a presented in Sect. 2.5.2 do not attain this lower bound [112].

Cramer–Rao Bound as Variance Estimator

Assuming that in many practical cases the bound (2.54) is tight, it is often used as an approximate value for estimators of variances, i.e.,

$$\text{Var}(\hat{p}) \simeq \frac{1}{I(p)}. \tag{2.55}$$

Parameter estimates are often obtained by numerical maximization of the likelihood function. In cases where no analytical formulae are available, the value of the Fisher information, $(\partial/\partial p) \log f(x,p)$, can be obtained numerically by resampling. By resampling, we mean averaging $[(\partial/\partial p) \log f(x,p)]^2$ by the use of stochastic simulations with a variant of the Markov chain Monte Carlo method, discussed later in this chapter.

Sufficient Statistics

A sufficient statistic has the property that it provides the same information as the whole sample. Sufficiency of statistics can be checked by the Fisher factorization criterion, which states that $t(x_1, x_2, ..., x_N)$ is a sufficient statistic for observations $x_1, x_2, ..., x_N$ if

$$f(x_1, x_2, ..., x_N, p) = g(t, p) h(x_1, x_2, ..., x_N) \tag{2.56}$$

for some functions g and h. By substituting (2.56) in (2.52), we can see that indeed

$$I_{X_1,X_2,...,X_N}(p) = I_{t(X_1,X_2,...,X_N)}(p).$$

Rao–Blackwell Theorem

This theorem shows how one can improve estimators of parameters by applying sufficient statistics. Denote by $\hat{p}$ any estimator of a parameter p, given observations $X_1, X_2, \ldots, X_N$, and define a new estimator $\hat{p}^{new}$ as the conditional expectation

$$\hat{p}^{new} = E[\hat{p}|t(X_1, X_2, \ldots, X_N)]$$

where $t(X_1, X_2, \ldots, X_N)$ is a sufficient statistic for p. The Rao–Blackwell theorem states that

$$E[(\hat{p}^{new} - p)^2] \leq E[(\hat{p} - p)^2].$$

Consider the uniform distribution examined earlier in Sect. 2.5.1. Knowing that $t(x_1, x_2, \ldots, x_N) = \max(x_1, x_2, \ldots, x_N)$ is a sufficient statistic for the parameter a, one can improve moment estimator $\hat{a}_{mom}$ by defining the following Rao–Blackwell estimator:

$$\hat{a}^{RB} = E[\frac{2}{N} \sum_{n=1}^{N} X_n | \max(X_1, X_2, \ldots, X_N)]. \tag{2.57}$$

2.6 The Expectation Maximization Method

For the majority of examples considered in the previous sections, computing the maximum likelihood estimates of the parameters was accomplished by means of analytical expressions. Also, in the examples analyzed, it was straightforward to prove that there existed unique maxima of the likelihood functions over the parameter spaces. However, in numerous problems of data analysis, employing the principle of maximum likelihood may lead to numerical computational problems of considerable complexity. Moreover, multiple extrema of the likelihood function often exist. Therefore, in many situations, ML estimates are computed by using numerical, static, dynamic, or combinatorial optimization. Some of these methods will be illustrated in later chapters.

A special and remarkable approach to the numerical recursive computation of ML estimates is the expectation maximization (EM) method [63, 190]. This approach is intended for the situation where the difficulty in obtaining ML estimates arises from the existence of missing (also called hidden or latent) variables. If the missing variables had been observed, the ML estimation would have been fairly simple. In such circumstances, the EM method proceeds recursively. Each of recursions consists of an E-step involving computing the conditional expectation with respect to the unknown data, given the available data, and an M-step, involving maximization with respect to parameters. The construction of the algorithm guarantees that each iteration increases the value of the likelihood function. Owing to its simplicity and robustness, the EM method is widely applied, and although it converges relatively slowly, many publications, which describe new, elegant, and useful possibilities of employing the EM idea are constantly appearing in the scientific literature.

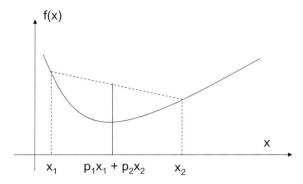

Fig. 2.4. Illustration of the convexity of $f(x)$. For convex $f(x)$, if $p_1 \geq 0$, $p_2 \geq 0$, and $p_1 + p_2 = 1$, then $f(p_1 x_1 + p_2 x_2) \leq p_1 f(x_1) + p_2 f(x_2)$

2.6.1 The Derivations of the Algorithm

The idea of EM recursions relies on an inequality for the conditional expectation of the log-likelihood of missing variables. Below, we show two methods of establishing this inequality used in the literature, by using Jensen's inequality and by using the Kullback–Leibler distance measure. We first briefly state the necessary supplementary results.

Jensen's Inequality

The definition of convexity for a function $g(x)$, illustrated in Fig. 2.4, is

$$g(p_1 x_1 + p_2 x_2) \leq p_1 g(x_1) + p_2 g(x_2), \ p_1 \geq 0, \ p_2 \geq 0, \ p_1 + p_2 = 1. \quad (2.58)$$

Using induction, we can prove that this implies an analogous inequality for any $n \geq 2$,

$$g(p_1 x_1 + p_2 x_2 + \cdots + p_n x_n) \leq p_1 g(x_1) + p_2 g(x_2) + \cdots + p_n g(x_n), \quad (2.59)$$

$p_i \geq 0$, $i = 1, 2, \ldots, n$, $p_1 + p_2 + \cdots + p_n = 1$. We can also move from a one-dimensional space of arguments $x \in R$ to a more general m-dimensional space, $x \in R^m$, and the inequality remains valid. Every convex function $g(x)$, $R^m \to R$ satisfies (2.59), and (2.59) is called the finite, discrete Jensen's inequality. We can let n pass to infinity and (2.59) remains valid.

It is also possible to replace the discrete probability distribution containing atoms $p_1, p_2, \ldots, p_n$ appearing in (2.59) with a continuous distribution $f(x)$, where $\int_{-\infty}^{+\infty} f(x) dx = 1$, and the inequality analogous to (2.59) is

$$g\left[\int_{-\infty}^{+\infty} x f(x)\right] \leq \int_{-\infty}^{+\infty} g(x) f(x) dx, \quad (2.60)$$

which can also be expressed, with the use of the expectation operator, as

$$g[E(X)] \leq E[g(X)]. \qquad (2.61)$$

In the above X is a random variable with a probability density function $f(x)$. The inequality (2.60) or (2.61), valid for every convex function $g(x)$, is called the continuous Jensen's inequality. Jensen's inequality can also be stated in a more general form, namely

$$g\left[\int_{-\infty}^{+\infty} h(x)f(x)\right] \leq \int_{-\infty}^{+\infty} g[h(x)]f(x)dx, \qquad (2.62)$$

or

$$g(E[h(X)]) \leq E(g[h(X)]), \qquad (2.63)$$

where again $g(x)$ is a convex function and $h(x)$ is any measurable function. Observe that (2.63) becomes equivalent to (2.61) when we substitute $Y = h(X)$.

Kullback–Leibler Distance

Consider two finite, discrete random variables X and Y, both assuming values $1, 2, \ldots, n$ with probabilities $p_1, p_2, \ldots, p_n$, $p_1 + p_2 + \cdots + p_n = 1$ for X, and $q_1, q_2, \ldots, q_n$, $q_1 + q_2 + \cdots + q_n = 1$ for Y. The Kullback–Leibler distance $K_{X,Y}$ between the distributions of X and Y is defined as

$$K_{X,Y} = -\sum_{i=1}^{n} q_i \ln \frac{p_i}{q_i}. \qquad (2.64)$$

It can be seen that $K_{X,Y} \geq 0$ and that

$$K_{X,Y} = 0 \Leftrightarrow p_i = q_i, \ i = 1, 2, \ldots, n \qquad (2.65)$$

(see Exercise 6). The Kullback–Leibler distance is also called the entropy of the distribution $p_1, p_2, \ldots, p_n$ relative to the distribution $q_1, q_2, \ldots, q_n$.

For continuous random variables X and Y, with corresponding probability density functions $f_X(z)$ and $f_Y(z)$, their Kullback–Leibler distance is defined as

$$K_{X,Y} = -\int_{-\infty}^{+\infty} f_Y(z) \ln \frac{f_X(z)}{f_Y(z)} dz, \qquad (2.66)$$

and, again,

$$K_{X,Y} \geq 0 \qquad (2.67)$$

and $K_{X,Y} = 0 \Leftrightarrow f_X(z) = f_Y(z)$ (possibly except for a set of measure zero).

EM Recursions

Let us assume that the available observation (or observations) is modeled by a random variable (or random vector) X and that the aim is to estimate the parameter (or parameter vector) p. Also, we assume there exist some missing observations X^m. By merging available and missing observations we obtain

$$X^c = (X^m, X)$$

called the complete observations. We shall show examples of situations where estimating p on the basis of the log-likelihood function with an available observation x,

$$\ln[f(x,p)]$$

leads to computational problems, yet maximization of log-likelihood function on the basis of the complete observations

$$\ln[f(x^c,p)],$$

is fairly straightforward.

Let us express the conditional distribution of missing observations given the available observations and parameters, $f(x^m|x,p)$, with the use of Bayes' formula:

$$f(x^m|x,p) = \frac{f(x^m,x,p)}{f(x,p)} = \frac{f(x^c,p)}{f(x,p)}. \tag{2.68}$$

We can invert the above and obtain

$$f(x,p) = \frac{f(x^c,p)}{f(x^m|x,p)}$$

and then take logarithms of both sides, which results in

$$\ln f(x,p) = \ln f(x^c,p) - \ln f(x^m|x,p). \tag{2.69}$$

We assume a guess for the parameters, denote it by p^{old}, and recall that x is known and fixed. The distribution of the unknown x^m given the available data x is $f(x^m|x,p^{old})$. We average (2.69) over the distribution of unknown data, or, in other words we compute the expectations of both sides of (2.69) with respect to $f(x^m|x,p^{old})$. Since $E[h(X)|X] = h(X)$ for every function $h(X)$, this can be written as follows:

$$\ln f(x,p) = E[\ln f(X^c,p)|x,p^{old}] - E[\ln f(X^m,p)|x,p^{old}]. \tag{2.70}$$

We introduce the notation

$$Q(p,p^{old}) = E[\ln f(X^c,p)|x,p^{old}] = \int f(x^m|x,p^{old}) \ln f(x^c,p) dx^m \tag{2.71}$$

and

$$H(p, p^{old}) = E[\ln f(X^m, p) | x, p^{old}] = \int f(x^m | x, p^{old}) \ln f(x^m | x, p) dx^m, \tag{2.72}$$

and so
$$\ln f(x, p) = Q(p, p^{old}) - H(p, p^{old}). \tag{2.73}$$

From (2.72), it follows that

$$H(p^{old}, p^{old}) - H(p, p^{old}) = -\int f(x^m | x, p^{old}) \ln \frac{f(x^m | x, p)}{f(x^m | x, p^{old})} dx^m.$$

To the right-hand side of (2.73) we can apply either Jensen's inequality (2.62), with the convex function $g(x^m) = -\ln(x^m)$ and the function $h(x^m) = f(x^m | x, p)/f(x^m | x, p^{old})$ or the inequality (2.67) for the Kullback–Leibler distance (2.66). Both will result in the conclusion that

$$H(p^{old}, p^{old}) - H(p, p^{old}) \geq 0. \tag{2.74}$$

If we are able to find a new estimate p^{new}, which has the property that $Q(p^{new}, p^{old}) > Q(p^{old}, p^{old})$, then from (2.73) and (2.74) we conclude that

$$\ln f(x, p^{new}) > \ln f(x, p^{old}),$$

and so we have been able to increase the log-likelihood. Typically p^{new} will be chosen by maximization of $Q(p, p^{old})$ with respect to p.

Summing up the above considerations leads to the following construction of the EM algorithm:

E-step. Compute $Q(p, p^{old})$ as defined in (2.71).
M-step. Compute $p^{new} = \arg\max_p Q(p, p^{old})$.

By repeating the E-and M-steps with successive substitutions $p^{old} = p^{new}$, we increase, iteratively, the value of the log-likelihood $\ln f(x, p^{old})$. In many cases such iterations will lead to a unique global maximum. However, EM recursions can also end up in local maxima, and, moreover, examples can be found where despite the step-by-step increase of $\ln f(x, p)$, successive estimates of p do not reach any local maximum.

2.6.2 Examples of Recursive Estimation of Parameters by Using the EM Algorithm

We now illustrate the use of the EM algorithm and its convergence with the aid of several examples.

Exponential Distribution with Censored Observations

Censoring often appears in survival studies [57], and in measurements when the range of a measurement device is not sufficient to cover the full scale of the variability of the variable observed. Here we consider an exponential random variable T, with a probability density function (2.36). The problem is to estimate the parameter a on the basis of N observations. However, here there is a censoring mechanism with a constant threshold C, which means that if a measurement of T is greater than C we do not know its exact value, but only the information that the threshold C has been exceeded. Assume that the observations $t_1, t_2, ..., t_k$ did not exceed the threshold C and that $t_{k+1}, ..., t_N$ were above C. So the available information is $t_1, t_2, ..., t_k$ and $[t_{k+1}, ..., t_N$ exceeded $C]$. The complete information would be $t^c = t_1, t_2, ..., t_k, t_{k+1}, ..., t_N$. In order to set up the EM recursion, we start from a parameter guess a^{old}. The expression for $Q(a, a^{old})$ with $f(t, a)$ given by (2.36) then reads

$$Q(a, a^{old}) = E\{\ln f(T^c, a) | t_1, t_2, ..., t_k, [t_{k+1}, ..., t_N \geq C], a^{old}\}$$

$$= \sum_{i=1}^{k} \ln[a \exp(-at_i)] + \sum_{i=k+1}^{N} E(\ln[a \exp(-at_i)] \mid t_i \geq C, a^{old})$$

$$= N \ln a - a \sum_{i=1}^{k} t_i - a(N-k) \frac{\int_{C}^{+\infty} t a^{old} \exp(a^{old} t) dt}{\int_{C}^{+\infty} a^{old} \exp(a^{old} t) dt}$$

$$= N \ln a - a \left[\sum_{i=1}^{k} t_i + (N-k)\left(C + \frac{1}{a^{old}}\right) \right].$$

In the transformations above we used

$$E(-at_i \mid t_i \geq C, a^{old}) = -a \frac{\int_{C}^{+\infty} t a^{old} \exp(a^{old} t) dt}{\int_{C}^{+\infty} a^{old} \exp(a^{old} t) dt} = -a\left(C + \frac{1}{a^{old}}\right).$$

From the above, the value a^{new} maximizing $Q(a, a^{old})$ with respect to a is

$$a^{new} = \frac{N}{\sum_{i=1}^{k} t_i + (N-k)(C + 1/a^{old})}. \quad (2.75)$$

We can index the recursions of the EM estimate of the parameter a by numbers $1, 2, ..., m, ...$, i.e., we write $a_m = a^{old}$ and $a_{m+1} = a^{new}$. From (2.75), we can compute the limit

$$\hat{a} = \lim_{m \to \infty} a_m = \frac{k}{\sum_{i=1}^{k} t_i + (N-k)C}. \quad (2.76)$$

By computing analytically the limit $\lim_{m \to \infty} a_m$ we finally obtain the ML estimate of a. One can also derive the same result (the ML estimate for an exponential distribution with censored observations) by writing down the appropriate log-likelihood function for this case (Exercise 7).

Mixture Distributions

Mixtures of distributions are often applied to model or investigate the underlying structure in experimental data [191]. Mixture distributions have the form

$$f^{mix}(x, \alpha_1, \ldots, \alpha_K, p_1, \ldots, p_K) = \sum_{k=1}^{K} \alpha_k f_k(x, p_k), \quad (2.77)$$

where $\alpha_1, \ldots, \alpha_K, p_1, \ldots, p_K$ are the parameters of the mixture distribution. The weights (probabilities) $\alpha_1, \ldots, \alpha_K$ are nonnegative and add up to one, i.e.,

$$\sum_{k=1}^{K} \alpha_k = 1, \quad (2.78)$$

and the $f_k(x, p_k)$ are probability density functions. A random variable X will have the mixed probability distribution given in (2.77) if it is obtained according to the following scheme: (1) generate a random integer number k from range $1, \ldots K$ with probabilities $\alpha_1, \ldots, \alpha_K$, and (2) Generate a number (or vector) x from the probability distribution $f_k(x, p_k)$. Most often the $f_k(x, p_k)$ are distributions of the same type, for example Gaussian or Poisson, with different parameters, but it is also possible that distributions of different types are mixed. We call $f_k(x, p_k), k = 1, 2, \ldots, K$ the component distributions.

Suppose that a random sample of size N is drawn from the mixture distribution (2.77). Computing ML estimates of parameters $\alpha_1, \ldots, \alpha_K, p_1, \ldots, p_K$ typically leads to problems of numerical optimization. However, there is a natural approach using the idea of the EM algorithm. Namely, we assume the complete information $x^c = k_1, k_2, \ldots, k_N, x_1, x_2, \ldots, x_N$; in other words we assume that we know the index k_n of the component distribution $f_{k_n}(x_n, p_{k_n})$ which generated observation x_n. Clearly, with this complete information, the ML estimation problem splits into separate problems, (a) estimation of the parameters $p_1, \ldots, p_M$ of the component distributions and (b) ML estimation of the weights $\alpha_1, \ldots, \alpha_K$. The latter can be solved on the basis of numbers of occurrences of the indices k_n. Owing to this decomposition, the log-likelihood function for the complete data assumes the form

$$\ln[f(x^c, p)] = \sum_{n=1}^{N} \ln \alpha_{k_n} + \sum_{n=1}^{N} \ln f_{k_n}(x_n, p_{k_n}), \quad (2.79)$$

where $x^c = k_1, k_2, \ldots, k_N, x_1, x_2, \ldots, x_N$ and $p = \alpha_1, \ldots, \alpha_K, p_1, \ldots, p_K$.

E-step. We now make a guess of the parameters $p^{old} = \alpha_1^{old}, \ldots, \alpha_K^{old}$, $p_1^{old}, \ldots, p_K^{old}$ and we write down an expression for $Q(p, p^{old})$, where, in accordance with our notation, the available information is $x = x_1, x_2, \ldots, x_N$ and the missing information is $x^m = k_1, k_2, \ldots, k_N$:

44   2 Probability and Statistics

$$\begin{aligned}Q(p,p^{old}) &= E[\ln f(X^c,p)|x,p^{old}] \\ &= E\left(\sum_{n=1}^{N} \ln \alpha_{k_n}|x,p^{old}\right) + E\left[\sum_{n=1}^{N} \ln f_{k_n}(x_n,p_{k_n})|x,p^{old}\right] \\ &= \sum_{n=1}^{N} E(\ln \alpha_{k_n}|x,p^{old}) + \sum_{n=1}^{N} E[\ln f_{k_n}(x_n,p_{k_n})|x,p^{old}] \\ &= \sum_{n=1}^{N}\sum_{k=1}^{K} p(k|x_n,p^{old})\ln \alpha_k + \sum_{n=1}^{N}\sum_{k=1}^{K} p(k|x_n,p^{old})\ln f_k(x_n,p_k).\end{aligned}$$
(2.80)

The distribution, $p(k|x_n,p^{old})$ of the missing data conditional on the available data and the parameter guess is given by Bayes' formula

$$p(k|x_n,p^{old}) = \frac{\alpha_k^{old} f_k(x_n,p^{old})}{\sum_{\kappa=1}^{K} \alpha_\kappa^{old} f_\kappa(x_n,p^{old})}. \tag{2.81}$$

M-step. The expression for $Q(p,p^{old})$ can be readily optimized with respect to the weights, $\alpha_1,\ldots,\alpha_K$. Taking into account the constraint (2.78), by computations similar to those in (2.41)–(2.43), we obtain

$$\alpha_k^{new} = \frac{\sum_{n=1}^{N} p(k|x_n,p^{old})}{N}. \tag{2.82}$$

The above recursion for the weights is valid regardless of the form of the component distributions. In order to derive recursions for EM estimates of the parameters of component distributions $p_1^{new},\ldots,p_K^{new}$, we now focus on two special cases.

Mixed Poisson Distribution

Assume that the kth component distribution in the nth experiment, $f_k(x_n,p_k)$, is a Poisson distribution with an intensity parameter $p_k = \lambda_k$:

$$f_k(x_n,\lambda_k) = \exp(-\lambda_k)\frac{\lambda_k^{x_n}}{x_n!}. \tag{2.83}$$

Now $p(k|x_n,p^{old})$ is given by (2.81) with the pdf $f_k(x_n,p^{old})$ replaced by the Poisson distribution (2.83) and with an initial parameter guess λ_k^{old}, $k = 1,2,\ldots,K$:

$$p(k|x_n,\lambda^{old}) = \frac{\alpha_k^{old}\exp(-\lambda_k^{old})(\lambda_k^{old})^{x_n}}{\sum_{\kappa=1}^{K}[\alpha_\kappa^{old}\exp(-\lambda_\kappa^{old})(\lambda_\kappa^{old})^{x_n}]}. \tag{2.84}$$

In the above, $\lambda^{old} = \lambda_1^{old},\ldots,\lambda_K^{old}$. Substituting (2.83) in (2.80) and maximizing with respect to λ_k yields the update λ_k^{new}:

$$\lambda_k^{new} = \frac{\sum_{n=1}^{N} x_n p(k|x_n,\lambda^{old})}{\sum_{n=1}^{N} p(k|x_n,\lambda^{old})}, \quad k=1,2,\ldots,K. \tag{2.85}$$

Mixed Normal Distribution

Here, all component distributions are normal with parameters μ_k, σ_k, $k = 1, 2, ..., K$. For the nth observation we have

$$f_k(x_n, \mu_k, \sigma_k) = \frac{1}{\sigma\sqrt{2\pi}} \exp\left[-\frac{(x_n - \mu_k)^2}{2\sigma_k^2}\right]. \tag{2.86}$$

With an initial parameter guess $\mu_k^{old}, \sigma_k^{old}$, $k = 1, 2, \ldots, K$, the expression for the missing data conditional on the available data and the parameter guess assumes the form

$$p(k|x_n, p^{old}) = \frac{\alpha_k^{old} \exp[-(x_n - \mu_k^{old})^2/[2(\sigma_k^{old})^2]]}{\sum_{\kappa=1}^{K} \alpha_\kappa^{old} \exp[-(x_n - \mu_\kappa^{old})^2/[2(\sigma_\kappa^{old})^2]]}. \tag{2.87}$$

In the above we have used the notation $p^{old} = \alpha_1^{old}, \ldots, \alpha_K^{old}, \mu_1^{old}, \ldots, \mu_K^{old}, \sigma_1^{old}, \ldots, \sigma_K^{old}$ for a vector composed of all estimated parameters. When (2.86) is substituted in (2.80), maximization with respect to μ_k, σ_k yields the following updates for the mean and for the dispersion parameter

$$\mu_k^{new} = \frac{\sum_{n=1}^{N} x_n p(k|x_n, p^{old})}{\sum_{n=1}^{N} p(k|x_n, p^{old})}, \quad k = 1, 2, ..., K, \tag{2.88}$$

and

$$(\sigma_k^{new})^2 = \frac{\sum_{n=1}^{N} (x_n - \mu_k^{new})^2 p(k|x_n, p^{old})}{\sum_{n=1}^{N} p(k|x_n, p^{old})}, \quad k = 1, 2, ..., K. \tag{2.89}$$

2.7 Statistical Tests

Testing statistical hypotheses is very important in the analysis of statistical data. There are many types of statistical tests suitable for many specific situations. Here we describe some facts necessary for understanding the material presented later. We present the main ideas and, instead of going through a detailed development and classification, we provide some examples.

2.7.1 The Idea

Suppose we wanted to verify whether a coin was symmetric and we tossed it 50 times. As a result we observed 50 heads. Although this result of the experiment is not inconceivable, we would definitely not believe in the symmetry of the coin. In other words, we would reject the hypothesis of a symmetry. But what if we saw 10 heads and 40 tails? Is there a premise for rejecting the hypothesis of symmetry of the coin?

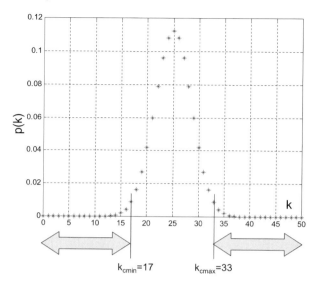

Fig. 2.5. Illustration of the construction of the critical region for the binomial test. In the plot above the probabilities $p(k)$ for the binomial distribution are marked by the asterisks. The critical region, $\mathcal{C} = \{k \in [0, 50] : k \leq k_{c\,\min} \text{ or } k \geq k_{c\,\max}\}$, is represented by the two sided, wide arrows.

Let us introduce the appropriate terminology. The hypothesis of symmetry is called the null hypothesis. Our experiment involved Bernoulli trials, and the null hypothesis can be expressed in terms of a value of the probability of success p for the binomial distribution, namely as $p = 0.5$. Therefore, the statistical test, called the binomial test, is parametric. The alternative to the null hypothesis is $p \neq 0.5$, which means that we accept both lower and higher alternatives. So our statistical test is two-sided.

We perform an experiment assuming that the null hypothesis is true. Then, we either reject the null hypothesis, if the observed result of the experiment is improbable, or, otherwise, we state that there is no premise for rejecting the null hypothesis. We must decide what probability value would correspond to "improbable". A typical value of the probability taken as a threshold is 0.05. This threshold value is called the significance level.

In order to compute whether the result of an experiment is improbable or not we construct the critical region of the test, bounded by a critical value or values. The critical region for the binomial test related to our experiment is denoted by $\mathcal{C}$; it is symmetric with respect to the most probable value of the number of successes $k = 25$, and is bounded by two critical values $k_{c\,\min} = 17$ and $k_{c\,\max} = 33$. That is $\mathcal{C} = \{k \in [0, 50] : k \leq k_{c\,\min} \text{ or } k \geq k_{c\,\max}\}$. The construction of the critical region $\mathcal{C}$ is presented in Fig. 2.5. We can readily compute $P[k \in \mathcal{C}] = 0.0328$.

So if, in the Bernoulli-trials experiment of tossing a coin 50 times, we observe k^{observed} successes (heads) and we have $k^{\text{observed}} \in \mathcal{C}$, we reject the null hypothesis of symmetry because the event that happened, $\mathcal{C}$, was improbable. Both values of k mentioned above, $k = 50$ and $k = 10$ would lead to the rejection of the null hypothesis. We also say that we reject the null hypothesis at the significance level $\alpha = 0.0328$. The value $\alpha = 0.0328$ stems from the construction of the critical set $\mathcal{C}$ described above.

In the above, we have introduced and used the terms "critical value" and "critical region". It is also very convenient to introduce and use the notion of the p-value of a statistical test. The p-value is the lowest (i.e., most favorable) significance level attainable given the observed value of the test statistic. The p-value depends on the result of the experiment (the value of the statistic related to the experiment). For our binomial test example, it is computed by defining the bound of the critical region equal to k^{observed}. So since $k^{\text{observed}} = 10 < 25$ then we set $k_{c\,\text{min}} = k^{\text{observed}} = 10$, and, using the symmetry, $k_{c\,\text{max}} = 50 - k^{\text{observed}} = 40$. This leads to the p-value $p = 2.3861 \times 10^{-5}$.

2.7.2 Parametric Tests

Parametric tests involve hypotheses concerning the values of parameters of probability distributions. One example is the binomial test discussed above, where the null hypothesis concerns the value of the parameter p of the binomial distribution. In this subsection we shall give other examples of parametric tests. When using parametric tests, we must have evidence that variables in the experiment follow the assumed classes of probability distributions.

The next example of an experiment leading to a parametric statistical test is the following scenario. A new medication for reducing hypertension was invented. In order to study whether the new medication was superior to the standard therapy, two groups of hypertension patients were compared, one was treated with the standard medication and the other with the new drug. Both groups were of equal size K. We assume that blood pressure in both groups can be modeled by normally distributed random variables X_1 and X_2. After the experiment, average blood pressures $\bar{X}_1$ and $\bar{X}_2$ were computed for the two groups. The null hypothesis, $\mu_1 = \mu_2$, concerns the mean values of these normal variables. Under the null hypothesis and the additional assumption that the variances of X_1 and X_2 are equal, the statistic

$$t = \frac{\bar{X}_1 - \bar{X}_2}{\sqrt{\frac{1}{K}(s_1^2 + s_2^2)}} \tag{2.90}$$

where s_1 and s_2 are the standard deviations for the two groups, follows the t-distribution with $2(K-1)$ degrees of freedom. This allows us to compute the p-value of the test and, as a consequence, to either reject the null hypothesis or not, depending on whether $p < \alpha$ or not, where α is the significance level desired. Rejecting the null hypothesis gives statistical proof of the efficiency of the new medicine.

The statistical test described above is parametric and is called the t-test. Depending on sizes of the groups, variances and other assumptions, different variants of the t-test can be constructed. Generally, tests belonging to the t family are used for comparing mean values of normally distributed variables. Other examples of parametric tests are, the ANOVA test for comparing means between multiple groups of measurements and the Bartlett test for homogeneity of variances.

2.7.3 Nonparametric Tests

In many situations involving analysis of statistical data, the assumption of known distributions of variables cannot be justified. Therefore many statistical tests have been developed, which allow for nonparametric statistical inference. Examples of such tests are Smirnov–Kolmogorov, Kruskal–Wallis, Man–Whitney and Wilcoxon tests [297]. Below, we briefly describe the construction of the Wilcoxon test.

Assume that in a class of K students the academic achievements of the students were unsatisfactory, as indicated by low test scores, and, because of this, an additional afternoon study program was enforced. After a half-year of additional afternoon classes, the tests were repeated. We are interested in whether the scores improved or not. One possibility is to compute average scores before and after the afternoon study program and use the t-test. However, we assume that we have evidence to believe that the distribution of test scores is not normal. In such a situation, the Wilcoxon sign-rank test can be applied. We need to compute the following Wilcoxon statistics. For each student, we compute the difference between the two sores, D_i. Then we take the absolute values $|D_1|, |D_2|, ..., |D_K|$, we order them from smallest to largest, and we assign them ranks from 1 to K, $r(|D_1|), r(|D_2|), ..., r(|D_K|)$. We also keep a record of the original signs of the differences, and we denote by I^+ the list of indices i for which the signs were positive and we denote by I^- the complementary list of negative signs. The Wilcoxon statistic T^+ is defined as

$$T^+ = \sum_{i \in I^+} r(|D_i|). \tag{2.91}$$

The null hypothesis is that the afternoon study program does not improve (change) the distribution of test scores. Under the null hypothesis, the distribution of the Wilcoxon statistic T^+ can be computed for a given K, and it can be proven that it does not depend on the distribution of test scores. For large K, the statistic T^+ converges to a normal distribution. On the basis of the distribution of the Wilcoxon statistic, the p-value of the test can be computed.

2.7.4 Type I and II statistical errors

As can be seen from the above examples, most often in statistics proving a statement means rejecting the null hypothesis that the converse is true. However, owing to the randomness in statistical experiments, statistical inference is exposed to the risk of committing errors. Two possible types of error in statistical inference are commonly called type I and type II statistical errors.

A type I error is rejecting the null hypothesis when in reality it is true. A type I statistical error is also called a false discovery, with an obvious interpretation.

A type II error is accepting (not rejecting) the null hypothesis in a situation when in reality it is false.

The parameters very often used when developing and analyzing statistical tests are the significance level α, p-value and the power of the test. All of them are related to type I and II statistical errors. When we say that a hypothesis is rejected at a significance level α this means that the probability of committing a type I statistical error is lower than α. When a procedure for computing a statistical test returns a p-value equal to p, this means that the related null hypothesis can be rejected at a significance level $\alpha \geq p$. Finally, the power of a test is one minus the probability of committing a type II statistical error.

2.8 Markov Chains

A stochastic process is a family of functions of a variable t, $\{X(t,\omega), t \in T, \omega \in \Omega\}$ (t is usually understood as a time), parametrized by random outcomes ω. For any fixed outcome ω, $X(.,\omega)$ is a function; for any fixed time t, $X(t,.)$ is a random variable.

Markov processes constitute the best-known and useful class of stochastic processes [78, 129]. A Markov process is a special case of a stochastic process in that it has a limited memory. Limited memory means that for a process $X(t,\omega)$ which has been running in the past ($t \leq t_0$), the future $\{X(t,\omega), t > t_0\}$ is characterized by the present, i.e., $X(t_0,\omega)$. This latter property is known as the Markov property.

A Markov chain is a Markov process for which $X(t,\omega) \in S$, where S is a discrete set. Usually the state space S is a subset of the integers. In other words, a Markov chain exhibits random transitions between discrete states. The theory presented here is focused on the case of a finite number of states, N, numbered $1, 2, \ldots, N$. Also, we discuss most systematically the case of discrete times $0, 1, 2, \ldots, k, \ldots$. However, we also add some facts about the case of continuous time. Most frequently, we write $X_k(\omega)$ or X_k instead of $X(k,\omega)$.

As already stated, the defining property of a Markov chain is that the future of the chain is determined by the present, i.e., X_k. This can be expressed by the following equation:

$$P(X_{k+1} = j \mid X_k = i, X_{k-1} = i_1, X_{k-2} = i_2, ...) = P(X_{k+1} = j \mid X_k = i). \tag{2.92}$$

The conditional probability $P(X_{k+1} = j \mid X_k = i)$ is called the transition probability from $X_k = i$ to $X_{k+1} = j$ and is denoted by p_{ij}, where

$$p_{ij} = P(X_{k+1} = j \mid X_k = i). \tag{2.93}$$

An important property of the Markov chains discussed here is their time homogeneity, which means that their transition probabilities p_{ij} do not depend on time.

The Markov property (2.92) has most important consequences for the analysis of Markov chains and allows us to derive recursive relations for probabilities related to X_k. In particular, the probability of the occurrence of the sequence of states $i_0, i_1, ..., i_K$ is given by the product of transition probabilities

$$P[i_0, i_1, ..., i_K] = \pi_{i_0} p_{i_0 i_1} ... p_{i_{K-1} i_K}, \tag{2.94}$$

where $\pi_{i_0} = P[X_0 = i_0]$. The above equation can be derived by using the chain rule (2.10) and the Markov property (2.92).

2.8.1 Transition Probability Matrix and State Transition Graph

The transition probabilities p_{ij} given in (2.93) can be represented by an $N \times N$ matrix, P, called the transition probability matrix of the chain,

$$P = \begin{bmatrix} p_{11} & p_{12} & \cdots & p_{1N} \\ p_{21} & p_{22} & \cdots & p_{2N} \\ \cdots & \cdots & \cdots & \cdots \\ p_{N1} & p_{N2} & \cdots & p_{NN} \end{bmatrix}. \tag{2.95}$$

State transitions and their probabilities can also be represented by a state transition graph, such as the one shown in Fig. 2.6. Here circles represent states and arrows represent state transitions. Each of the arrows is labeled by the transition probability. State transition graphs give an intuitive understanding of the properties of Markov chains, because of their graphical form and the fact that arrows are placed in them only for transitions that have nonzero probabilities. The representations of a chain by the state transition graph and by a transition probability matrix are equivalent. The transition probability matrix corresponding to the graph in Fig. 2.6 is

$$P = \begin{bmatrix} 0.5 & 0.5 & 0 & 0 \\ 0 & 0 & 0.8 & 0.2 \\ 0 & 0 & 0.1 & 0.9 \\ 0 & 0 & 0 & 1 \end{bmatrix}. \tag{2.96}$$

Probabilities of transitions from state i to all other states add up to one, i.e.,

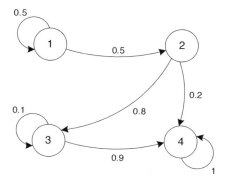

Fig. 2.6. Graph of state transitions for the Markov chain represented in (2.96).

$$\sum_{j=1}^{N} p_{ij} = 1. \qquad (2.97)$$

A matrix P, which has the property (2.97) is called a stochastic matrix. The corresponding property of the state transition graph is "The weights of the transition arrows pointing out from a state i add up to 1".

It often happens that the transition probability matrix is a sparse matrix, where many of the state transitions have a probability of zero, in which case the graph representation becomes both more comprehensive and more efficient.

2.8.2 Time Evolution of Probability Distributions of States

Having specified the transition probabilities by means of a transition probability matrix or a state transition graph, and given an initial probability distribution of the states, one can compute the evolution of the probability distribution of the states with time. Let us assume that at time 0, the probability distribution of the states is

$$P[X_0 = i] = \pi_i(0), \qquad (2.98)$$

and so $\sum_{i=1}^{N} \pi_i(0) = 1$. Using the law of total probability (2.3), we can compute the probability distribution of the states in the next step:

$$P[X_1 = j] = \pi_j(1) = \sum_{i=1}^{N} \pi_i(0) p_{ij}. \qquad (2.99)$$

Introducing a row vector notation for the probabilities of states at a time instant k,

$$\pi(k) = [\pi_1(k), \pi_2(k), ..., \pi_N(k)], \qquad (2.100)$$

we can represent (2.99) by using matrix multiplication, as

$$\pi(1) = \pi(0)P. \tag{2.101}$$

By repeatedly applying (2.101), we obtain

$$\pi(k) = \pi(0)P^k. \tag{2.102}$$

2.8.3 Classification of States

The classification of Markov chain states and the related classification of Markov chains is important for understanding the theory and applications of Markov chains. Below, we present this classification, illustrated by properties of state transition graphs.

Irreducibility

A Markov chain is irreducible if and only if its state transition graph has the property that every state can be reached from every other state. The Markov chain whose state transition graph is shown in the upper plot in Fig. 2.7 is irreducible. If a Markov chain is not irreducible, as in the case of the one in the lower plot in Fig. 2.7 then, by renumbering its states, its transition probability matrix can be transformed to the block matrix form

$$P = \begin{bmatrix} Q & 0 \\ U & V \end{bmatrix}, \tag{2.103}$$

where the upper right block consists of zeros and Q is a square matrix corresponding to an irreducible Markov sub-chain.

The transition probability matrix P of an irreducible Markov chain has the property that $P^k > 0$ for some k. By $P^k > 0$, we mean that all entries are strictly positive.

Persistent and Transient States

A state i is persistent if a Markov chain starting from i returns to i with probability 1. In other words, in the infinite sequence of states of the Markov chain starting from state i, state i occurs an infinite number of times. A state which is not persistent is called transient. It occurs in this sequence only a finite number of times, In the Markov chain whose state transition graph is shown in the upper plot in Fig. 2.7 all states are persistent; for the chain in the lower plot of figure 2.7 states 3 and 5 are transient and states 1, 2, and 4 are persistent.

Let us define

$$f_i^{(k)} = \text{Prob}[\text{Chain starting from } i \text{ has its first return to } i \text{ after } k \text{ steps}], \tag{2.104}$$

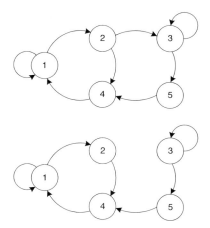

Fig. 2.7. *Upper plot*: a graph of state transitions for an irreducible Markov chain. *Lower plot*: Markov chain given by this state transition graph is not irreducible. In both plots, arrows represent transitions of nonzero probability.

with the convention $f_i^{(0)} = 0$, and

$$f_i = \sum_{k=1}^{\infty} f_i^{(k)}. \quad (2.105)$$

Since the events in (2.104) are exclusive, the sum of their probabilities cannot exceed one, i.e., $f_i \leq 1$. Using f_i we can give another condition for transient and persistent states: a state i is transient if $f_i < 1$, and persistent if $f_i = 1$.

The probabilities f_i can be computed on the basis of the entries of matrices $P, P^2, \ldots, P^k, \ldots$. We define

$$p_{ii}^{(k)} = \text{Prob}[\text{Chain starting from } i \text{ returns to } i \text{ after } k \text{ steps}] \quad (2.106)$$

and we adopt the convention $p_{ii}^{(0)} = 1$. The events in (2.106) are not exclusive. We also see that $p_{ii}^{(k)}$ is the i,i entry of the matrix P^k. Using the law of total probability (2.3), for the events in (2.106) and (2.104) we have

$$p_{ii}^{(k)} = f_i^{(1)} p_{ii}^{(k-1)} + f_i^{(2)} p_{ii}^{(k-2)} + \ldots + f_i^{(k)} p_{ii}^{(0)}. \quad (2.107)$$

Writing the above for $k = 1, 2, \ldots$ gives a system of linear equations, which allows us to solve for $f_i^{(k)}$.

Using the probabilities $p_{ii}^{(k)}$, we can state one more condition. If $\sum_{k=0}^{\infty} p_{ii}^{(k)} = \infty$, then state i is persistent. If $\sum_{k=0}^{\infty} p_{ii}^{(k)} < \infty$, then state i is transient. This dichotomy can be proved by applying the method of generating functions to equations (2.107) (see Exercise 16).

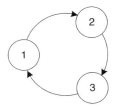

Fig. 2.8. Graph of state transitions corresponding to periodic states

If the state i is persistent, we can ask what is the expected waiting time μ_i for the recurrence of i. On the basis of (2.104), μ_i can be computed as follows:

$$\mu_i = \sum_{k=1}^{\infty} k f_i^{(k)}. \tag{2.108}$$

Periodic States

In Fig. 2.8 there is an example of a transition graph corresponding to periodic states. The states 1, 2 and 3 are periodic with period three. Generally, a state i of a Markov chain is periodic if $p_{ii}^{(k)} \neq 0$ only for $k = \nu t$, $t = 0, 1, \ldots$ and an integer $\nu > 1$. The largest such ν is called the period of the state i. Periodicity is rarely encountered in applications of Markov chains. It is, rather, a theoretical possibility, which has to be excluded when formulating precise definitions and proving theorems. A state i is aperiodic if no $\nu > 1$ satisfies the property stated above.

2.8.4 Ergodicity

A state i is called ergodic if it is aperiodic and persistent. A Markov chain is called ergodic if all its states are ergodic. For Markov chains with a finite number of states, ergodicity is implied by irreducibility and aperiodicity.

2.8.5 Stationary Distribution

The stationary (or invariant) distribution of a Markov chain is defined as the π_S (a row vector) such that
$$\pi_S = \pi_S P,$$
whenever it exists. In general, π_S does not have to be unique. For example, if

$$P = \begin{bmatrix} P_1 & 0 \\ 0 & P_2 \end{bmatrix}$$

and $\pi_{S1} = \pi_{S1} P_1$ and $\pi_{S2} = \pi_{S2} P_2$, then for any $\alpha \in [0, 1]$, $\pi_S = \alpha[0\ \pi_{S1}] + (1-\alpha)[\pi_{S2}\ 0]$ is a stationary distribution. Stationary distributions are related to limit distributions, defined by $\pi(\infty) = \lim_{k\to\infty} \pi(k)$. If the distribution exists in the limit, it is stationary. If a Markov chain is ergodic, then the limit of $\pi(k)$ exists and does not depend on the initial distribution $\pi(0)$, i.e.,

$$\lim_{k\to\infty} \pi(k) = \pi_S. \qquad (2.109)$$

In such a case, the stationary distribution is unique. Moreover, the limit of P^k also exists and

$$\lim_{k\to\infty} P^k = \mathbf{1}\pi_S. \qquad (2.110)$$

In the above, $\mathbf{1}$ is an N-element column vector with all entries equal to one. The ith column of the limiting matrix in (2.110) consists of identical elements, equal to π_{Si}; the ith element of the vector π_S. We can also demonstrate (Exercise 17) that

$$\pi_{Si} = \frac{1}{\mu_i}. \qquad (2.111)$$

We call a Markov chain stationary if its initial distribution is its stationary distribution

$$\pi(0) = \pi_S. \qquad (2.112)$$

In such a chain, by the definition of π_S, $\pi(k) = \pi_S$ for each k. In other words, the Markov chain evolves in accordance with its stationary distribution.

2.8.6 Reversible Markov Chains

Here we consider a Markov chain in the reversed order, $\{X_k, X_{k-1}, X_{k-2}, ...\}$. It can be proven that the process X_k, X_{k-1}, X_{k-2}, ... again has the Markov property. By using the Bayes' rule (2.2), we can compute the transition probability from state i to state j in reversed time,

$$p_{ij}^{reversed} = P[X_{k-1} = j | X_k = i] \qquad (2.113)$$
$$= \frac{P[X_{k-1} = j]P[X_k = i | X_{k-1} = j]}{P[X_k = i]} = \frac{\pi_j(k-1)p_{ji}}{\pi_i(k)}.$$

There is an inconsistency in the notation in expression (2.113), since $p_{ij}^{reversed}$ depends on the time instant k. For simplicity of notation, we suppress index k. Nevertheless, we learn from (2.113) that the Markov chain with reversed time becomes inhomogeneous.

In most applications, it is important to analyze the reversed Markov chain under the additional assumption of stationarity (2.112). In that case the Markov chain with reversed time becomes homogeneous. We have $P[X_{k-1} = j] = \pi_{Sj}$ and $P[X_k = i] = \pi_{Si}$, and (2.113) becomes

$$p_{ij}^{reversed} = \frac{\pi_{Sj} p_{ji}}{\pi_{Si}}. \qquad (2.114)$$

We call a Markov chain reversible if it satisfies

$$p_{ij}^{reversed} = p_{ij}. \tag{2.115}$$

It is interesting that reversibility implies stationarity of both the forward and the reversed chain. Indeed, if

$$p_{ij} = \frac{\pi_j(k-1)p_{ji}}{\pi_i(k)}$$

for all i,j then if we set $i=j$ we have $\pi_i(k-1)/\pi_i(k) = 1$.

From the definition (2.115) we can understand that when we observe (or record) states of a reversible Markov chain, we cannot tell whether it is proceeding forward or backward. Combining (2.114) and (2.115), we obtain the following condition for the reversibility of a Markov chain:

$$p_{ij}\pi_{Si} = \pi_{Sj}p_{ji}. \tag{2.116}$$

It is also called the local balance condition, or the detailed balance condition, owing to the following interpretation. Assume that we are recording events in a Markov chain. The average number of transitions from state i to j, per recorded event, is $p_{ij}\pi_{Si}$. Analogously, for transitions from state j to i the average number per event is $\pi_{Sj}p_{ji}$. By the condition (2.116), in a reversible Markov chain these numbers are equal.

2.8.7 Time-Continuous Markov Chains

In the above we have assumed that transitions between states could only happen at discrete times $0, 1, 2, \ldots, k, \ldots$. Now, we assume that transitions between discrete states $1, 2, \ldots, N$ can occur at any time t, which is a real number. We denote the resulting stochastic process by $X(t)$ and introduce the $N \times N$ transition matrix $P(t-s)$, with entries

$$p_{ij}(t-s) = P[X(t) = j | X(s) = i]. \tag{2.117}$$

The Markov property of $X(t)$ is equivalent to the Chapman–Kolmogorov equation

$$p_{ij}(s+t) = \sum_{n=1}^{N} p_{in}(s)p_{nj}(t). \tag{2.118}$$

Using the matrix notation $P(t)$, (2.118) can be written as

$$P(s+t) = P(s)P(t). \tag{2.119}$$

In the above, $s \geq 0$, $t \geq 0$, and

$$P(0) = I, \tag{2.120}$$

where I means the identity matrix. $P(t)$ is differentiable [129], and by computing the derivative, from (2.119) we obtain

$$\frac{d}{dt}P(t) = P'(t) = QP(t), \qquad (2.121)$$

where the matrix Q called the intensity matrix of the time - continuous Markov chain $X(t)$, is given by the limit of the derivative at zero,

$$Q = \lim_{t \to 0^+} \frac{dP(t)}{dt}.$$

The constructions of Markov processes $X(t)$ used in practical applications, for example in the nucleotide substitution models described in Chap. 6, start with defining the intensity matrix Q. Such an approach is the most natural. Given intensity matrix Q, state transition matrix $P(t)$ can be obtained by solving (2.121) with the initial condition (2.120). The solution is

$$P(t) = \exp(Qt) = \sum_{m=1}^{\infty} \frac{(Qt)^m}{m!}. \qquad (2.122)$$

For each $t \geq 0$, $P(t)$ is a stochastic matrix, and given an initial probability distribution $\pi(0)$ of states $1, 2, \ldots, N$, we can compute the distribution at time t, from

$$\pi(t) = \pi(0)P(t). \qquad (2.123)$$

The construction of the process using intensities implies that for any state i, the probability of a transition $i \to j$ in the interval $(t, t + \Delta t)$ is equal to $q_{ij}\Delta t + o(\Delta t)$, i.e.,

$$P[X(t + \Delta t) = j | X(t) = i] = q_{ij}\Delta t + o(\Delta t). \qquad (2.124)$$

For the diagonal elements of the intensity matrix Q, we define

$$q_{ii} = -\sum_{j \neq i} q_{ij}. \qquad (2.125)$$

It is possible to derive (2.121), (2.122) using (2.124) and (2.125).

2.9 Markov Chain Monte Carlo (MCMC) Methods

Monte Carlo methods, based on random number generators, allow one to perform a variety of tasks, including stochastic simulations, computing integrals in high dimensions, and optimizing functions and functionals. The Markov chain Monte Carlo approach is, additionally, based on using Markov chains for performing these tasks. An important tool in Markov chain Monte Carlo methods is the Metropolis–Hastings algorithm, [195, 116]. It was originally developed for computing (or estimating) integrals in high-dimensional state spaces in molecular physics, but subsequently found many different applications. Here we describe this algorithm and present applications of it to sampling and stochastic optimization.

The Metropolis–Hastings method gives a solution to the following problem: construct an ergodic Markov chain with states 1, 2, ..., N and with a prescribed stationary distribution, given by a vector π_S. By constructing a Markov chain, we mean defining its state transition probabilities. Clearly, there are an infinite number of Markov chains with a stationary distribution π_S. Given transition probabilities we can compute the stationary distribution π_S, but there is no explicit formula for the inverse relation. Metropolis–Hastings method provides a solution to this problem by starting from any ergodic Markov chain with states 1, 2, ..., N and then modifying its transition probabilities in such a way that the local balance condition (2.116) is enforced. Therefore the modified Markov chain becomes reversible and has the desired stationary distribution π_S.

Employing this idea, let us assume that we have defined an irreducible, aperiodic Markov chain with states 1, 2, ..., N and transition probabilities q_{ij}. In the next step, we modify these probabilities by multiplying them by factors a_{ij}, which leads to a new Markov chain with transition probabilities

$$p_{ij} = a_{ij} q_{ij}. \qquad (2.126)$$

We want to choose the factors a_{ij} such that transition probabilities p_{ij} satisfy the local balance condition (2.116). Substituting (2.126) in (2.116), we obtain

$$a_{ij} q_{ij} \pi_{Si} = a_{ji} q_{ji} \pi_{Sj}. \qquad (2.127)$$

There are two variables and one equation here, so again an infinite number of solutions is possible. A simple solution is to assume that one of the factors a_{ij} and a_{ji} is equal to one. There are two possibilities. However, we should take into account the condition that multiplying factors should satisfy $a_{ij} \leq 1$ for all i, j. This condition stems from the fact that the scaling in (2.126) must not lead to probabilities out of the range $(0, 1]$. This, finally, leads to the solution

$$a_{ij} = \min\left(1, \frac{q_{ji} \pi_{Sj}}{q_{ij} \pi_{Si}}\right). \qquad (2.128)$$

Equation (2.126), with a_{ij} specified in (2.128), allows us to compute the transition probabilities p_{ij} for all $i \neq j$. For the probabilities p_{ii}, we use the formula

$$p_{ii} = 1 - \sum_{j \neq i} p_{ij}, \qquad (2.129)$$

following from (2.97).

As seen from the rule (2.128), the expression for a_{ij} does not depend on the absolute values of π_{Si} but only on their ratios. This means that it is enough to know π_S up to a proportionality constant. This is an important feature, which allows one to simulate distributions for which a norming constant is difficult to find.

2.9.1 Acceptance–Rejection Rule

The Metropolis–Hastings method for modifying transition probabilities (2.126)–(2.129) can be formulated in the terms of the acceptance–rejection rule, very useful in practical implementations. Let us assume that we have (i) defined an irreducible, aperiodic Markov chain with states $1, 2, \ldots, N$ and transition probabilities q_{ij}, and (ii) developed a program for simulating transitions between its states. The modification of the transition probabilities q_{ij} described in (2.126)–(2.129) is equivalent to adding the following acceptance–rejection rule to the program for simulating transitions between states of Markov chain q_{ij}. When a transition $i \to j$ is encountered, compute a_{ij} according to (2.128). If $a_{ij} = 1$, do not intervene (move to state j). If $a_{ij} < 1$, then, with probability a_{ij}, move to state j and, with probability $1 - a_{ij}$, cancel the transition $i \to j$ (stay in the state i).

2.9.2 Applications of the Metropolis–Hastings Algorithm

By using the Metropolis–Hastings algorithm we can perform random sampling from arbitrary distributions. This is very useful, for example for estimating shapes or parameters of complicated posterior distributions. Another important application of the Metropolis–Hastings algorithm is stochastic optimization. An example is a search for the most likely tree given data (see Chap. 7). For each tree, we compute the corresponding probability (likelihood), but owing to the huge number of all possible trees, one cannot go through all of them and pick out the one with the highest probability. Instead, we can construct a Markov chain such that different trees correspond to its states. Applying the Metropolis–Hastings algorithm, we visit (sample) trees with frequencies corresponding to their probabilities. Trees with higher probabilities are visited more frequently, whereas trees with low probabilities are unlikely to be visited at all. Subsequently, we can limit the search for the most likely tree to trees visited in the Metropolis–Hastings sampling procedure.

2.9.3 Simulated Annealing and MC3

Is it possible to use the idea of the Metropolis–Hastings algorithm for optimization, over the argument space, of any function $f(x)$, not necessarily a likelihood? The challenge is that $f(x)$ may assume both positive and negative values and may not have a probabilistic interpretation.

Consider the transformation

$$p(x) = \exp\left[\frac{f(x)}{T}\right], \qquad (2.130)$$

based on the idea of the Boltzmann energy distribution. The function $p(x)$ is always strictly positive and assumes its maximum at the same argument value $x_{\max}$ as that for which $f(x)$ does. This function does not necessarily correspond to a probability distribution, since its integral is generally not equal to 1. However, only strict positivity is important here, since, as already noted, (2.126)–(2.129) depend only on ratios of elements of the vector π_S. It is then possible to program an algorithm for searching for maximum of $p(x)$ by using Metropolis–Hastings sampling, based on $p(x)$. If the space of arguments x is continuous, it is discretized before applying the Metropolis–Hastings algorithm.

Equation (2.130) contains a free parameter T. By the analogy to the Boltzmann energy distribution, this parameter is interpreted as a "temperature". Changing its value influences the properties of the sampling algorithm. Increasing the temperature makes the browsing through the argument space more intensive, since the transitions from a higher to a lower $p(x)$ become more likely. Decreasing the temperature makes transitions less likely. In the method of simulated annealing [152], the temperature is changed, according to some schedule, along with browsing through the argument space. Simulated-annealing algorithms start browsing with a high temperature and then, gradually, the temperature is lowered as iterations come close to the neighborhood of the maximum.

Another very useful idea, named MC3, is to perform the search through the argument space using several (often three) Metropolis–Hastings samplers with different temperatures [184]. The sampling algorithms operate in parallel and can exchange their states depending on the values of the likelihoods.

2.10 Hidden Markov Models

In the preceding sections, when deriving properties of Markov chains, we assumed that the sequences of states were observable. However, this assumption is often not satisfied in applications of Markov chain models. Hidden Markov models (HMM) [69, 236, 157] are frequently applied in such situations. A hidden Markov model is a Markov chain whose states are not observable. Only a sequence of symbols emitted by the states is recorded.

More specifically, let us consider a Markov chain with states $1, 2, \ldots, N$, over a discrete time interval $0, 1, 2, \ldots, k, k+1, \ldots, K$. Additionally, there are M possible symbols denoted by $o_1, o_2, \ldots, o_m, o_{m+1}, \ldots, o_M$, called emissions. Each of the states has an associated probability distribution of emissions

$$b_{im} = \text{Prob}[\text{state } i \text{ emits } o_m]. \tag{2.131}$$

2.10.1 Probability of Occurrence of a Sequence of Symbols

From (2.94) and (2.131), we conclude that the probability of occurrence of states $i_0, i_1, \ldots, i_K$ and symbols $o_{j_0}, o_{j_1}, \ldots, o_{j_K}$ is

$$P[i_0, o_{j_0} i_1, o_{j_1} \ldots, i_K, o_{j_K}] = \pi_{i_0} b_{i_0 j_0} p_{i_0 i_1} b_{i_1 j_1} \ldots p_{i_{K-1} i_K} b_{i_K j_K}. \tag{2.132}$$

2.10 Hidden Markov Models   61

The probability of recording a sequence of symbols $o_{j_0}, o_{j_1}, ..., o_{j_K}$ is obtained by summing (2.132) over all possible sequences $i_0, i_1, ..., i_K$, which leads to

$$P[o_{j_0}, o_{j_1}, ..., o_{j_K}] = \sum_{i_0=1}^{N} \pi_{i_0} b_{i_0 j_0} \sum_{i_1=1}^{N} p_{i_0 i_1} b_{i_1 j_1} \cdots \sum_{i_K=1}^{N} p_{i_{K-1} i_K} b_{i_K j_K}. \quad (2.133)$$

When using the above expression in practical computations, we arrange the summation in a recursive manner. There are two possibilities, leading to a backward or a forward algorithm.

2.10.2 Backward Algorithm.

We can organize the recursive computation of (2.133) starting from the last sum. We denote the last sum by

$$B_{K-1}(i_{K-1}) = \sum_{i_K=1}^{N} p_{i_{K-1} i_K} b_{i_K j_K} \quad (2.134)$$

and we see that for $B_k(i_k)$, defined as

$$B_k(i_k) = \sum_{i_{k+1}=1}^{N} p_{i_k i_{k+1}} b_{i_{k+1} j_{k+1}} \cdots \sum_{i_K=1}^{N} p_{i_{K-1} i_K} b_{i_K j_K}, \quad (2.135)$$

there holds a recurrence relation

$$B_k(i_k) = \sum_{i_{k+1}=1}^{N} p_{i_k i_{k+1}} b_{i_{k+1} j_{k+1}} B_{k+1}(i_{k+1}), \quad (2.136)$$

valid for $k = 0, 1, ..., K - 2$. Finally,

$$P[o_{j_0}, o_{j_1}, ..., o_{j_K}] = \sum_{i_0=1}^{N} \pi_{i_0} b_{i_0 j_0} B_0(i_0). \quad (2.137)$$

The defined recurrence defined above involves storing N-dimensional arrays and summations over one index.

2.10.3 Forward Algorithm.

Another possibility is to start from the first sum. in (2.133) Defining

$$F_k(i_k) = \sum_{i_0=1}^{N} \pi_{i_0} b_{i_0 j_0} \cdots \sum_{i_{k-1}=1}^{N} p_{i_{k-2} i_{k-1}} b_{i_{k-1} j_{k-1}} p_{i_{k-1} i_k}, \quad (2.138)$$

we realize that $F_k(i_k)$, $k = 1, ..., K-1$ can be computed by using the following recursion:

$$F_{k+1}(i_{k+1}) = \sum_{i_k=1}^{N} F_k(i_k) b_{i_k j_k} p_{i_k i_{k+1}}. \tag{2.139}$$

Now $P[o_{j_0}, o_{j_1}, ..., o_{j_K}]$ is given by

$$P[o_{j_0}, o_{j_1}, ..., o_{j_K}] = \sum_{i_K=1}^{N} F_K(i_K) b_{i_K j_K}. \tag{2.140}$$

Similarly to the backward algorithm, the forward algorithm requires storing N-dimensional arrays and summations over one index.

2.10.4 Viterbi Algorithm

The Viterbi algorithm solves the following problem: given a sequence of symbols $o_{j_0}, o_{j_1}, \ldots, o_{j_K}$, find the most probable sequence of states $i_0, i_1, \ldots, i_K$. In other words we wish to compute the sequence of states that maximizes the conditional probability

$$P[i_0, i_1, ..., i_K | o_{j_0}, o_{j_1}, ..., o_{j_K}] = \frac{P[i_0, o_{j_0} i_1, o_{j_1} ..., i_K, o_{j_K}]}{P[o_{j_0}, o_{j_1}, ..., o_{j_K}]}. \tag{2.141}$$

Since $P[o_{j_0}, o_{j_1}, ..., o_{j_K}]$ is only a scaling factor here then maximizing the conditional probability (2.141) is equivalent to maximizing the joint probability (2.132) over all sequences of states $i_0, i_1, ..., i_K$. Taking the natural logarithm of both sides of (2.132) and defining

$$L(i_0, i_1, ..., i_K) = \ln P[i_0, o_{j_0} i_1, o_{j_1} ..., i_K, o_{j_K}],$$

we obtain

$$L(i_0, i_1, ..., i_K) = \ln \pi_{i_0} + \sum_{k=0}^{K-1} (\ln b_{i_k j_k} + \ln p_{i_k i_{k+1}}) \tag{2.142}$$

and the maximization problem becomes

$$\max_{i_0, i_1, ..., i_K} L(i_0, i_1, ..., i_K) \tag{2.143}$$

This maximization problem can be solved with the use of dynamic programming (Chap. 5) since decisions to be made come in sequential order and one can define partial scores related to each stage of the decision-making process, namely

$$L_0(i_0, i_1, ..., i_K) = L(i_0, i_1, ..., i_K) = \ln \pi_{i_0} + \sum_{k=0}^{K-1} (\ln b_{i_k j_k} + \ln p_{i_k i_{k+1}}) \tag{2.144}$$

and

$$L_m(i_m, i_{m+1}, ..., i_K) = \sum_{k=m}^{K-1} (\ln b_{i_k j_k} + \ln p_{i_k i_{k+1}}). \quad (2.145)$$

On the basis of (2.144) and (2.145) we can derive a Bellman equation for updating arrays of optimal partial scores,

$$\hat{L}_{K-1}(i_{K-1}) = \max_{i_K}(\ln b_{i_{K-1} j_{K-1}} + \ln p_{i_{K-1} i_K}) \quad (2.146)$$

and

$$\hat{L}_m(i_m) = \max_{i_{m+1}}[\ln b_{i_m j_m} + \ln p_{i_m i_{m+1}} + \hat{L}_{m+1}(i_{m+1})]. \quad (2.147)$$

By solving the above Bellman recursion, we can compute the solution to the maximization problem (2.143).

2.10.5 The Baum–Welch algorithm

One more problem often considered in the area of HMM models is to estimate the transition probabilities of a Markov chain, given a sequence of symbols $o_{j_0}, o_{j_1}, ..., o_{j_K}$. The maximum likelihood solution to this problem is to maximize probability in (2.133) over the entries p_{ij} of the Markov chain transition probability matrix. However, since this is an optimization problem in a high dimensionality, using some special approach seems desirable. One of the approaches is the Baum–Welch algorithm. We mention this algorithm here because it applies the idea of the EM recursions presented in this chapter, in Sect. 2.6. The parameters to be estimated are, initial probabilities of states π_i and transition probabilities p_{ij}. The observed variables are the symbols $o_{j_0}, o_{j_1}, ..., o_{j_K}$. The hidden variables are the states $i_0, i_1, ..., i_K$. Using these assumptions and denoting the vector including all estimated parameters by p, we can specify $Q(p, p^{old})$ defined in (2.71) as follows:

$$Q(p, p^{old}) = \sum_{i_0=1}^{N} ... \sum_{i_K=1}^{N} \left[\ln \pi_{i_0} + \sum_{k=0}^{K-1} (\ln b_{i_k j_k} + \ln p_{i_k i_{k+1}}) \right]$$
$$\times \pi_{i_0}^{old} b_{i_0 j_0} p_{i_0 i_1}^{old} b_{i_1 j_1} ... p_{i_{K-1} i_K}^{old} b_{i_K j_K}. \quad (2.148)$$

The above expression is the E-step. The M-step involves maximization of $Q(p, p^{old})$ over parameters, $\pi_i, p_{ij}, i, j = 1, ..., N$. We are omitting details of the computations. The issues arising in the practical construction of the appropriate algorithm are discussed in detail in, for example, [157].

2.11 Exercises

1. Derive expressions for the maximum likelihood estimators of the distributions described in Sects. 2.3.1–2.4.6.

2. Derive expressions for the expectations and variances of the ML and the moment estimators described in Sect. 2.5.1.
3. Verify that the estimator in (2.48) has the property $\lim_{k\to\infty} \hat{a}_{mom,k} = \max_{1\leq n\leq N} x_n$.
4. Compare the ML and moment estimators for a random variable distributed uniformly over the interval $a_{\min}, a_{\max}$.
5. Write a computer program for the ML estimate of the parameter a of a Cauchy distribution by solving (2.51) numerically. Estimate numerically its variance.
6. Verify the fact, used in (2.57), that $\max(x_1, x_2, ..., x_N)$ is a sufficient statistic for the parameter a.
7. Compute the variance of the estimator $\hat{a}^{RB}$ in (2.57).
8. Prove the assertion (2.65). Hint: use

$$-\sum_{i=1}^{n} q_i \ln \frac{p_i}{q_i} = \sum_{i=1}^{n} (p_i - q_i) - \sum_{i=1}^{n} q_i \ln \frac{p_i}{q_i}.$$

9. Write down the log-likelihood function corresponding to an exponential distribution with right-censored observations and compute the ML estimate of the parameter a. Compare it to (2.76).
10. Write a computer program to generate random variables described by distributions that are mixtures.
11. Write a computer program for iterating updates of parameter estimates for mixed Poisson distributions (2.83)–(2.85) and mixed-normal distribution (2.86)–(2.89). By use of repeated simulations, study the problem of the existence of local maxima and the convergence of EM iterations to local maxima.
12. Develop a method for computing the approximate variances of estimated mixture parameters by using Cramer–Rao approximation (2.55) for the variance .
13. Analyze the problem of estimating parameters by EM iterations when the observations come from a mixture of one normal and one uniform distribution.
14. Derive EM recursive estimates, analogous to (2.87)–(2.89), for mixtures of multivariable normal distributions (see, e.g., [191]).
15. Find the state transition matrices for the Markov chains presented in Figs. 2.7 and 2.8.
16. By using probabilities given in (2.104) and (2.106) we can define the generating functions

$$F_i(z) = \sum_{k=0}^{\infty} z^k f_i^{(k)}$$

and

$$P_i(z) = \sum_{k=0}^{\infty} z^k p_{ii}^{(k)}.$$

Prove that
$$P_i(z) = \frac{1}{1 - F_i(z)}.$$
Use this to prove the condition $\sum_{k=0}^{\infty} p_{ii}^{(k)} = \infty$ for persistence of state i.
17. Prove the assertion (2.111).
18. Assume that we are observing sequential states of a Markov chain X_k, $k = 0, 1, 2, \ldots,$. Develop a method for estimating the entries of the state transition matrix P in (2.95).
19. Develop a computer program for random simulations of state transitions in a Markov chain with the discrete time.
20. Develop a computer program for random simulations of state transitions in a Markov process described by the intensity matrix Q, in (2.122).
21. Develop a computer program for simulating transitions between states and emitting symbols in an HMM.
22. Study the following problem: what is the most probable sequence of symbols $o_{j_0}, o_{j_1}, \ldots, o_{j_K}$ for a given HMM? Develop an algorithm for solving this problem.

3
Computer Science Algorithms

Computer science is one of the basic technologies behind bioinformatics. Bioinformatic databases must be constructed, organized, maintained, and developed with the use of computers. For many scientists, however, when they are using bioinformatic databases in their research, the computer science algorithms involved remain hidden in a black box.

In this chapter we present some of these algorithms in detail, while for some others we discuss only their underlying ideas. Some methods can be practically coded on the basis of reading this text but others require further research and consulting more literature.

The algorithms that we present were developed in the course of the evolution of scientific ideas and they deserve interest for the ideas and concepts contained in them. Our rationale when presenting them here is more utilitarian. Research in bioinformatics relies very heavily on these algorithms. Two situations are common. (1) We want pursue a research project, for example on the sizes of chromosomes of different eukaryotic organisms, which requires access to many different database resources. Manual browsing may be difficult, and the research can be greatly improved by applying algorithmic automated information retrieval with a reasonable level of human supervision. (2) The data downloaded from a bioinformatic resource has a very large volume, such that it requires a dedicated-information processing algorithm. Both situations call for software to be developed along with pursuing research in bioinformatics. There are ready-made algorithms available in software packages, but there are still issues in the design of a desired algorithm which require knowledge of the formulations of the ready-made algorithms and, more generally, knowledge of computer science and information-processing methods.

3.1 Algorithms

We discuss several algorithms, and study and compare some of their properties, with emphasis on their complexity. So we start with a few words on what algorithms are and how their complexity is measured. An algorithm is a well-defined sequence of operations applicable to some initial data. An algorithm must have a rule for terminating its execution, when some conditions are satisfied. The result of applying an algorithm is some output data, related to the task for which the algorithm was created. Algorithms can be defined by a verbal description or in more formal ways, graphically, by flow charts, or by writing a computer code. A convenient method of formalize an algorithm is the use of a pseudocode, which lists operations in an algorithm in a computer-style format, but aims at a demonstrative presentation while avoiding the technicalities of software-specific implementations. One formalized mathematical model for algorithms is the Turing machine, which consists of

(1) a doubly infinite tape of symbols, which contains both input and output data;
(2) a printing/reading head;
(3) a list of possible states;
(4) a program, which actually specifies the steps of the algorithm, given the data.

The Turing machine is not a practical tool for developing algorithms, but, rather, it serves for proving mathematical theorems, evaluating the algorithmic complexity of problems, comparing different algorithms, etc.

An algorithm is typically applicable to various sets of data and the time for its execution depends on the length (size) of this input data. The rule which relates the execution time of an algorithm to its input data length is called the complexity or the computational time of the algorithm. Clearly, we are interested in developing algorithms with lowest possible complexity.

Also, in the course of its execution, an algorithm produces intermediate data, which must be stored in computer memory. The memory storage capacity (occupancy) required by an algorithm is the second parameter characterizing the efficiency of an algorithm. Again, the volume of intermediate data is related to the size of the input data, and the memory occupancy efficiency is described by the relation between these two quantities.

3.2 Sorting and Quicksort

Words, strings, numbers, vectors, and so forth can be compared by size or by lexical order of their letters. Putting lists of elements into an ascending or descending order is called sorting. It is one of the basic algorithms in computer science. Sorting employs pairwise comparisons between elements of a list. The number of comparisons required by an algorithm to perform its task is a measure of its efficiency. We shall describe two algorithms for sorting, "simple sort" and "quicksort". They solve the same problem but differ in the number of steps they need to complete the task. We assume that X is a list of numbers with K elements, $X(1), X(2), \ldots, X(K)$, and the aim of sorting is rearranging X in decreasing order, $X(1) \geq X(2) \geq \ldots \geq X(K)$.

3.2.1 Simple Sort

This algorithm uses the operations of comparing two numbers and swapping $X(i) \leftrightarrows X(j)$, and proceeds by the most obvious steps, going through all possible comparisons. Using a variant of a pseudocode inspired by the syntax of Delphi, it can be written as follows:

Program Simple Sort
for $k = 1$ to K
    for $j = k + 1$ to K
        if $X(k) < X(j)$
            swap $X(i) \leftrightarrows X(j)$
        endif
    end
end

The above procedure executes two nested loops. Since all possible pairwise comparisons are always applied, the number of comparisons does not depend on the initial ordering of entries in the data vector X and is always equal to $K(K-1)/2$. So the complexity of the simple sort algorithm, measured by number of comparisons given length of data vector K, is polynomial of degree 2, which we represent as $[O(K^2)]$.

3.2.2 Quicksort

By using a clever approach [123], we can significantly reduce the number of comparisons necessary to sort a list of K elements. Assume that our input list of elements has a structure

$$X = [X_1 \ X_2], \qquad (3.1)$$

with two sublists X_1 and X_2, and that any element in X_1 is greater than or equal to any element in X_2. Clearly in such situation we would sort X_1 and X_2 separately, which would lead to saving execution time of the algorithm. However, there is no guarantee that X has this decomposed structure. Can we therefore transform (permute) X into $[X_1 \ X_2]$ of the form described above?

The solution is as follows. Pick a random element $x^s \in X$, call it a "splitter", and, by doing $K - 1$ comparisons between the splitter x^s and remaining elements of X rearrange it such that elements of X which are greater than or equal to x^s are moved to "front" (obtain indexed lower than the index of x^s), and elements X which are smaller than or equal to x^s are moved to "back" (obtain indexes higher than the index of x^s). These operations form the first step of the algorithm, represented symbolically below.

Step 1.

$$X \to [X_1 \; x^s \; X_2]. \tag{3.2}$$

All elements in X_1 are greater than or equal to x^s and x^s is greater than or equal to all elements in X_2. Now the result of permuting elements of X, (3.2), has the required structure (3.1) but we do not terminate the algorithm, rather we repeat the above step 1 for both X_1 and X_2. We pick up randomly splitters $x^{s1} \in X_1$ and $x^{s2} \in X_2$ and by doing comparisons *separately* for X_1 and X_2 we move to the next

Step 2.

$$[X_1 \; x^s \; X_2] \to [X_{11} \; x^{s1} \; X_{12} \; x^s \; X_{21} \; x^{s2} \; X_{22}].$$

Successive steps apply the same idea to sublists X_{11}, X_{12}, X_{21} and X_{22}, and so forth.

What is the computational time of the Quicksort algorithm? First, the computational time has an element of randomness, since the splitters are chosen randomly from X. In order to estimate the average time, we observe that, on average, the splitter x^s is close to the middle of the list X. If splitter divides a list into roughly equal sublists, the number of steps will be proportional to the number of successive divisions of X into halves, which is proportional to $\log_2 K$. Each step of the algorithm requires fewer than K comparisons, which leads to the final estimate

$$O(K \log_2 K).$$

If, by extreme bad luck, splitters were always chosen to be the biggest or smallest elements of the lists, then sorting by quicksort would require the worst case of $[O(K)^2]$ comparisons, same as simple sort. This is, however, very improbable.

When the length of the list of elements to be sorted is short, the sorting method applied does not make much difference, but when lists are very long, the efficiency of the method has a serious impact on the computational time.

3.3 String Searches. Fast Search

Strings are sequences of symbols defined over some alphabet, for example *abbbaa* is a string over the alphabet $\Sigma = \{a, b\}$. An often encountered problem is that of searching for occurrences of one string of symbols (or equivalently, characters, or letters), which we shall denote by P (the pattern string) inside another string of symbols, which we shall denote by S (the search string or text). Typically, by P we understand an item such as word (or sentence), shorter than S, which we imagine as a text. Again, we shall describe two different algorithms, an easy one with a longer execution time and a cleverer one, which saves computational expense.

3.3.1 Easy Search

This algorithm is obvious and goes through sequential comparisons of letters of P and S and sliding P along S. We denote the lengths of the strings P and S by K^P and K^S, respectively, and by $P(i)$ and $S(j)$, ith and jth symbol of P and S, $i = 1, 2, \ldots, K^P$, $j = 1, 2, \ldots, K^S$. The pseudocode for the "easy search" algorithm is

Program Easy Search
for $j = 1$ to $K^S - K^P$
    $i = j$
    while symbol__compare$[P(i), S(j)] == 1$
        $i = i + 1$
    endwhile
    if $i == j + K^P$
        break the for loop and report (first) string match at $i = j$
    endif
end
report no match

The above program uses a function symbol__compare(a, b), which returns 1 if the symbols a and b are the same, and 0 if not. What is the computational complexity of easy search? Its computational time depends not only on the length of the data, but also on the data itself, that is, on the order of symbols in S and P, since number of symbol comparisons depends on how many times the while loop in the program is iterated before it stops owing to mismatch of characters. It is easy to evaluate, intuitively, the worst-case computational time. Assume that $P = aaab$ and

$$S = aaac\_aaac\_aaac \ldots aaac\_aaab \tag{3.3}$$

From this example we can see that, by arranging artificially created data as in the string above we can make the number of necessary symbol comparisons (the number of calls of the function symbol__compare) of the order of $K^P K^S$. However, in typical data the situation (3.3) generally does not happen, and, practically the complexity of the algorithm is of the order of K^S. We note that at least K^S character comparisons must be done to make sure that P does not occur in S.

3.3.2 Fast Search

The algorithm above can be improved by more sophisticated plans of comparisons between characters of P and S [39, 155]. Advanced algorithms for string

72   3 Computer Science Algorithms

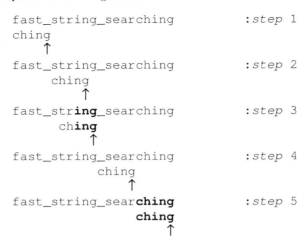

Fig. 3.1. Steps of Boyer–Moore fast string search

searches have a data-dependent structure in the sense that the order of operations performed depends on the characters in the pattern string P and on the results of comparisons between the characters in P and S. We present the idea behind the construction of the algorithm developed by Boyer and Moore [39]. Again we assume that we are comparing characters of a pattern P and a string S and that, in the course of algorithm we slide P along S from left to right. The interesting observation is that when we are comparing strings from left to right, it is most informative to start the character comparisons from the character of S facing the last character $P(K^P)$. Instead of formally defining all steps of the algorithm, we shall illustrate it using the following example, where the pattern is $P = ching$ and the string is $S = fast\_string\_searching$. The consecutive steps of the algorithm are presented in Fig. 3.1. The algorithm has a pointer (a number), which specifies the position in the string S which the last character of P is currently facing. This pointer is represented by ↑ in Fig. 3.1. Each step of the algorithm has the following structure.

One Step of the Boyer–Moore Algorithm

(1) Compare strings S and P, starting from ↑ in the backwards direction (right to left).
(2) Depending on the sequence of matching characters found and depending on the known structure of the string P,
   - either report "match found" and terminate,
   - or move ↑ to the right by an appropriate number of positions and go to (1).

Construction of the Boyer–Moore Algorithm

Let us discuss the steps in Fig. 3.1. *step 1*. At the beginning the strings P and S are aligned at their leftmost positions and the character in S corresponding to $P(K^P)$ is "_". Since this character does not appear in P there is no possibility of finding a match until we move pointer to the right by $K^P = 5$ positions. So the pointer is moved to the right by K^P positions. *step 2*. The character in S corresponding to $P(K^P)$ (pointed to by the pointer ↑) is now "n". This character appears in P at the position one before last and we move the pointer by one position right, such that the two characters "n" coincide. *step 3*. Comparing backwards, we find a matching substring "ing". This substring does not appear in P at any position except at the end. Therefore, analogously to step 1, we can again move the pointer to the right by K^P positions, since there is no possibility of obtaining a match in a move by a lower number of positions. *step 4*. This is analogous to step 1. The character "r" does not appear in P so we move the pointer by K^P positions to the right. *step 5*. Match.

In the example above, the algorithm needed only four steps and 11 character comparisons between P and S to find a string match. On the other hand it requires more complicated operations to be done during its operation than were necessary in easy search, namely inquiring whether and at which position the character "n" or the substring "ing" appeared in the string P. These operations are again string search problems, but owing to their repetitive nature in fast search, they can be coded more efficiently, by indexing the string P (overviewed in the next section) at the beginning of the procedure.

One can see that the speed of execution of the Boyer–Moore fast search algorithm, on average, increases with increasing the length of the string P. Its average execution time is cK^S, where c is less than 1. For this reason, this method is called a sublinear string search algorithm.

3.4 Index Structures for Strings. Search Tries. Suffix Trees

The algorithms described in the previous section involved the situation where both the pattern string P and the search string S were supplied as the initial data. However, a situation encountered very often is one where the search string S remains the same and multiple inquiries are made about its contents. In such a situation, it is reasonable to derive some indexing structures that should speed up access to elements in S. Here we describe some of the approaches to doing this [154, 268]. Before discussing memory structures for representing strings and improving searching tasks, we should highlight three aspects (parameters) that can be used for grading the efficiency of any proposed approach: (1) the construction time, the computational load related to

74   3 Computer Science Algorithms

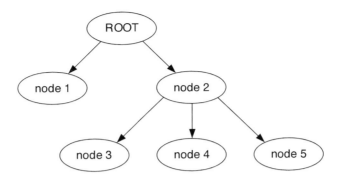

Fig. 3.2. A treelike structure, which can be used to store data in the memory of a computer

creating the indexing structure in the memory of the computer; (2) the memory capacity necessary to store the data created; and (3) the access time, i.e., how fast we can perform string search inquiries with the use of the structure. As for (3), the access time for all indexed structures is typically of the order of the length of the pattern string. However, (1) and (2) may be different for different approaches, as we will see later in this section.

3.4.1 A Treelike Structure in Computer Memory

Let us describe the key idea behind organizing data in computer memory in a treelike structure. This will help in going through subsequent algorithms. We shall discuss this using the example in Fig. 3.2. The tree has nodes and branches. We assume that branches are directed arcs and that they express relations between parent and child nodes. Each of the nodes is located in some area in memory. A node (or, more precisely, the memory area occupied by a node) may contain data that is significant for the searching tasks but, importantly, parent nodes also contain information about the memory locations (addresses) of their child nodes. For example, the data stored in node 2 in Fig. 3.2 will hold memory addresses of nodes 3, 4, and 5. The addresses written in the nodes describe the topology of the tree. ROOT is a distinguished node, which allows one to address the whole tree. For building treelike structures, computer languages with dynamic memory addressing are most efficient; good examples are C and C++. Using the topology of the tree and all the other information written in the nodes, tree search algorithms can explore this data structure very efficiently.

An important index which characterizes a tree structure is its size, which we assume is proportional to the number of its nodes. Why is the size of a tree proportional to the number of nodes and why can we disregard the number of branches? In a tree, each child node has exactly one parent; we can assign

3.4 Index Structures for Strings. Search Tries. Suffix Trees

each branch to the child node that it points to and therefore the number of branches is always equal to the number of nodes.

3.4.2 Search Tries

To develop the ideas of indexing we start from search tries. The term trie is derived from "re*trie*val". These are memory structures for representing efficiently lists of words, for example *trie, search, sea, string,* and *seal*. Assume that, given a pattern string P we are required to answer the question whether P appears among items (words) in the list. A naive and inefficient approach would be successive comparisons between P and words in the list. A much better approach mimics a method which everybody uses when looking up a word in a dictionary. When looking for the word "*pattern*" we start from the section for the letter "*p*" then we move to the pages for "*pa*", and so on. In order to build a search trie using this idea, we put our list of words into lexicographical (alphabetical) order and add an artificial terminating symbol, such as $, at the end of each word:

$$sea\$$$
$$seal\$$$
$$search\$$$
$$string\$$$
$$trie\$ \qquad (3.4)$$

The terminating symbol $ is necessary because, when searching the trie which we shall construct for the above list, we must know whether we have encountered the end of a string (the terminating symbol $) or whether our pattern is only a substring of some other word. The search trie based on (3.4) is presented in Fig. 3.3. Each of the nodes relates to one of the characters in the list of words. The ROOT branches into two nodes because there are two possible first letters in the list (3.4), "*s*" and "*t*". The node "*a*" branches into three because, starting with the three-character prefix "*sea*", there are three possible choices for the fourth character "$", "*l*" and "*r*" for the words in the list (3.4), and so forth.

If the list (3.4) is stored in computer memory as the trie structure shown in figure 3.3, then an appropriate tree search algorithm will establish whether any pattern P belongs to (3.4), using a number of character comparisons proportional to the length of P. The construction of such algorithms is quite straightforward, (see exercises at the end of this chapter). Note that at the terminal nodes of the tree in Fig. 3.3 we can put pointers to memory locations containing explanations of these words, translations of these words to another language etc.

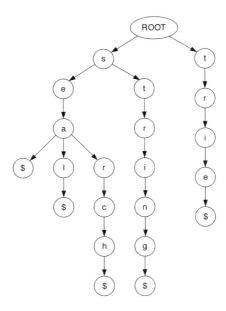

Fig. 3.3. A search trie for the list of words in (3.4)

3.4.3 Compact Search Tries

Since, as mentioned above, the size of a treelike structure is proportional to the number of nodes, the trie in Fig. 3.3 might seem somewhat inefficient owing to the necessity of traversing long sequences of nodes without any branchings. A more efficient structure with respect to the number of nodes is a compact search trie, or Patricia trie. The abbreviation Patricia stands for Practical Algorithm To Retrieve Information Coded in Alphanumeric [199]. The idea is to merge nodes if there are no branchings between them; and the compact search trie corresponding the list of words in (3.4) and to search trie presented in Fig. 3.3 is shown in Fig. 3.4. We observe that using the idea of merging, we have reduced number of nodes from 22 in the trie in Fig. 3.3, to nine in the compact trie in Fig. 3.4.

We should also mention that writing a computer program code for searching through compact tries is a little more complicated, since, unlike the trie in Fig. 3.3 where comparisons were done always between single characters, now (for the compact trie in Fig. 3.4) the items to be analyzed are both characters and substrings. Nevertheless, the savings in trie size are typically sufficient that it is worth developing a more complicated program.

3.4 Index Structures for Strings. Search Tries. Suffix Trees   77

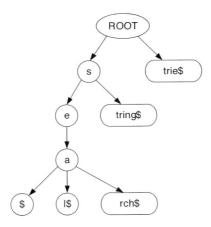

Fig. 3.4. The compact search trie (Patricia trie) corresponding to the strings listed in (3.4)

3.4.4 Suffix Tries and Suffix Trees.

Fast searching for words in vocabularies, as described above, is one possible string-searching problems. In the context of bioinformatics such searching can be applied, for example in algorithms for determining open reading frames (ORFs) (see Sect. 8.6, and Exercise 3 in this chapter). However, many other string-searching problems can be formulated, such as checking fast whether a pattern P is a substring of a string S, counting the numbers and positions of occurrences of patterns in strings, searching for the longest repeating substrings, and searching for substrings shared by two strings. Some of the possible string searching issues are listed as exercises at the end of this chapter. Effective, fast algorithms for solving these problems are of great utility in the analysis of biological sequence data. String-indexing structures convenient for addressing these problems are suffix tries and suffix trees.

A suffix trie (or suffix tree) for a string S is a search trie (or compact search trie) as in Fig. 3.3 or 3.4, constructed for all suffixes of the string S. A suffix of S is a trailing part of S. If we employ more a expanded notation for string of length n, namely $S(1:n)$, where the range of the indices of the characters of the string, $1:n$, is included, then a suffix of S is every substring of the form $S(i:n)$, $1 \leq i \leq n$. Consider a string

$$S = CACTAACTGA \qquad (3.5)$$

defined over the alphabet of letters A, C, G, T, which can symbolize the nucleotides in DNA. Below we show the set of all suffixes of S,

$$CACTAACTGA\$$$
$$ACTAACTGA\$$$
$$CTAACTGA\$$$
$$TAACTGA\$$$
$$AACTGA\$$$
$$ACTGA\$$$
$$CTGA\$$$
$$TGA\$$$
$$GA\$$$
$$A\$, \quad (3.6)$$

and the same set ordered alphabetically is

$$A\$$$
$$AACTGA\$$$
$$ACTAACTGA\$$$
$$ACTGA\$$$
$$CACTAACTGA\$$$
$$CTAACTGA\$$$
$$CTGA\$$$
$$GA\$$$
$$TAACTGA\$$$
$$TGA\$ \quad (3.7)$$

The terminating artificial symbol "$\$$" has been added for reasons analogous to those already discussed. Using (3.7), we can easily construct a search trie and a compact (Patricia) search trie for list of suffixes. These are called suffix trie, shown in Fig. 3.5 and suffix tree, shown in Fig. 3.6. The number of nodes in the trie in Fig. 3.5 is 58, while the suffix tree in Fig. 3.6 has only 18 nodes. More generally, if a string has length n, then number of nodes of its suffix trie is proportional to n^2, which we denote $O(n^2)$, while, on the average, the number of nodes of the suffix tree is of the order of $O(n)$. This becomes important for long strings. Looking at the tree in Fig. 3.6 we might have doubts concerning the true saving in memory use, since there are fewer nodes but they are occupied by longer substrings. However, in practice, the suffix tree for the string S in (3.5) will look more like that shown in Fig. 3.7. Here the nodes are not occupied by strings, but instead they contain ranges of characters in S given by pairs of indices of (pointers to) characters in S. Therefore, the memory capacity used by the information written in the nodes is only that necessary to hold two indices.

3.4 Index Structures for Strings. Search Tries. Suffix Trees   79

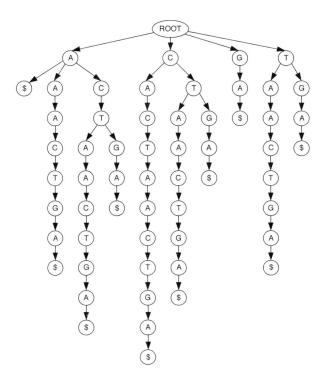

Fig. 3.5. Suffix trie for the string $S = CACTAACTGA$

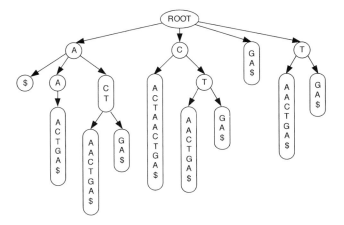

Fig. 3.6. Suffix tree for the string $S = CACTAACTGA$

80   3 Computer Science Algorithms

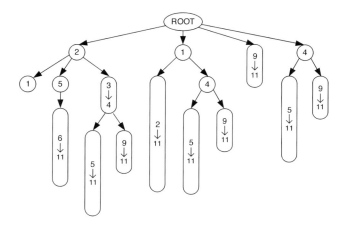

Fig. 3.7. In memory of the computer, the nodes of suffix trees contain indices to the text rather than the text itself. This figure shows the same suffix tree as in Fig. 3.6. However, strings in the nodes have been replaced by indices. The indices of the characters of the string are explained in the main text

3.4.5 Suffix Arrays

Suffix arrays [185] are indexed structures for strings that are more memory-efficient than suffix trees. A suffix array is a lexicographically ordered array of suffixes of the text. So, for the string S, if we assign numbers $1, 2, \ldots, 10$ to its suffixes listed in (3.6) (1 corresponds to the longest and 10 to shortest suffix), then, on the basis of (3.7), the suffix array for S will be

$$10\ 5\ 2\ 6\ 1\ 3\ 7\ 9\ 4\ 8.$$

The memory space required for suffix array corresponding to a string of length n is again $O(n)$; however, it is lower than memory occupancy of suffix trees in the sense that proportionality coefficients are different. If the memory requirement for suffix trees is C_1 and for suffix arrays it is $C_2 n$, then $C_2 < C_1$.

3.4.6 Algorithms for Searching Tries

In the above we have described indexed structures for efficient representation of strings, namely suffix tries, suffix trees and suffix arrays. In our presentation of searching algorithms below, we shall limit ourselves to searching through searching through suffix tries, which are less efficient, but easier to browse through. The algorithms, which we describe below are quite easily extendible from tries to trees. However, using them for suffix arrays would require more effort and, probably, consulting more specialized literature.

What is important is that string-indexed structures enable not only fast pattern matching but also many other very interesting and useful inquiries

Fig. 3.8. Results of searching for two patterns, $P_1 = ACT$ and $P_2 = CT$, in the string $S = CACTAACTGA$, on the basis of the suffix trie for the string S presented in Fig. 3.5. Both P_1 and P_2 are found, and their locations in the suffix trie are marked by shaded nodes

about strings. Below, we list some possibilities for using suffix tries, and we sketch the construction of the appropriate algorithms.

(1) *Pattern occurrence.* In Fig. 3.8, we show the results of searching for two patterns, $P_1 = ACT$ and $P_2 = CT$, in the string S, on the basis on the suffix trie for the string S presented in Fig. 3.5. Not surprisingly, both P_1 and P_2 are found; their locations in the suffix trie are marked by shaded nodes.

Below we show a pseudocode for matching occurrences of a pattern P of length K in a string S, on the basis of the suffix trie for S:

Program trie search for pattern
set $node = ROOT$
for $k = 1$ to K
    $c\_list \leftarrow \text{get\_children}(node)$
    if $P(k) \in c\_list$
        $node \leftarrow \text{child\_index}[P(k)]$
    elseif

```
            return(NO_MATCH)
            break
        endif
    endfor
return(MATCH)
```

The above program uses two functions, get_children, which returns a list of characters stored in the child nodes of a given node, and child_index, which returns the node index of the child whose character matches the character $P(k)$. The number of children of any node in the suffix trie is equal to or smaller than the number of symbols in the alphabet (which is constant). Therefore the computational complexity of the above algorithm is $O(K)$, of the order of the length of the pattern string. Note the saving in execution time in this algorithm, compared with the fast search described in Sect. 3.3. The latter algorithm needed $O(N)$, where N is the length of the string (text) S.

For simplicity, we have skipped one condition, sometimes quite important, when writing the above program code. Namely, before we invoke the function get_children(*node*), we should make sure that *node* is not a terminal symbol $.

(2) *Numbers of occurrences and positions of patterns.* The trie search program above verifies that the patterns $P_1 = ACT$ and $P_2 = CT$, occur in the string S, as seen in Fig. 3.8. Both of these patterns appear twice in string S, which can also be detected, by an appropriate method for searching the suffix trie. The number of occurrences of a pattern P in the string S is equal to the number of suffixes of S starting with P (more precisely, sharing P as their common prefix). In the suffix trie, this can be verified by the number of terminal symbols $ among the descendants (children, grandchildren, great-grandchildren, etc.) of the pattern P. Also, the positions of the occurrences of pattern P in the string S can be computed from the lengths of suffixes starting with P.

(3) *Longest repeating pattern.* The longest repeating pattern in the string S is the longest path in the associated suffix trie, from root down to a branching into at least two children. The path through a suffix tree with this property can be found by an appropriate search through the trie. Since the trie size is $O(n^2)$, the search will not take longer than that. For the suffix trie in Fig. 3.5 the longest repeating pattern is ACT.

(4) *Longest pattern shared by two strings.* This is an important issue in the analysis of DNA sequences. It can also have applications in other areas, such as analysis of scientific texts. Assume that two strings S, and Q are given, and form a new string as follows:

$$T = S\$Q\#.$$

Two artificial terminating/separating symbols have been added, $ and #. Construct a suffix trie for T and search it for the longest path P such that (a) it goes from the root down to a branching into at least two children, and (b) among descendants of P, we find both $ and #. This will give us the solution to the problem.

(5) *Palindromes*. Palindromes are strings, which read the same both from left to right and from right to left, for example *abba* or *abbcbba*. In the analysis of DNA sequences searching for palindromes and analyzing their frequencies is motivated by their supposed evolutionary importance. DNA palindromes are defined in a somewhat more complicated way, clarified in Chap. 8. For a string S, we define by $\bar{S} = \text{reverse}(S)$, the string resulting from reversing the order of characters. Now, by forming

$$T = S\$\bar{S}\#$$

and searching as in (4) above, we can find (the longest or all) palindromes in S.

(6) *Approximate matches*. Searching for approximate matches is also an important task in the analysis of texts and sequences. Two problems of approximate string matches are: (a) for a given pattern P, determine whether there is a substring Q in a text S such that the distance between P and Q is less then k; and (b) for two texts S and T, find the longest substring $Q \in S$ such that there is a substring $R \in T$ with a distance between R and Q less than k. By the distance between patterns, we mean the number of mismatching characters. Searching for approximate matches is more involved than searching for strict (regular-expression) matches, and requires more operations. Nevertheless, algorithms can be developed by extending and developing some ideas in regular-expression matching [111, 268, 263].

3.4.7 Building Tries

Before we can perform trie search operations, we must first have the trie constructed in the memory of the computer. We present below an efficient, recursive method for trie construction. This algorithm is given for example in Sect. 4.2.1 of [268], where it is called "a brute force method". It can be 'easily implemented and, despite some inefficiency resulting from its simplicity, can be useful for many tasks. When building tries and trees in the examples already described, we started from ordering strings alphabetically. In the algorithm below, this is not necessary; it will only slow down the computational time. An example of building a suffix trie, in six steps, for the string $CATCA$ is given in Fig. 3.9. The algorithm proceeds, starting from the root, by sequentially adding new suffixes

$$CATCA\$$$
$$ATCA\$$$
$$ATCA\$$$
$$TCA\$$$
$$CA\$$$
$$A\$$$

84   3 Computer Science Algorithms

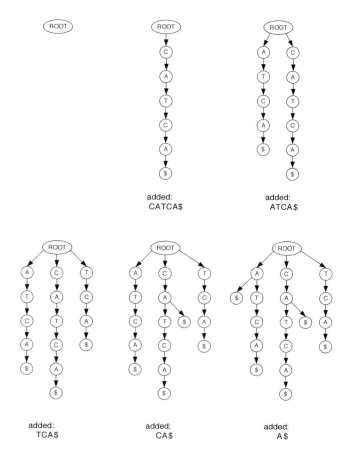

Fig. 3.9. Presentation of the steps in building a suffix trie for the sequence $CATCA$

to the existing trie. Each step of the algorithm includes two main operations.

One Step of the Trie-Building Algorithm

(1). Find the node of the present trie at which a suffix should be added. Add the suffix at the node found in (1), take the next suffix, and go back to (1).

Operation (1) of searching for the node at which the new suffix should be added can be accomplished by a (slightly modified) algorithm that performs a trie search for a pattern, as already described. If a suffix does not have any prefix in common with those already added to the trie, then it is added at the root of the trie, as in steps 2, 3, and 4 in Fig. 3.9. Otherwise, the suffix tail is added at the node, as shown in steps 5 and 6 in Fig. 3.9.

3.4.8 Remarks on the Efficiency of the Algorithms

The algorithms for building and searching tries presented in the above are quite simple, yet they enable one to pursue interesting projects involving string searches, and sequence and text analyses, provided that the sequences analyzed are not very long, for example they consist of texts of length of order of hundreds of characters. Some of their possibilities can be enhanced in ways illustrated by the exercises at the end of this chapter. When we try to apply these procedures to longer and longer strings, however, we encounter problems related mainly to prohibitive size of the suffix tries. So, in the next step in the development of software dedicated to string searches, we move to the more efficient data structure, suffix trees, shown in Figs. 3.6 and 3.7. This will easily move the string sizes from hundreds to thousands.

However, one can do better than that. More sophisticated algorithms [111, 268, 263] can both construct and search through trees (arrays) at lower computational expense, than can those presented here. Manbers and Myers [185] have given an algorithm for sorting suffixes of a string of length N, which performs the task in $O(N \log N)$ time. With the most advanced algorithms, one can perform string searches on very large texts or sequences, such as data in chromosomes or entire genomes (see Chap. 8). Some of the software for string search analyses can be downloaded from Internet sites such as [341].

3.5 The Burrows–Wheeler Transform

As mentioned above, pattern searches and string comparisons are performed on extremely large sets of genomic data; some results of such searches are presented in Chap. 8. Suffix trees and suffix arrays, covered in the previous section, provide very efficient techniques, but recently a method based on the Burrows–Wheeler (BW) transform, which both compresses the sequence data and allows for fast searches, has been developed. The Burrows–Wheeler method was initially aimed at creating an effective, lossless compression tool for long data strings by using the idea of transforming (permuting) the initial data string to an easily compressible form [47]. However, it was soon recognized that the BW transform cat itself be a very fast memory-occupancy-effective search tool for substrings of any length [84, 119, 180]. In this section we present the construction of the BW transform, the BW inverse transform, and a description of how the BW transform can be employed as a search engine.

Let us assume the following string over the alphabet A, C, G, T:

$$S = CACTAACTGA. \qquad (3.8)$$

The BW transform of the string S (in our example, of length $n = 10$) is constructed as follows. We first build an array $Z(S)$ such that the rows of the array $Z(S)$, numbered from 0 to $n-1$, are consecutive left-to-right cyclic

$$Z(S) = \begin{matrix} & & & & & & & & & & \text{Row} \\ \text{C} & \text{A} & \text{C} & \text{T} & \text{A} & \text{A} & \text{C} & \text{T} & \text{G} & \text{A} & 0 \\ \text{A} & \text{C} & \text{A} & \text{C} & \text{T} & \text{A} & \text{A} & \text{C} & \text{T} & \text{G} & 1 \\ \text{G} & \text{A} & \text{C} & \text{A} & \text{C} & \text{T} & \text{A} & \text{A} & \text{C} & \text{T} & 2 \\ \text{T} & \text{G} & \text{A} & \text{C} & \text{A} & \text{C} & \text{T} & \text{A} & \text{A} & \text{C} & 3 \\ \text{C} & \text{T} & \text{G} & \text{A} & \text{C} & \text{A} & \text{C} & \text{T} & \text{A} & \text{A} & 4 \\ \text{A} & \text{C} & \text{T} & \text{G} & \text{A} & \text{C} & \text{A} & \text{C} & \text{T} & \text{A} & 5 \\ \text{A} & \text{A} & \text{C} & \text{T} & \text{G} & \text{A} & \text{C} & \text{A} & \text{C} & \text{T} & 6 \\ \text{T} & \text{A} & \text{A} & \text{C} & \text{T} & \text{G} & \text{A} & \text{C} & \text{A} & \text{C} & 7 \\ \text{C} & \text{T} & \text{A} & \text{A} & \text{C} & \text{T} & \text{G} & \text{A} & \text{C} & \text{A} & 8 \\ \text{A} & \text{C} & \text{T} & \text{A} & \text{A} & \text{C} & \text{T} & \text{G} & \text{A} & \text{C} & 9 \end{matrix}$$

Fig. 3.10. The array $Z(S)$ resulting from circular shifts of the string $S = CACTAACTGA$

rotations (shifts) of S, as shown in Fig. 3.10. In the next step, we sort the rows of $Z(S)$ in lexicographic order. This leads to the array $Z_1(S)$ shown in Fig. 3.11. The last column of the array $Z_1(S)$ is the Burrows–Wheeler transform $BW(S)$ of S. For the string S in (3.8), its BW transform is

$$BW(S) = TGCAAAATCC.$$

Can the BW transform be inverted? Clearly, the BW transform of all cyclic rotations of S is the same as the string given above. In this sense, the transform presented in Fig. 3.11 is not invertible. But if we assume that we have data that allows us to find the correct phase of S then the answer is yes. One possible to define the phase of the string S is to inspect the rows of the array $Z_1(S)$. Since the rows of $Z_1(S)$ are all cyclic shifts of S, then at least one of them must be equal to S. We denote the index of the first such row by r; in Fig. 3.11, we have $r = 4$. This information allows the correct inversion of $BW(S)$. Another, more practical possible way to define the phase is to add one unique, artificial letter, $, to the alphabet and to label the end of the string with it; $ appears only once in the string and we assume that it is lowest in lexicographic order. Instead of S given in (3.8) we have $CACTAACTGA$$. The position of $ in $BW(S)$ will then take account of the row index r, since we know that it occurs as the last character of S.

3.5.1 Inverse transform.

We create one more array, $Z_2(S)$, which is obtained by moving the last column of the array $Z_1(S)$ to the front. Both $Z_1(S)$ and $Z_2(S)$ are shown in Fig. 3.12. Showing these two arrays together as in the figure is our starting point for describing the idea behind inverting the BW transform. The BW transform $BW(S)$ of S is simultaneously the last column of $Z_1(S)$ and the first column of $Z_2(S)$. The first column of $Z_1(S)$, which is also, simultaneously, the second

3.5 The Burrows–Wheeler Transform   87

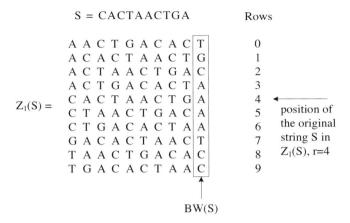

Fig. 3.11. The array $Z_1(S)$ is obtained by sorting the rows of the array $Z(S)$ in lexicographic order. The last column of $Z_1(S)$ is the Burrows–Wheeler transform $BW(S)$ of the string S

column of $Z_2(S)$, can easily be obtained from $BW(S)$ by ordering its characters alphabetically. This column will, therefore, be denoted by $SORT(S)$. The remaining columns of $Z_1(S)$ and $Z_2(S)$ are not known at the moment when we start the inverse transform, and they are shaded gray to depict this fact. By their construction, both arrays $Z_1(S)$ and $Z_2(S)$ contain all cyclic shifts of S as their rows. However, they appear in different orders in the two arrays. We ask, "Can we establish the correspondence between the rows of the matrices $Z_1(S)$ and $Z_2(S)$?" Consider all rows in $Z_2(S)$ which begin with the letter A. We can notice that these rows come in lexicographic order. Since the rows in the array $Z_1(S)$ which begin with the letter A also come in lexicographic order, then there is an obvious one-to-one correspondence between them. The same can be repeated for the letters C, G, and T, which leads to the one-to-one correspondence between all rows of the arrays $Z_1(S)$ and $Z_2(S)$. This correspondence is depicted by arrows in Fig. 3.12. We denote the transformation of the row numbers i which follows from this correspondence by $\varUpsilon(i)$.

From the description above we can see that the transformation $\varUpsilon(i)$ has the following meaning: a left-to-right circular shift of row i of the array $Z_1(S)$ by one position changes it to row $j = \varUpsilon(i)$ of the the array $Z_1(S)$. The transform inverse to $\varUpsilon(i)$ can be described as follows: right-to-left circular shift of row no i of the array $Z_1(S)$ by one position changes it to row $j = \varUpsilon^{-1}(i)$. The transformation $\varUpsilon(i)$ is depicted by the arrows in Fig. 3.12; for example, $\varUpsilon(0) = 3$, $\varUpsilon^{-1}(0) = 8$.

We can reconstruct S, either from left to right, by applying successively $\varUpsilon(i), \varUpsilon(\varUpsilon(i)), \ldots$ to the letters of $SORT(S)$, starting from the position $i = 4$,

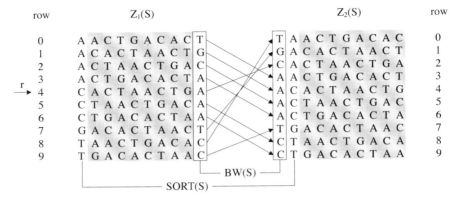

Fig. 3.12. Inverting the Burrows–Wheeler transform (see the text). The algorithm for inverse transform can be obtained by using the observation that there is a one-to-one correspondence between rows of arrays $Z_1(S)$ and $Z_2(S)$

or from right to left, by applying $\Upsilon^{-1}(i)$, $\Upsilon^{-1}(\Upsilon^{-1}(i))$, ... to the letters of $BW(S)$, again starting from $i = 4$. This is shown in Fig. 3.13.

3.5.2 BW Transform as a Compression Tool

The length of $BW(S)$ is of course equal to the length of S. However, if S is a text written in a natural language, with some structure, the sequence $BW(S)$ will be easy to compress. Typically, $BW(S)$ is a sequence such as

$$xxxxxxaaaaaaabbbbbbbbb\ldots, \qquad (3.9)$$

with a strong repetitive character and can compressed simply by storing successive letters and numbers of counts of them, for example, the sequence (3.9) can be stored as $x-6$, $a-6$, $b-9$, ... Why does the BW transform transform "natural" strings to very compressible forms? The answer is as follows [47]. Assume that S is an English text with many repetitions of the word "the". When $BW(S)$ is computed, "the" occurs many times at the beginning of successive shifts of S. In the next shift of a sequence with "the" at the leading position, we shall see "he" at the front, and "t" as the last character. All sequences with "he" at the front will most probably be arranged one after another, owing to the alphabetical ordering performed when $BW(S)$ is computed. So, for example, 100 occurrences of the word "the" in the text S will typically result in a block of 100 letters "t" in $BW(S)$. In [119], the following estimate appears. The suffix array [185] for the human genome constitutes approximately 12 gigabytes (3 billion 4-byte integers) of RAM. However, the

3.5 The Burrows–Wheeler Transform   89

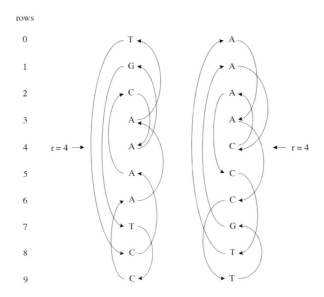

Fig. 3.13. Inverse BW transform. Reconstruction of the original string $S = CACTAACTGA$. In the *right plot*, starting from row $r = 4$ the reconstruction goes from left to right by traversing $SORT(S) = AAAACCCGTT$, following the arrows. In the *left plot* the reconstruction goes from right to left by traversing $BW(S) = TGCAAATCC$, following the arrows. Here the direction is reversed relative to the right plot

BW string alone, which is sufficient to determine word counts, as explained in the next subsection, can be compressed to about 1 gigabyte of RAM.

An interesting property of BW compression, which is intuitively clear from the above explanation, is that for natural-language texts the compression rate improves with increasing length of the text. This makes it a desirable tool for many applications.

3.5.3 BW Transform as a Search Tool for Patterns

Let us add \$ at the end of S in (3.8), i.e., define $S\$ = CACTAACTGA\$$. As mentioned, we use this to label the end of the string. The BW transform of $S\$$ is $BW(S\$) = TCAG\$AATCCA$, and the transformation $\Upsilon(i)$, which can be obtained by comparing $BW(S\$)$ and $SORT(S\$)$, is as follows:

i	= 0	1	2	3	4	5	6	7	8	9	10
$\Upsilon(i) =$	2	5	6	10	1	8	9	3	0	7	4

.

90   3 Computer Science Algorithms

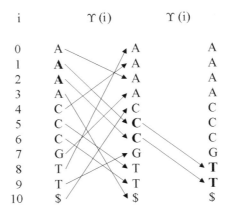

Fig. 3.14. Illustration of use of the BW transform to count occurrences of the substring ACT in the string $S = CACTAACTGA$

We shall use the above to illustrate the application of the BW transform to counting occurrences of patterns in S. How many times does the pattern ACT occur in S? Suppose we are performing an inverse BW transform by applying $\Upsilon(i)$ to elements of $SORT(S\$)$. We apply $\Upsilon(i)$ to all occurrences of letter A to $SORT(S\$)$ see how many of the $\Upsilon(A)$ end in entries of $SORT(S\$)$ containing the letter C, and then, in the next step, see how many of the AC have T as the next letter. This is illustrated in Fig. 3.14 where the transformation $\Upsilon(i)$ is represented graphically by arrows. By "following arrows" we count that the number of occurrences of the substring ACT in $S\$$ equals 2.

Given the compressed form of the BW transform for a given text S, one can perform pattern searches and compute numbers of occurrences of P in S, in a time proportional to the length of P [119].

3.5.4 BW Transform as an Associative, Compressed Memory

BW transform technique can also be understood as a kind of associative, compressed memory. We can decompress all the text at once, but we are often interested in decompressing only a fragment of a text. As seen from the above, decompressing goes together with reading from the beginning of the fragment. If we know that the text fragment to decompress starts with, for example, *"go to the specific location ..."*, then, using the search procedure presented in Fig. 3.14 we can retrieve the appropriate paragraph, of given length, from the compressed text.

3.5.5 Computational Complexity of BW Transform

Writing a computer code for performing a BW transform, as defined in fig. 3.11 is quite straightforward (Exercise 15), involves cyclic rotations and arranging rows of an array in alphabetical order. However, both the memory capacity requirement and computational time in this simplest approach will be of the order of $O(N^2)$. For applications to large strings, more efficient methods are desired. Burrows and Wheeler [47] have described a method that performs the operation, on average, in $O(N \log N)$ time, using $O(N)$ space in computer memory.

3.6 Hashing

Hashing is a technique of recording (bookkeeping) the occurrences of some data objects by using a randomized method to address a data structure called a hash table or accumulator array, to bookkeep the process [91]. This data structure is then used to obtain useful information about the object under study. The idea of hashing and hash tables originally arose among computer programmers, specifically developers of interpreter languages, to create an easily addressable memory storage structure to keep and update lists of words, labels, etc. In the course of time hashing techniques became very widely used, not only in computer sciences but also in other areas, such as computer vision, pattern analysis, and bioinformatics.

3.6.1 Hashing functions for addressing variables

An interpreter language, like Basic, executes a list of commands. It may encounter a line such as

$$\vdots$$
$$WIDTH = 7$$
$$\vdots$$

which means that a variable named $WIDTH$ should be created and given a value 7. A problem arises: at which specific memory locations should the name of the variable and the value of the variable be placed? One possibility is to (a) go through the text of the program and find all variables names, (b) arrange them alphabetically, (c) store in memory on the basis of alphabetical order of the names. If we wanted to keep the list of variable names in memory and search through this list to determine which memory address a variable occupies, it would be rather time-consuming. An idea which can be very helpful is to construct a hashing function, which will point, rather randomly, to some memory address. For example, assume that the array for storing the values of variables in the computer memory is planned to have 100 entries; we define a hashing function as follows

$$\text{memory location of variable} = \text{product of ASCII codes of characters in its name } mod(100).$$
(3.10)

Since the ASCII codes of the letters in our variable name $WIDTH$ are $(W)87$, $(I)73$, $(D)68$, $(T)84$, and $(H)72$, then its memory location will be

$$\text{memory location of } WIDTH = (87 \cdot 73 \cdot 68 \cdot 84 \cdot 72) mod(100) = 64.$$

The next variable encountered in the program can be "$HEIGHT$" and it will get memory address $(72 \cdot 69 \cdot 73 \cdot 71 \cdot 72 \cdot 84) mod(100) = 12$. Note that executing the hashing function for a given variable name is much faster than searching for it through the list of variable names.

It is rather difficult to predict where our hashing function is going to go with our variable, but this does not pose a problem. The problem which one meets when using hashing functions is collisions, namely it is possible that a hashing function will give us the same value for two different names. This problem will be discussed in the next subsection. Defining hashing functions is an art in computer programming. One can imagine many choices of hashing function for the problem of addressing variables presented here that will perform comparably well. A good hashing function (1) spreads addresses approximately uniformly over the address space and (2) is likely to give different values if the objects differ in an intuitive sense. The second requirement is maybe not completely clearly stated here, but can we explain it using the example of the function defined "ad hoc" in (3.10). Intuitively, a weak point of the hashing function (3.10) is that it is insensitive to permutations of characters in variable names; for example, if one defines variables "ABC", "BAC" and "CAB", for some computations concerning triangle ABC, they will all get the same memory address.

3.6.2 Collisions

We have reserved space for 100 variables and now variables are appearing with more or less random names, created by a programmer. No matter what hashing function we use, it is very unlikely that this hashing function called for 100 different names will point different numbers in the range 1, ..., 100. For example, as already said ABC, BAC, and CAB will all get the same address, 43, also GH will point to the same address, 12 as $HEIGHT$. The occurrence of such phenomenon is called a collision. At first sight it may seem that collisions kill the idea of hashing, but fortunately there are many methods for collision resolution. Collision resolution procedures search through the data array for a free address. There may be several different strategies, such as:

- *Linear probing.* If the address in the data array is busy try the next in row position until a free address is found. At the end of the data array wrap to the front.
- *Double hashing (or multiple hashing).* If a memory location pointed to by the current hashing function is already occupied, try the next hashing function.
- *Hashing with buckets.* In this variant of the collision resolution, more than one record can be stored at the position pointed by a hashing address.

3.6.3 Statistics of Memory Access Time with Hashing

There are three parameters, which interfere with each other when data is being stored in computer memory by the use of a hashing method, N, the number of memory addresses, R, the number of variables (records) to be stored and T_A, the access time. Clearly, the average of the access time, $E(T_A)$, should increase with increasing packing density R/N in the computer memory [154, p. 539]. However, it is useful to develop some intuition about the quantitative aspects of this relation. So, let us derive an appropriate expression. Assume multiple hashing model with all hashing functions independent, and ideal in the sense that they spread addresses over N locations in memory with a uniform, discrete distribution, each address having a probability $1/N$ of being pointed to. Assume also that the operation of calling a hashing function for a record (variable) name and the operation of one memory access, together take one unit of computer time. Imagine the process of successive storing of R records in N locations. If $r - 1$ records have already been stored, then the probability that the hashing function will point to a free address is

$$p_r^{free} = \frac{N - r + 1}{N}.$$

If we denote by T_A^r a discrete random variable modeling the number of computer time units necessary to place the rth record in a free memory location, then $T_A^r = 1$ with probability p_r^{free}, $T_A^r = 2$ with probability $(1 - p_r^{free})p_r^{free}$, $T_A^r = 3$ with probability $(1 - p_r^{free})^2 p_r^{free}$, and so on. Distribution of T_A^r is geometric (chapter on probability and statistics) and it is easy to derive

$$E(T_A^r) = \frac{N}{N - r + 1}.$$

Then we have that on average, a memory access operation for one record will take

$$E(T_A) = \frac{1}{R} \sum_{r=1}^{R} E(T_A^r) = \frac{N}{R} \sum_{r=1}^{R} \frac{1}{N - r + 1} \qquad (3.11)$$

units of computer's time. Note that when N and R are large numbers, the expectation above can be approximated well by

$$E(T_A) = \ln(1 - x) \qquad (3.12)$$

where $x = R/N$.

94   3 Computer Science Algorithms

3.6.4 Inquiring About Repetitive Structure of Sequences, Comparing Sequences and Detecting Sequence Overlap by Hashing

From the perspective of bioinformatics a process more interesting than just memory access by hashing is using hash tables to inquire the structure of biological sequences. We shall describe some of the techniques for doing this in this subsection. In the applications in the title of this subsection, the idea of hashing is composed with the related method of accumulator arrays, which we discuss first.

Accumulator Arrays

Accumulator arrays serve for the purpose of registering occurrences of items in data. For example, assume we have a sequence of DNA, composed of the symbols for nucleotides, a, c, g, and t. The following example sequence

```
tgagtttgta cattactttt cgtatttcta taaacaaaaa aaagaagtat aaagcatctg
catagcaatt aataaaaagg tgaccatccc atatatataa cactcaaatt tgatggatcc
gtggcttgct gaatcaaatc ttgtacgcta gactctacac ttagtccatt acccataagc
ttctcttcta cacctttaag ggccctataa gactcttggt tttcgttcct ..........
```

is a front fragment of the gene TEL1 on chromosome II of the organism, baker's yeast (Saccharomyces cerevisiae). The sequence of nucleotides was downloaded from the NCBI Gene Bank page [326]. This gene codes for one of the protein kinases. It is a homolog of the human ATM gene, which has many functions in the human genome. For the purpose of this example, however, the information about the functions of the is not important. The fragment above is rather short compared with the length of the whole gene TEL1, which is about 9000 base pairs long. We know that amino acids and their orders are coded by triplets of nucleotides (see Chap. 8). There are $4^3 = 64$ possible triplets, and we might be interested in how often each of the triplets appear in the DNA sequence analyzed. So we build an array addressed by 64 possible triples of nucleotides, called accumulator array, and reading the sequence triplet by triplet, tga, gag, agt, gtt, ... we increment the corresponding entry of the accumulator array at each step. Note that in order to perform the inquiry about frequencies of nucleotide triplets, we only need 64 memory locations, no matter how long the DNA sequence analyzed is.

Hash tables

What if we want to pursue a similar inquiry to that described above, but instead of triplets we now want to analyze patterns of length 20; in DNA sequences, these are called 20-mers or 20-tuples. Clearly the size, 4^{20} of an array of all possible 20-mers is prohibitive in terms of computer memory. The number of different 20-mers in our DNA sequence is of course far less than 4^{20}, but we cannot predict which ones will appear in the sequence. What helps is to derive a hashing function for assigning memory addresses to 20-mers, as described in Sect. 3.6.1. Once we have an engine to store them in memory, we can record occurrences of 20-mers in the analogous way to that described for triplets.

Because it is simple and efficient at the same time, this idea is used very intensively in the analysis of biological sequences. The items whose occurrences we register can be of different types and can come from many different biological sequences. Note that in this memory structure, called a hash table, we can record not only the occurrence of an item, but also its type, source, etc. This gives us a lot of flexibility in elaborating algorithms and creates an extremely broad range of applications [68, 117, 179, 288]. We list some of them below.

(1) *Repetitive structure of a sequence.* We construct a hash table for the sequence, using for example 20-mers again, and by inspecting the entries of the hash table which are incremented to values greater or equal to 2, we obtain knowledge about the repetitive structure of the sequence.
(2) *Comparing a biological sequence with a database.* Suppose we have downloaded a large database of biological sequences and we want to compare our sequence against it. So we create a hash table and we record the occurrences of items (e.g., 20-mers) both in the sequences in the database and in our sequence. We record not only the occurrence but also the source, i.e., the database or our sequence. Then by looking through the hash table for entries activated by both items from the database and items from our sequence, we learn about their similarities [68, 179, 288].
(3) *Sequence overlap.* A similar technique can be used in DNA assembly. Assume we have a large collection of DNA sequences, each of a length of the order of 1000 base pairs. Our aim is to detect all pairwise overlaps in these sequences. We define two sequences to overlap if they share at least two different 20-mers. By using a hash table, we can quickly build a graph of their overlap structure [117].

3.7 Exercises

1. Write a computer program for the quicksort algorithm described in Sect. 3.2. By using randomly generated data, study the computational time of this algorithm.
2. Write a computer program for the easy search algorithm described in Sect. 3.3.

3. Draw a search trie for the three-letter codons of the genetic code, listed in Table 8.1. Use this trie for developing a program for translating triples of nucleotides into codes of amino acids.
4. Develop a computer program for building a search trie, given a list of words.
5. Develop a computer program for building a suffix trie for a given string.
6. Write a computer program for searching for a pattern in a search trie.
7. Write a computer program for the following:
   a) Searching for the longest repeating pattern in a string,
   b) Searching for the longest pattern shared by two strings.
   c) Searching for palindromes in a string.
8. Develop appropriate algorithm and write a computer code for searching for shortest non repeating patterns in strings.
9. Elaborate an algorithm and write a computer program for searching for a substring in a given string S which approximately matches a pattern P.
10. Try to generalize programs in problems 7.1–7.3 by replacing matching by approximate matching.
11. Optimize the programs developed in the previous problems by replacing less effective structure of a suffix trie by the more memory-efficient suffix trees.
12. Compute the BW transforms of the following strings:

$$abababababababab\$$$

$$abcdabcdabcdabcd\$$$

Discuss the form of the results obtained.
13. Write a computer program for performing the BW transform for a given string S.
14. Write a computer program for performing the inverse BW transform.
15. Develop a computer program for searching for a pattern P in a string S by using the BW transform.
16. Write a computer program for storing memory items with by use of hashing. Experiment with different hashing functions and with different methods of collision resolution.
17. a)
    b) Compute the variance of the random variable T_A whose expectation was computed in equation (3.11).
    c) Derive probability distribution of the random variable T_A.
18. Write a computer program for recording occurrences of l-mers in a DNA sequence with the use of a hash accumulator array. Experiment with using this program for the tasks mentioned in section 3.6.4.
19. Describe, how we can use the idea of hash arrays to find all occurrences of palindromes of length 10 in a given text. Write an appropriate computer program.
20. Derive the approximation obtained in (3.12).

4
Pattern Analysis

A pattern is a general concept and means a form, a template, a model, or, more abstractly, a set of rules or a data structure. Pattern recognition is the detection of underlying patterns in data. By pattern analysis we mean analysis of data on the basis of patterns, involving pattern recognition, classification, modeling and statistics.

An important class of patterns is those related to images. Images are an interesting form of biomedical data, and looking for patterns in images can give useful information. Also, the analysis of images provide nice, comprehensive examples of pattern analysis algorithms.

In this chapter we cover some pattern analysis algorithms. Pattern analysis is a broad field, with many applications and strong links to information processing, computer science, biometrics and biostatistics. The size of bioinformatic data and databases excludes most manual operations on this data, and the successful extraction of useful information relies heavily on the effectiveness of automatic browsing, searching, and linking. Combining of browsing bioinformatic data files with pattern analysis algorithms, such as automatic classification, has great potential and can lead to very interesting findings.

4.1 Feature Extraction

A feature is a mapping from a pattern space or image space to a feature space i.e., a space of numbers or vectors. Examples of features are the positions of image fragments, the areas of parts of images, lengths of contours of objects in images, the numbers of occurrences of certain strings in sequences, and the coefficients of series expansions (Taylor, Fourier, etc.) of functions associated to images or patterns related to experiments performed.

Feature extraction is one of the initial steps of pattern analysis. The definition of features for a given situation is a crucial element in the art of construction of pattern analysis algorithms. If the features defined correspond well to the type of information one is looking for, then it is likely that the whole pattern analysis system will perform satisfactorily. However, it can also happen that the relations between the defined features and patterns we are after are so ambiguous that the pattern analysis system will eventually fail.

In the systems of measurements that are performed in molecular biology, biochemistry or genetics, it often happens that experiments and the analysis of them provide hundreds, thousands, or even more features, and the problem is to reduce their dimensionality or to look for a hierarchy in the data.

4.2 Classification

The classification problem for patterns involves setting discrimination rules, based on the features and a knowledge of the classes of the patterns. In the sequel, we will treat terms "classes" and "states of an experiment" as synonymous. The simplest formulation of the classification task is as follows. There are two possible states of the experiment coded, for example, as 0 for normal and 1 for disease. The feature extraction system has already been designed and we have $x_1, x_2, \ldots, x_n$, which are feature vectors known to correspond to state 0, and $x_{n+1}, x_{n+2}, \ldots, x_{n+m}$, which are feature vectors known to correspond to state 1. The feature space is k-dimensional, i.e., $x \in R^k$. The problem is to find a scalar function (a classifier) $f(x)$ defined on the feature space such that $f(x_i) = 0$ for all $i = 1, \ldots, n$ and $f(x_i) = 1$ for all $i = n+1, \ldots, n+m$. If we succeed in the construction of the classifier, it will allow automated classification of experimental states on the basis of extracted features.

A more general formulation of the classification problem involves more than two classes. Also, it may be either impossible or unsuitable to obtain a perfect discrimination. As a result of noise in the data, the knowledge about the classes may be erroneous, and so it can be more reasonable to allow for some classification errors.

4.2.1 Linear Classifiers

Let us code the states of the experiment by -1 and 1 instead of 0 and 1. This change of coding is motivated by our aim of using a "sign" function, which returns values -1 and 1. By a linear classifier function or linear discriminant function, we mean a function

$$f(x) = \text{sign}(w^T x + w_0). \tag{4.1}$$

In the above "sign" the a sign function, which returns -1 or 1 depending on whether the argument is negative or positive; x is a feature vector, which is a k-dimensional, column vector, w^T is a k-dimensional row vector of weights, w_0 is a scalar, and the superscript T denotes, as usual, vector transposition. The function (4.1) could also be called an affine classifier owing to the occurrence of the offset term w_0. The geometric locus in the space $x \in R^k$

$$L = \{x : w^T x + w_0 = 0\} \tag{4.2}$$

is called a separating hyperplane.

With the data $x_1, x_2, \ldots, x_n$ and $x_{n+1}, x_{n+2}, \ldots, x_{n+m}$, belonging to two classes as described above, the problem of the construction of the linear classifier (4.1) can be formulated as follows. Find a vector $w \in R^k$ and a scalar w_0 such that $w^T x_i + w_0 < 0$ for all $i = 1, \ldots, n$ and $w^T x_i + w_0 > 0$ for all $i = n+1, \ldots, n+m$. Using matrix and vector notation, these $n+m$ inequalities can be written as

$$M \begin{bmatrix} w \\ w_0 \end{bmatrix} < 0 \tag{4.3}$$

where $[w^T \ w_0]^T$ is a $k+1$-dimensional column vector, and M is a matrix defined as follows

$$M = \begin{bmatrix} x_1^T & 1 \\ \vdots & \vdots \\ x_n^T & 1 \\ -x_{n+1}^T & -1 \\ \vdots & \vdots \\ -x_{n+m}^T & -1 \end{bmatrix}. \tag{4.4}$$

As we can see, the problem of constructing a linear classifier reduces to the problem of looking for a vector $[w^T \ w_0]^T$ which satisfies the system of linear inequalities (4.3). The system (4.3) is homogeneous, which means that if $[w^T \ w_0]^T$ solves it, then any other vector $\alpha [w^T \ w_0]^T$ obtained by multiplication by a positive constant $\alpha > 0$ is also a solution. So, equivalently to (4.3) we can analyze

$$M \begin{bmatrix} w \\ w_0 \end{bmatrix} \leq -\mathbf{1}, \tag{4.5}$$

where $\mathbf{1}$ means a vector with all entries equal to one.

One method for solving (4.3) or (4.5) with respect to $[w^T \ w_0]^T$ is by using the linear programming algorithm mentioned in Chap. 5. One can easily define a linear programming problem with the property that its solution solves the system of inequalities (4.3). One possibility is as follows:

$$\max z \tag{4.6}$$

subject to

$$M \begin{bmatrix} w \\ w_0 \end{bmatrix} \leq -\mathbf{1} z \tag{4.7}$$

and

$$0 \leq z \leq 1. \tag{4.8}$$

In the above formulation, the variables of the linear programming problem are w, w_0, and z, (one scalar variable z has been added). If the optimal solution

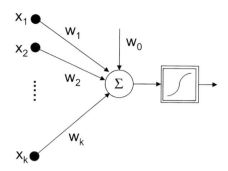

Fig. 4.1. Graphical representation of a model of an artificial neuron

to (4.6)–(4.8) is $z_{\text{opt}} = 1$, then a solution to (4.3) exists. If $z_{\text{opt}} = 0$, then the system (4.3) is infeasible.

The method of choosing weights $w_1, \ldots, w_k$ by solving linear programming problem, as shown above, is very efficient. However, many other methods are also in use. One group of approaches to adjusting the weights in classifiers uses iterative procedures [67], where weights are modified step by step and the procedure stops when solution to (4.3) is achieved. Such procedures are called training of the classifier. For single classifiers (4.1), they are only toy algorithms, but become important when classifiers are organized in larger structures, namely artificial neural networks, as we outline below.

4.2.2 Linear Classifier Functions and Artificial Neurons

The linear classifier functions (4.1) are closely related to artificial neurons. A model of an artificial neuron can be represented graphically as shown in Fig. 4.1. Signals (the elements of the vector x) $x_1, \ldots, x_k$ are multiplied by weights $w_1, \ldots, w_k$, summed up with offset w_0. This step of signal transformation is the same as in the linear classifier (4.1). The threshold element "sign" which appears in (4.1) is, in the neuron in Fig. 4.1 replaced by a smooth function, for example a sigmoid [67]. The function in the output block of the artificial neuron is called the neuron activation function. This function is chosen to be a smooth approximation of a thresholding element, i.e., a sigmoid, logistic, arctan function, etc. The smoothness makes neuron activation functions more physically sound and, more importantly, makes it possible to construct training algorithms based on derivatives.

4.2.3 Artificial Neural Networks

As discussed above, linear classifiers or single neurons can perform linear discrimination; in other words the separation can only be done by means

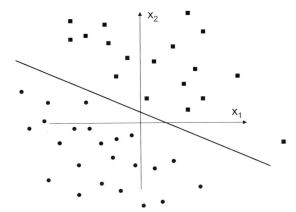

Fig. 4.2. An example of two classes which can be separated linearly. The feature vectors are two-dimensional. The two classes are marked by circles and squares

of lines, planes, or hyperplanes. An example of a linear separation of feature vectors belonging to two classes, depicted by circles and squares, is shown in Fig. 4.2.

However, one may wish to design discriminant systems which allow more complicated boundaries between classes. This aim can be achieved by combining several neurons into a network, as shown in Fig. 4.3. The neural network presented in the upper part of Fig. 4.3 is called a multilayer perceptron, or hidden-layer perceptron. This is a simple example, where the input vector x has two components x_1, x_2 and the total number of neurons in the network is three. This neural network is organized into three layers. The first, input layer is built from the input signals x_1, x_2. The second, hidden layer contains two neurons, which, as their inputs, take sums of the input signals with different weights, w_{11}^1 and w_{12}^1 with an offset w_{10}^1 for the first neuron, and w_{11}^1 and w_{12}^1 with an offset w_{20}^1 for the second neuron. The superscript 1 indexes the first layer. The outputs from the neurons in the second layer are fed into the last, third layer, which has only one neuron, with an output signal y. In the lower plot in Fig. 4.3 we present the shape of a separation line which can be obtained with the use of the neural net in the upper plot. This line separates two different classes determined by states of the experiment, marked by circles and squares. Such a shape of the separation line cannot be obtained with a single-neuron classifier.

Artificial neural networks of the type shown in the upper part of Fig. 4.3 can have more than one hidden layer, as well as more neurons in each of the layers. The crucial task is the training algorithms for artificial neural networks. A well-known recursive algorithm for adjusting the values of the weights is called back propagation [67].

102   4 Pattern Analysis

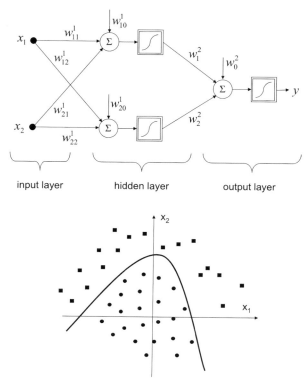

Fig. 4.3. *Upper plot*: a neural net with three layers, two input signals x_1 and x_2, and one output signal y. *Lower plot*: the shape of a separation line, which can be obtained with the use of neural net from the upper plot. The line separates two classes marked by circles and squares

4.2.4 Support Vector Machines

We now return to the problem of the construction of linear classification functions (4.1). In this subsection, we introduce an approach using supporting vector machines (SVMs) (SVM). The idea of SVMs involves designing a linear classifier which is optimal in the sense of its distances to points belonging to separated classes. Before explaining this idea in more detail, let us recall the fact from the multidimensional analytical geometry, namely that the distance between a hyperplane $H = \{x : w^T x + w_0 = 0\}$ and a point $y \in R^k$ is given by the formula

$$d(H, y) = \frac{|w^T y + w_0|}{\|w\|}. \tag{4.9}$$

In Fig. 4.4 we have shown a graphical representation of the situation where the feature space is two-dimensional, i.e., the vector x has two components x_1, x_2 and there are two classes, marked by circles and squares. This figure shows two plots, with identical locations of some feature vectors corresponding to two classes. There are infinitely many possible linear discriminant functions for separating the two classes. The left and right plots present two separating lines related to two different linear discriminant functions. In the left-hand plot we presented a "randomly chosen" separating line, obtained, for example, by some recursive procedure for modification of the weights. In the right-hand plot, we present a separating line related to a special discriminant function $f^*(x) = \text{sign}(w^{*T}x + w_0^*)$. Denoting the feature vectors in Fig. 4.4 by $x_1, x_2, \ldots, x_n$ (those corresponding to circles), and $x_{n+1}, x_{n+2}, \ldots, x_{n+m}$ (corresponding to squares), the special property of this separating line can be explained as follows. The separating line in the right-hand plot, $L^* = \{x : w^{*T}x + w_0^* = 0\}$, has the property that (i) it separates the two classes and (ii) it maximizes the minimal distance $d(L^*, x_k)$ between L^* and the points $x_1, x_2, \ldots, x_{n+m}$:

$$w^*, w_0^* \leftarrow \max_{w,w_0} \min_{1 \leq k \leq n+m} d(L, x_k). \tag{4.10}$$

The conditions (i) and (ii) determine uniquely the parameters w^*, w_0^*. Using (4.9) and (4.10) and recalling the idea of construction of the matrix M in (4.4), one can derive that the parameters w^*, w_0^* can be obtained from the solution to the quadratic programming problem

$$\min w^T w, \tag{4.11}$$

subject to the constraints

$$M \begin{bmatrix} w \\ w_0 \end{bmatrix} \leq -\mathbf{1}. \tag{4.12}$$

The symbols M and $\mathbf{1}$ have the same meaning as in (4.3)–(4.5). The quadratic programming problem is also mentioned in Chap. 5 in (5.56) and (5.57).

There are many examples where the optimal discriminant function

$$f^*(x) = \text{sign}(w^{*T}x + w_0^*)$$

shown in the right-hand plot in Fig. 4.4 has better properties than a "randomly chosen" discriminant function, such as the one in the left hand plot in Fig. 4.4. One may also wish to extend the method to discriminating to more than two classes and to more complicated shapes of the separating lines or surfaces. An appropriate methodology can be designed by developing the ideas sketched above, [45, 46]. Classifiers based on an optimal separation of the kind shown in the right plot in Fig. 4.4 and described in (4.9)–(4.12) are called supporting vector machines.

4.3 Clustering

Clustering involves a situation where there is a need to identify classes solely on the basis of feature vectors. We infer classes by using the hypothesis that

104   4 Pattern Analysis

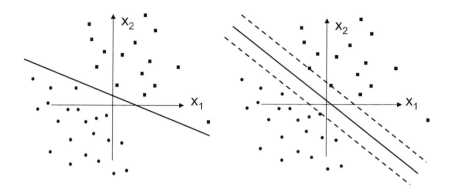

Fig. 4.4. In the *left-hand plot* a "randomly chosen" separating line is presented. In the *right-hand plot* we show the separating line which maximizes the minimal distance $d(L^*, x_k)$ between L^* and points x_1, x_2, ..., x_{n+m}

separate classes correspond to regions where data points occur with increased density. We call such regions of increased density clusters. An example is shown in Fig. 4.5. In the plot in figure 4.5 we can see clearly that the data points, representing some features or patterns, tend to be concentrated around two points, forming two data clusters. So it may be reasonable to hypothesize that these two clusters are related to two different classes in the data.

We shall present two algorithms for clustering, the K-means algorithm and the hierarchical clustering algorithm. Both of these approaches are related to methods presented also in other chapters of this book. The K-means algorithm can be interpreted in terms of analyzing mixtures by using EM methods (Chap. 2), and hierarchical clustering is closely related to inferring trees (Chap. 7). In order to decide whether the data points are densely or sparsely located, one needs to use some distance measure. The most natural is the Euclidean distance, but other distances can also be used.

4.3.1 K-means Clustering

The idea behind the algorithm for K-means clustering is very simple, and similar to the idea the EM algorithm for estimating the parameters of mixtures of distributions.

Concerning the construction of the algorithm, we do the following:

(I) We assume that the number K of clusters is known, and we make two more assumptions, as follows.

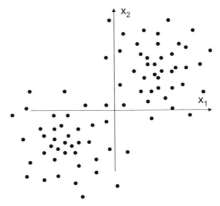

Fig. 4.5. Example of a pattern of feature vectors. The classes are not predefined. The hyporthesis of the existence of two classes is made based on the fact that points tend to be concentrated around two centers, forming two data clusters

(II) Each of the clusters has a center point x_i^C, $i = 1, \ldots, K$, with the coordinates equal to the mean of the coordinates of the data points that belong to that cluster.

(III) For each of the data points x, we decide which of the clusters does it belong to by computing the distances between that point and centers of all clusters $d(x, x_i^C)$, $i = 1, \ldots, K$. We take the index i of x_i^C that minimizes the distance $d(x, x_i^C)$ to indicate the cluster containing x.

On the basis of the assumptions (I)–(III), the following design of the clustering algorithm is quite obvious. We choose randomly some initial values for the centers of the clusters x_i^C, $i = 1, \ldots, K$, and then iterate the following two steps until convergence is obtained:

Step 1. Assign each data point to a cluster, on the basis of the criterion in (III).

Step 2. On the basis of the assignment in step 1, update values for centers of clusters as defined in (II).

The above algorithm is both simple and an efficient tool for searching for clusters.

4.3.2 Hierarchical Clustering

The drawback of the K-means clustering algorithm is the need to know the number of clusters in advance. There are several methods to overcome this difficulty. The natural approach is to perform clustering for different numbers of clusters and try to estimate number of clusters by some method of assessing the quality of the clustering.

106   4 Pattern Analysis

It is also worthwhile to consider another approach, hierarchical clustering. In this approach, a tree or a treelike structure is constructed on the basis of pairwise distances between feature vectors. Clusters are obtained by "cutting" the tree at some level, and the number of clusters is controlled by deciding at which level of the hierarchy of the tree the splitting is performed. The construction of the tree is based on neighbor joining, an idea described also in Chap. 7. There are several variants of the hierarchical-clustering algorithm [67]. Here we shall describe the basic idea and some of the possible modifications.

Assume that the data points (feature vectors) are $x_1, x_2, \ldots, x_n$ and define a matrix D^0 of distances between them

$$D^0 = [d(x_i, x_j)]. \tag{4.13}$$

In the above, $d(x_i, x_j)$ denotes the Euclidean distance between the feature vectors x_i and x_j. The idea of neighbor joining is (i) to find a pair of feature vectors x_{i*} and x_{j*} with a minimal distance

$$i^*, j^* \leftarrow \min_{i,j} d(x_i, x_j), \tag{4.14}$$

and (ii) to join x_{i*} and x_{j*}. Joining x_{i*} and x_{j*} is often realized by replacing x_{i*} and x_{j*} by their mean:

$$x_{i*}, x_{j*} \rightarrow \frac{1}{2}(x_{i*} + x_{j*}) = y. \tag{4.15}$$

After joining x_{i*} and x_{j*} we update the matrix of distances between feature vectors, i.e., $D^0 \rightarrow D^1$; the entries of D^1 which need updating are the distances $d(x_i, y)$ between the new vector y and the feature vectors x_i that were not involved in the joining operation. Performing the above-described operations sequentially leads to the construction of a neighbor-joining tree for the vectors $x_1, x_2, \ldots, x_n$.

Sequential joining of vectors, as defined in (4.15), leads to the formation of clusters. By keeping track of the indexes of the vectors that have been joined to each other, we know which vectors belong to which cluster. Since we are focused on clusters rather than on a tree, we may make some modifications to the algorithm described by (4.13)–(4.15). The sequential application of the replacement rule (4.15) leads to defining vectors y, which can be interpreted as centers of clusters. Then, one can use a modification of the rule (4.15), defined by

$$x_{i*}, \text{cluster}(x_{j1}, \ldots, x_{jm}) \rightarrow \text{cluster}(x_{j1}, \ldots, x_{jm}, x_{i*}) \tag{4.16}$$

and

$$y = \text{cluster\_center}(x_{j1}, \ldots, x_{jm}, x_{i*}) = \frac{x_{j1} + \ldots + x_{jm} + x_{i*}}{m+1}, \tag{4.17}$$

for merging x_{i*} with cluster$(x_{j1},...,x_{jm})$ and for computing the center of the cluster. When deciding about merging vectors with clusters, we can use the distances between the vectors and the centers of clusters.

Other variants of the hierarchical-clustering algorithm are also possible, for example one may use using other definitions of the distance function or other rules for merging vectors with existing clusters [235, 67]. Some possible definitions of distances are the euclidean, correlation, Pearson or Spearman, and Manhattan distances. The rules used most often for defining clusters are single-linkage clustering, where the distance between two clusters i and j is the minimum of distances between members of clusters i and j; complete-linkage clustering, where the distance between two clusters is the maximum of the distances between their members; and average-linkage clustering, where the distance between two clusters is the mean value of the distances between members of the clusters.

4.4 Dimensionality Reduction, Principal Component Analysis

A need for dimensionality reduction arises when the number of features is large. Experimental results in molecular biology and biochemistry often lead to the creation of a large number of measurement data points. Examples are gene expression intensities in DNA microarrays, proteomic spectra, and data concerning conformations of large molecules such as proteins. In such situations the number of features (measurements) obtained in each experiment is much bigger than the number of experiments. One expects that only some of the measurements will be correlated with the state of the experiment under study.

Two cases are possible. The first possibility is that the state of the experiment (e.g. diseased versus healthy) is known. A related problem is to select the subset of features most suitable for differentiating between experimental states. Some aspects of this problem are discussed in this book, in Chap. 11. The second possibility is that inference must be done solely on the basis of the set of feature vectors, without any knowledge about the underlying structure. A well-established methodology for this problem is principal component analysis [131]. The searching for principal components in the data is based on analysis of the variances along different directions in the feature space. The method of principal component analysis (PCA) can also be applied to the situation where the classes of experimental states are known. Below we present some of the computational aspects of these applications of principal component analysis.

4.4.1 Singular-Value Decomposition (SVD)

We start from a theorem on SVD of a real matrix. Let us define a real $m \times n$ matrix

$$A = \begin{bmatrix} a_{11} & a_{12} & \cdots & a_{1n} \\ a_{21} & a_{22} & \cdots & a_{2n} \\ \vdots & \vdots & \ddots & \vdots \\ a_{m1} & x_{m2} & \cdots & x_{mn} \end{bmatrix}.$$

The SVD theorem, [103, 131], states that A can always be represented as follows:

$$A = U \Sigma V^T \qquad (4.18)$$

where U and V^T are (nonsingular) real orthogonal transformation matrices, of dimensions $m \times m$ and $n \times n$ respectively and the superscript T represents matrix transposition. The $m \times n$-dimensional matrix Σ is composed of the following blocks:

$$\Sigma = \begin{bmatrix} \Xi_{r \times r} & O_{r \times (n-r)} \\ O_{(m-r) \times r} & O_{(m-r) \times (n-r)} \end{bmatrix},$$

where $O_{k \times l}$ denotes $k \times l$-dimensional matrix with all entries equal to zero, and $\Xi_{r \times r}$ is a diagonal matrix

$$\Xi_{r \times r} = \begin{bmatrix} \sigma_1 & 0 & \cdots & 0 \\ 0 & \sigma_2 & \cdots & 0 \\ \vdots & \vdots & \ddots & \vdots \\ 0 & 0 & \cdots & \sigma_r \end{bmatrix},$$

with real elements $\sigma_1 \geq \sigma_2 \geq ... \geq \sigma_r > 0$, and where $r = \mathrm{rank}(A)$. Clearly, $r \leq \min(n, m)$.

The numbers $\sigma_1, \sigma_2, ..., \sigma_r$ are called the singular values of the matrix A and the first r columns of the matrix U are called the principal directions of the matrix A. More precisely, the first r columns of U are the principal directions for the columns of the matrix A and the first r rows of the matrix V^T are the principal directions for the rows of the matrix A. The singular values and orthogonal matrices U and V^T are related to eigenvalues and eigenvectors of the Grammian matrices AA^T and A^TA, which can be clearly seen from (4.18). Recalling that the orthogonality of U and V implies that $U^TU = I_{m \times m}$ and $V^TV = I_{n \times n}$, where $I_{k \times k}$ denotes the $k \times k$ identity matrix, we have the following equalities for the Grammian matrices AA^T and A^TA

$$AA^T = U\Sigma \Sigma^T U^T = U \Xi^2_{m \times m} U^T \qquad (4.19)$$

$$A^TA = V \Sigma^T \Sigma V^T = V \Xi^2_{n \times n} V^T. \qquad (4.20)$$

In the above expressions, $\Xi^2_{k \times k}$ (where k equals either m or n) stands the diagonal matrix

4.4 Dimensionality Reduction, Principal Component Analysis

$$\Xi_{k \times k}^2 = \begin{bmatrix} \Xi_{r \times r}^2 & O_{r \times (k-r)} \\ O_{(k-r) \times r} & O_{(k-r) \times (k-r)} \end{bmatrix},$$

where

$$\Xi_{r \times r}^2 = \begin{bmatrix} \sigma_1^2 & 0 & \cdots & 0 \\ 0 & \sigma_2^2 & \cdots & 0 \\ \vdots & \vdots & \ddots & \vdots \\ 0 & 0 & \cdots & \sigma_r^2 \end{bmatrix}.$$

Since the expressions on the right-hand sides of (4.19) and (4.20) are Jordan canonical forms, the columns of the orthogonal matrix U are eigenvectors of the Grammian matrix AA^T and the columns of the orthogonal matrix V are eigenvectors of the Grammian matrix $A^T A$. One can also see that nonzero eigenvalues of both AA^T and $A^T A$ are equal to the squares of the singular values $\sigma_1^2, \sigma_2^2, \ldots, \sigma_r^2$ of the matrix A.

4.4.2 Geometric Interpretation of SVD

The representation (4.18) has several interesting geometric interpretations. One geometric interpretation is as follows. Let us understand A as a linear operator mapping n-dimensional vectors $x \in R^n$ to m-dimensional vectors $y = Ax$, $y \in R^m$. The representation (4.18) implies that for every linear operator of rank r, one can find two orthogonal bases, in the domain and image spaces R^n and R^m, respectively, such that the first r vectors of the orthogonal basis in the domain space R^n are mapped to first r vectors of the orthogonal basis in the image space R^m. The orthogonal basis in the domain space is given by the rows of the matrix V^T, and the orthogonal basis in the image space by the columns of the matrix U.

Another important geometric interpretation of the decomposition (4.18), which will be used in this book in several contexts, is related to expressing the principal directions of the matrix in terms of solutions to optimization problems and to computing projections onto subspaces. Let us interpret the matrix A as a set of n column vectors, each belonging to the space R^m:

$$A = [a_1 \; a_2 \; \ldots \; a_n], \quad (4.21)$$

$$a_k = \begin{bmatrix} a_{1k} \\ a_{2k} \\ \vdots \\ a_{mk} \end{bmatrix} \in R^m, k = 1, 2, \ldots, n.$$

We now ask a somewhat imprecise question: Which direction in the space R^m is most representative for the vectors a_k, $k = 1, 2, \ldots, n$? Precisely, we call a vector $c \in R^m$ the most representative for the set of vectors a_k, $k = 1, 2, \ldots, n$, if:

(1) c is a linear combination of vectors a_k,

$$c = \sum_{k=1}^{n} \beta_k a_k, \qquad (4.22)$$

with scalar coefficients β_k;
(2) the coefficients β_k, $k = 1, 2, ..., n$ are normalized to 1, i.e., $\sum_{k=1}^{n} \beta_k^2 = 1$; and
(3) the vector c is the longest possible under conditions (1) and (2).

By "longest possible" we mean the one with the largest Euclidean norm. Conditions (1)–(3) lead to the following maximization problem:

$$\max \|c\|$$

under the constraints

$$c = \sum_{k=1}^{n} \beta_k a_k, \quad \sum_{k=1}^{n} \beta_k^2 = 1.$$

Noting that $\max \|c\|$ is equivalent to $\max \|c\|^2$ and introducing the vector of coefficients

$$b = \begin{bmatrix} \beta_1 \\ \beta_2 \\ \vdots \\ \beta_n \end{bmatrix}, \qquad (4.23)$$

we can write the above maximization problem as

$$\max b^T A^T A b \qquad (4.24)$$

under the constraints

$$b^T b = 1. \qquad (4.25)$$

A necessary conditions for optimality for the constrained optimization problem (4.24), (4.25) (see Chap. 5) are that (b, λ) is the stationary point of the Lagrange functional

$$L(b, \lambda) = b^T A^T A b + \lambda (1 - b^T b). \qquad (4.26)$$

In the above λ is a scalar Lagrange multiplier. Stationarity is verified by comparing the gradients of $L(b, \lambda)$ with respect to λ and b, with zero and a zero vector. The condition

$$\frac{\partial L}{\partial \lambda} = 0 \qquad (4.27)$$

is equivalent to (4.25), and

$$\frac{\partial L}{\partial b} = 0 \qquad (4.28)$$

leads to
$$A^T A b - \lambda b = 0, \qquad (4.29)$$
or
$$\lambda b = A^T A b. \qquad (4.30)$$

This is an eigenvalue–eigenvector problem. So the problem of maximizing (4.24) with the constraint (4.25) leads to computing the eigenvector b and the eigenvalue λ of the symmetric matrix $A^T A$. The solution is nonunique since in general there are n (nonnegative, real) eigenvalues $\lambda_1, \lambda_2, \ldots, \lambda_n$ of $A^T A$. Nonuniqueness is a consequence of using only necessary optimality conditions. However, among those satisfying (4.30), the optimal (b, λ) can easily be identified. Substituting (4.30) in (4.24) results in

$$\max b^T A^T A b = \max \lambda b^T b = \max \lambda = \lambda_{\max}. \qquad (4.31)$$

So the solution (λ, b) is the maximal eigenvalue $\lambda_{\max}(A^T A)$ and the corresponding eigenvector b. From (4.30), (4.20), and (4.18) we now see that the eigenvector b corresponding to $\lambda_{\max}(A^T A)$ is the first column of the matrix V.

Let us return to the representative direction c. The relation (4.22) can be represented in vector notation as

$$c = Ab.$$

Multiplying both sides of the above equation by AA^T and recalling (4.30) and (4.25) we obtain

$$\lambda_{\max} c = A A^T c.$$

The conclusion is that the representative direction c, defined by conditions (1)–(3) above is, up to some scaling factor, the first principal direction of the matrix A. It turns out that the first principal direction (the first column of the matrix U) of the matrix A has the interpretation given by conditions (1)–(3) above. Repeating the above with the matrix A understood as a set of m rows, rather than a set of n columns as in (4.21), we obtain an interpretation of the first row of the matrix V^T as the first principal direction for the rows of the matrix A.

After establishing the meaning of the first principal direction of the matrix A, one can ask about other principal directions (the other columns of the matrix U). The second, third, and further principal directions of the matrix A, can again be interpreted as representative directions determined by some sets of vectors, in the following sense. Let us represent all vectors a_k, $k = 1, 2, \ldots, n$ as sums of two components, parallel and orthogonal to c:

$$a_k = \hat{a}_k + \tilde{a}_k. \qquad (4.32)$$

In the above $\hat{a}_k$ is parallel to c, which means that $\hat{a}_k = \rho_k c$ for some scalar value ρ_k, and $\tilde{a}_k$ is orthogonal to c, which means that their scalar product equals zero, i.e., $c^T \tilde{a}_k = 0$. Taking the scalar products of both sides of (4.32) with c, we obtain $\rho_k = c^T a_k / c^T c$, and consequently

$$\hat{a}_k = \frac{c^T a_k}{c^T c} c$$

and

$$\tilde{a}_k = a_k - \frac{c^T a_k}{c^T c} c. \tag{4.33}$$

By defining the row vector

$$\rho^T = [\rho_1 \ \rho_2 \ \dots \ \rho_n]$$

we can express the relation (4.33) in matrix–vector notation, as follows:

$$\tilde{A} = A - c\rho^T \tag{4.34}$$

where $\tilde{A}$ is defined as

$$\tilde{A} = [\tilde{a}_1 \ \tilde{a}_2 \ \dots \ \tilde{a}_n]. \tag{4.35}$$

The matrix $\tilde{A}$ is composed of the residual vectors $\tilde{a}_k$ (4.33), and it can be verified (we give it as Exercise 6) that, if the matrix A has a set of singular values $\sigma_1 > \sigma_2 \ \dots \ \sigma_r > 0$ then matrix $\tilde{A}$ will have singular values $\sigma_2 > \sigma_3 \ \dots \ \sigma_r > 0$, or in other words the largest singular value $\sigma_1 = \sigma_{\max}$ is replaced by zero. Now, solving the problem (4.24)–(4.25) with A replaced by $\tilde{A}$ will lead to computing the second singular value and the second principal direction of A, and so forth. This leads to the following representation of the matrix A:

$$A = \sum_{k=1}^{r} c_k \rho_k^T$$

where two sets of orthogonal vectors c_k and ρ_k^T, $k = 1, 2, \dots, r$ are called loadings and scores, respectively. The above representation also follows directly from the form of the singular-value decomposition (4.18).

The principal components and the singular-value decomposition also have a very important statistical interpretation in terms of variances of random variables. Let us consider a set of n random variables $X_1, X_2, \dots, X_n$. For each of them, m realizations are given (i.e., measured), which are denoted as follows: $x_{11}, x_{21}, \dots, x_{m1}, x_{12}, x_{22}, \dots, x_{m2}, \dots, x_{1n}, x_{2n}, \dots, x_{mn}$. We form the matrix of data (measurements) X,

$$X = \begin{bmatrix} x_{11} & x_{12} & \cdots & x_{1n} \\ x_{21} & x_{22} & \cdots & x_{2n} \\ \vdots & \vdots & \ddots & \vdots \\ x_{m1} & x_{m2} & \cdots & x_{mn} \end{bmatrix}. \tag{4.36}$$

Assume that the realizations are centered, which means that for each column k we have $\sum_{i=1}^{m} x_{ik} = 0$. The total sampling variance $\mathrm{Var}(X)$ of the data (4.36) is then defined as

4.4 Dimensionality Reduction, Principal Component Analysis

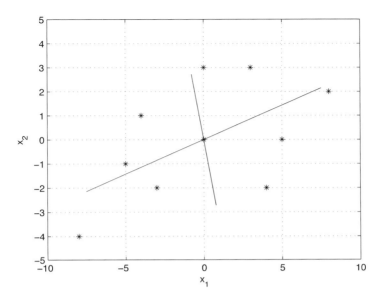

Fig. 4.6. Scatterplot of ten pairs of realizations of random vectors x_1, x_2. The vectors v_1^T and v_2^T, the principal directions of the matrix X, are marked by line segments. The longer segment is for the first principal direction, and the shorter segment is for the second principal direction

$$\text{Var}(X) = \frac{1}{m-1}\text{trace}(X^T X), \quad (4.37)$$

where trace(A) means the trace (sum of diagonal elements) of the matrix A. We can see that the sampling variance as defined by (4.37) is invariant with respect to orthogonal transformations of the data, that is, if $Y = XW^T$, where W^T is an $m \times m$-dimensional orthogonal matrix, then $\text{Var}(Y) = \text{Var}(X)$. From (4.20) it follows that the total sampling variance can be expressed in terms of the squares of the singular values of matrix X, $\sigma_1, \sigma_2, \ldots, \sigma_r$, where $r = \text{rank}(A)$:

$$\text{Var}(X) = \frac{1}{m-1}\sum_{k=1}^{r} \sigma_k^2.$$

Let us consider the SVD decomposition of the matrix X

$$X = U\Sigma V^T$$

and its equivalent form

$$XV = \Sigma U. \quad (4.38)$$

Denoting by U_r the matrix composed of the first r columns of U, and by V_r the matrix composed of the first r columns of V we can express the relation (4.38) as follows:

114   4 Pattern Analysis

$$XV_r = \Sigma U_r = [\sigma_1 u_1 \; \sigma_2 u_2 \; \ldots \; \sigma_r u_r],$$

where $u_1, u_2, \ldots, u_r$ are the columns of the matrix U_r (the principal directions of X). Since $\text{Var}(XV_r) = \text{Var}(\Sigma U_r) = \text{Var}(X)$ and the columns of U_r are orthogonal, we can characterize each of principal directions $u_1, u_2, \ldots, u_r$ in terms of how much of total variance they include. It is common to say that the first principal direction or component captures $(\sigma_1^2 / \sum_{k=1}^{r} \sigma_k^2) \cdot 100\%$ of total variance, the first two principal components capture $(\sigma_1^2 + \sigma_2^2)/\sum_{k=1}^{r} \sigma_k^2 \cdot 100\%$, and the first j principal components capture

$$\frac{\sum_{k=1}^{j} \sigma_k^2}{\sum_{k=1}^{r} \sigma_k^2} \cdot 100\% \qquad (4.39)$$

of the total variance.

As an example, let us consider the following data matrix

$$X = \begin{bmatrix} 3 & -3 & -8 & 4 & 0 & -4 & 5 & 8 & 0 & -5 \\ 2 & -2 & -4 & -2 & 0 & 1 & 0 & 2 & 3 & -1 \end{bmatrix}^T, \qquad (4.40)$$

consisting of ten realizations of two random variables X_1 and X_2, written as two columns of the matrix X. The superscript T stands for transposition. Columns of X are mean-centered. A scatterplot for pairs of realizations of X_1 and X_2 is shown by asterisks in Fig. 4.6. The SVD decomposition of X is

$$X = U\Sigma V^T, \qquad (4.41)$$

where

$$U = \begin{bmatrix} -0.237 & 0.364 & 0.559 & -0.168 & 0.000 & 0.196 & -0.279 & -0.503 & -0.084 & 0.307 \\ 0.220 & -0.194 & -0.269 & -0.567 & 0.000 & 0.392 & -0.270 & -0.082 & 0.527 & 0.095 \\ 0.563 & -0.291 & 0.726 & -0.017 & 0.000 & -0.021 & 0.075 & 0.197 & 0.116 & -0.113 \\ -0.211 & -0.534 & 0.032 & 0.655 & 0.000 & 0.262 & -0.225 & -0.196 & 0.246 & 0.143 \\ 0.000 & 0.000 & 0.000 & 0.000 & 1.000 & 0.000 & 0.000 & 0.000 & 0.000 & 0.000 \\ 0.229 & 0.364 & -0.058 & 0.256 & 0.000 & 0.801 & 0.178 & 0.172 & -0.170 & -0.122 \\ -0.308 & -0.243 & 0.106 & -0.210 & 0.000 & 0.171 & 0.836 & -0.184 & 0.118 & 0.125 \\ -0.528 & -0.049 & 0.222 & -0.159 & 0.000 & 0.147 & -0.169 & 0.754 & 0.036 & 0.157 \\ -0.053 & 0.510 & 0.079 & 0.265 & 0.000 & -0.189 & 0.141 & 0.073 & 0.772 & -0.065 \end{bmatrix}
\qquad (4.42)$$

$$\Sigma = \begin{bmatrix} 15.621 & 0 & 0 & 0 & 0 & 0 & 0 & 0 & 0 \\ 0 & 5.657 & 0 & 0 & 0 & 0 & 0 & 0 & 0 \end{bmatrix}^T, \qquad (4.43)$$

and

$$V^T = \begin{bmatrix} -0.962 & -0.275 \\ 0.962 & -0.274 \end{bmatrix}.$$

In the above $r = n$. The nonzero entries of Σ are the singular values of X, and the rows of V^T,

$$v_1^T = [-0.962 \; -0.275]$$

4.4 Dimensionality Reduction, Principal Component Analysis

and
$$v_2^T = [0.962 \ -0.274],$$

are two principal directions of the rows of X. The vectors given by the principal directions, with lengths scaled by the corresponding singular values, are also shown in Fig. 4.6. For rectangular matrices, such as X in (4.40), the matrix of singular values Σ always contains zero rows or columns, as does Σ in (4.43). So, instead of the decomposition (4.18) it may be reasonable to use an "economy" SVD, where zero columns or rows of the matrix Σ are skipped and the corresponding rows or columns of the matrix U or V^T are removed. For example, we have the following economy-size decomposition for (4.41):

$$X = U^0 \Sigma^0 V^T, \tag{4.44}$$

where

$$U^0 = \begin{bmatrix} -0.237 & 0.364 \\ 0.220 & -0.194 \\ 0.563 & -0.291 \\ -0.211 & -0.534 \\ 0.000 & 0.000 \\ 0.229 & 0.364 \\ -0.308 & -0.243 \\ -0.528 & -0.049 \\ -0.053 & 0.510 \end{bmatrix},$$

and

$$\Sigma^0 = \begin{bmatrix} 15.621 & 0 \\ 0 & 5.657 \end{bmatrix}.$$

The columns removed from the matrix U have no influence on the product representation of the matrix X.

In cases where data sets to be analyzed are large, using economy-size SVD can save a lot of computational time and memory space.

4.4.3 Partial-Least-Squares (PLS) Method

Here we assume that the input measurement data (also called the explaining or predictor variables) given by (4.36) are accompanied by measurements of a scalar output (also called the dependent variable) Y. So our data structure is now

$$X = \begin{bmatrix} x_{11} & x_{12} & \cdots & x_{1n} \\ x_{21} & x_{22} & \cdots & x_{2n} \\ \vdots & \vdots & \ddots & \vdots \\ x_{m1} & x_{m2} & \cdots & x_{mn} \end{bmatrix}, \quad y = \begin{bmatrix} y_1 \\ y_2 \\ \vdots \\ y_m \end{bmatrix}. \tag{4.45}$$

In a general setting we would assume a vector-valued output. However, here we shall confine the presentation to a scalar output, as given in (4.45). Now we try to form a linear combination of columns of the matrix X (the vectors x_1, ..., x_n), with coefficients β_k, $k = 1, 2, \ldots, n$, normalized to 1, $\sum_{k=1}^{n} \beta_k^2 = 1$, such that the resulting vector

$$c = \sum_{k=1}^{n} \beta_k x_k$$

maximizes the covariance or, equivalently, the scaled scalar product $c^T y/(m-1)$. Using a vector notation analogous to (4.24) and (4.25), we can state this maximization problem as

$$\max \frac{1}{m-1} y^T X b$$

with the constraint

$$b^T b = 1,$$

where b is a vector of parameters β_k as in (4.23). Using the technique of constrained optimization (Chap. 5), we obtain, analogously as to (4.26)–(4.29), the following optimal vector,

$$b = \frac{X^T y}{\sqrt{y^T X X^T y}}$$

and the first PLS direction (component),

$$c = \frac{X X^T y}{\sqrt{y^T X X^T y}}.$$

The second, third, and further PLS components are obtained by projections onto the direction given by c in the above expression and analyzing the residual vectors, analogously to (4.33)–(4.35).

4.5 Parametric Transformations

Transformations are workhorses in all areas of applied mathematics. Some examples already shown are generating functions and characteristic functions, discussed in Chap. 2. The characteristic function is actually the Fourier transform of the probability density function of a random variable. One- and two-dimensional Fourier transforms are also widely applied in pattern analysis, for example for noise reduction and extraction of image features. Fourier transformation is also used in bioinformatics, for example for analysis of repetitive structure of sequences. A DNA sequence is changed to numerical symbols by some method, then a Fourier transformation is applied to the numerical sequence obtained, and the resulting spectrum is used to search for special geometric-like or repetitive patterns. There are numerous textbooks (e.g., [66]) devoted to transforms including the Fourier, Laplace, Laurent, Hankel, and Hilbert transforms.

However, in this section, we focus on transformations which are not as widely known as the above, but underline interesting relations between computer-science algorithms and pattern analysis methods. They come under different names, such as parametric transforms, Hough transforms, and geometric hashing, but share the idea of using a process of pattern scanning in conjunction with addressing and operating on a data structure to record occurrences of data objects. The contents of this data structure can be then used to obtain useful information about the objects under study. Clearly, this idea has some similarity to the method of hashing and hash tables, presented in Chap 3.

In this section we discuss some of these approaches from the pattern analysis perspective. Owing to their flexibility they have large potential to serve in numerous procedures for browsing databases for correlations, similarities of different types, etc. Later we also show some applications of these methods in genomics and in protein docking.

4.5.1 Hough Transform

The Hough transform [132] provides a method for detecting parametric curves in images and estimating the values of their parameters. Most often, Hough transforms use contours in a binary as input data and apply a duality between points on the curve and the parameters of the curve. The Hough transform can also be understood as a feature extraction technique based on interpreting the contents of a digital image by using a feature space. The basic example is the detection of straight lines in images, as presented in Fig. 4.7. The task is to detect occurrences of straight lines in the image. We assume that the image to be analyzed is binary, as shown in the left plot in Fig. 4.7, and so it contains a number of discrete image points. The equation of a straight line in the image space x, y is

$$y = ax + b. \tag{4.46}$$

To accomplish the aim of detecting straight lines in the image, we create a parameter space (a plane) with coordinates a, b, as shown in the right-hand part of Fig. 4.7. For each of the points in the image x_i, y_i, we draw a corresponding line in the parameter space a, b,

$$y_i = ax_i + b. \tag{4.47}$$

Because all points in the image space x, y are collinear, all lines in the parameters space a, b intersect in one point. The occurrence of the point of intersection, a^*, b^* of many lines in the parameter space a, b indicates detection of a line in the image space $y = a^*x + b^*$.

In practical situations, the parameter space is discretized and consists of a finite number of pixels. The discretized parameter space is called an

118   4 Pattern Analysis

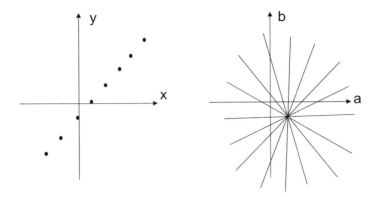

Fig. 4.7. The idea of the Hough transform. For each point x_i, y_i in the image space x, y, we draw a corresponding line $y_i = ax_i + b$ in the parameter space a, b. If the points in the image space belong to a line, as shown in the *left-hand plot*, their correponding lines in the parameter space intersect at one point, as depicted in the *right-hand plot*

accumulator array. Drawing lines corresponding to points found in the image is equivalent to incrementing memory locations in the accumulator array. The procedure of incrementing entries of accumulator array is often called voting. Detecting lines can be accomplished by browsing through the accumulator array and searching for local maxima.

The idea described above can also be used for detecting other parametric curves in images, for example circles and ellipses.

4.5.2 Generalized Hough Transforms

Generalized Hough transforms extend the idea described above to the non-parametric curves. The most straightforward generalization is as follows. Assume we are searching for occurrences of a shape, such as the one shown by the dashed curve in the left plot in Fig. 4.8. The binary image to be analyzed consists of points, also depicted in the left plot in Fig. 4.8. The problem is, does the shape occur in the image? We are not allowing rotations of the target shape, so it is natural to define a parameter space with translations Δx and Δy along the axes as coordinates. The procedure for updating the accumulator array associated with this parameter space is very similar to the one described in the previous subsection. We browse through the image and, after detecting a point with coordinates x_i, y_i, we draw the target shape in the parameter space translated by the vector $\Delta x = x_i$, $\Delta y = y_i$. This is shown in the right plot in Fig. 4.8. Again, the intersection of many of the drawn shapes at one point indicates the occurrence of the target shape in the image.

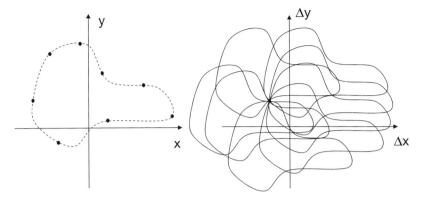

Fig. 4.8. The idea of the generalized Hough transform. For each point x_i, y_i in the image space (left plot) we draw the target shape (dashed curve) in the parameter space translated by the vector $\Delta x = x_i$, $\Delta y = y_i$. This is shown in the right image The intersection of many of drawn shapes at one point indicates the occurrence of the target shape in the image.

The algorithm described here does not allow us to detect rotated or rescaled target shapes. However, extensions that overcome this limitation have been proposed in several papers [17, 215, 228].

4.5.3 Geometric Hashing

Ideas similar to the above has been used to construct another algorithm called geometric hashing. [120, 292] Let us assume that we aim to search for occurrences in images of a pattern of points, $x_1, x_2, \ldots, x_n$, $x_i \in R^2$. The first step of the algorithm is to compute a signature that is invariant under translations, rotations and scale changes. The signature is the set of points in R^2 obtained by the following procedure. Go through all pairs of points x_i, x_j, $1 \leq i,j \leq n$. For each pair x_i, x_j, (I) find a transformation T that maps $x_i \to T(x_i) = (-1,0)$ and $x_j \to T(x_j) = (1,0)$, and (II) add all transformed points $T(x_m)$, $m \neq i$, $m \neq j$ to the signature.

Let us analyze a binary image given by another set of points $y_1, y_2, \ldots, y_m$, $y_i \in R^2$. From the above definition it is clear that if the set $\{y_1, y_2, \ldots, y_m\}$ contains $\{x_1, x_2, \ldots, x_n\}$ possibly translated, rotated and rescaled, then the signature of $\{y_1, y_2, \ldots, y_m\}$ contains the signature of $\{x_1, x_2, \ldots, x_n\}$. In the programs developed in practice the coordinates of the vectors of the signatures are discretized and stored with by use of data structures that are of the form of accumulator arrays.

4.6 Exercises

1. Assume the data points given in Table 4.1.

a) Assume that the points with numbers 1–5 correspond to class 1 and those with numbers 6–10 to class 2. Find a linear discriminant function for classes 1 and 2 using the method of linear programming as described in (4.6)–(4.8). A linear-programming algorithm can be found in many software packages. Draw the data from the Table 4.1 in the plane x, y. Draw the separating line obtained by solving (4.6)–(4.8).

b) For the same data and the same assumption that points with numbers 1–5 correspond to class 1 and those with numbers 6–10 to class 2, find the optimal linear discriminant function for classes 1 and 2 by solving the quadratic programming problem (4.11)–(4.12). Draw the data and the optimal separating line. Again, a quadratic programming algorithm can be found in many software packages.

Table 4.1. Table of data points to be used in exercises

No.	x	y
1	1	1.5
2	1.5	3
3	3	1
4	3	2
5	3.5	2
6	−0.5	−0.5
7	−0.5	2
8	−1	0.5
9	1	−1
10	2	−1

2. Separate the classes 1 and 2 defined in the previous exercise by using the artificial neural network shown in Fig. 4.3. There are many software packages that support the designing and training and artificial neural networks. One of them can be used to solve this exercise.
3. Decompose the data set from Table 4.1 into two classes using the K-means algorithm. Decompose the data set from Table 4.1 into three classes using the K-means algorithm.
4. Construct a neighbor-joining tree for the data set from Table 4.1 by using rules (4.13) and (4.15).
5. Build a hierarchical clustering tree by using an algorithm with the rules (4.16)–(4.17).
6. Prove that the largest singular value of the matrix $\tilde{A}$ in (4.35) is the second largest singular value of the matrix A in (4.21).
7. The transformation $T([x_i, y_i])$ mentioned in Sect. 4.5.3, is defined by $T([x_1, y_1]) = [-1, 0]$ and $T([x_2, y_2]) = [1, 0]$, where $[x_1, y_1]$ and $[x_2, y_2]$ are given in Table 4.1. Compute $T([x_i y_i])$ for all points in Table 4.1.
8. Develop a computer program for geometric hashing.

9. Study the problem of extending the geometric hashing algorithm described in Sect. 4.5.3 to the case of three-dimensional feature space [211].

5
Optimization

Optimization involves computing parameter values or making decisions such that certain performance or quality indices are maximized, or, conversely, some penalty or loss functions are minimized. Optimization is an area of scientific research, and also has extensive applications in technology, engineering design, etc. Optimization problems involve a large variety of situations, such as optimizing the parameters of engineering structures, developing optimal policies in decision-making problems, and optimizing controls in dynamical systems. The application of optimization techniques can also involve fitting models to data by optimizing the model parameters in such a way that output of the developed mathematical model is as close as possible to the measurements.

Optimization methods are traditionally divided into static and dynamic methods. Static optimization is the computing of extremal points of functions of one or several variables [296, 182]. This involves formulating appropriate optimality conditions and then proposing algorithms, often iterative, for satisfying these conditions. Dynamic optimization is the computing of optimal control functions or making optimal decisions in a sequential order, following the dynamics of some system. Dynamic optimization can involve continuous-time or discrete-time dynamics, and there are again two main approaches. One approach is to use variational type conditions leading to two-point boundary value problems. The other uses Bellman's optimality principle [26], which in discrete cases leads to a recursion for the cumulative score index, and its continuous-time limit leads to partial differential equation formulation called the Bellman–Jacobi equation [183]. Here we focus on discrete dynamic optimization and dynamic programming [26, 183]. There are some very well known examples of application of dynamic programming in bioinformatics, namely the Needleman–Wunsch and Smith–Waterman algorithms for DNA alignment. They will be described in Chaps. 8 and 9. Other examples of the application of dynamic programming will also be discussed.

We also include a section on combinatorial optimization in this chapter. By combinatorial optimization we mean problems which involve optimizing over paths in the plane or space, topologies of trees, graphs etc. Combinatorial optimization is related to theory of algorithms and their complexity, computer science and operation research. Some of issues regarding combinatorial optimization were mentioned in Chap. 3. Combinatorial optimization is of great importance in bioinformatics because it involves a lot of non-numerical data and the need to perform various operations on these data.

5.1 Static Optimization

The simplest example of static optimization is maximizing a function of one variable, as shown in Fig. 5.1. The function $f(x)$ has its maximal value $f_{\max}$ at $x = x_{\max}$. A characteristic of extremal points is that, provided the function is smooth, the derivative becomes zero, i.e.,

$$\frac{df(x)}{dx} = 0, \tag{5.1}$$

which can be used for computing $x_{\max}$ by solving the algebraic equation (5.1). At points where the derivative is negative the function decreases, and at points where the derivative is positive the function increases. This can be used for constructing recursive estimators for extremal points. The condition (5.1) applies for both maximal and minimal points. Resolving the distinction between a maximum and a minimum can be done by use of the second derivative (if the function is twice differentiable); namely if (5.1) holds at $x = x^*$ and

$$\frac{d^2 f(x)}{dx^2} > 0, \tag{5.2}$$

then x^* corresponds to a local minimum, and if the inequality sign in (5.2) is opposite then x^* corresponds to a local maximum.

The elementary conditions discussed above can be generalized to multi-dimensional functions $f : R^n \to R$. In the two-dimensional case, where the function is $f(x_1, x_2)$, the condition for an extremum analogous to (5.1), is

$$\frac{\partial f(x_1, x_2)}{\partial x_1} = 0, \quad \frac{\partial f(x_1, x_2)}{\partial x_2} = 0. \tag{5.3}$$

This condition means that the plane tangent to the graph of $f(x_1, x_2)$ at an extremal point is horizontal. The vector composed of partial derivatives in (5.3) is called gradient of f and denoted ∇f:

$$\nabla f(x_1, x_2) = \begin{bmatrix} \dfrac{\partial f(x_1, x_2)}{\partial x_1} \\ \dfrac{\partial f(x_1, x_2)}{\partial x_2} \end{bmatrix}. \tag{5.4}$$

The notation $\nabla f(x) = \partial f / \partial x$ is often also used, for the gradient vector, where x is a vector argument

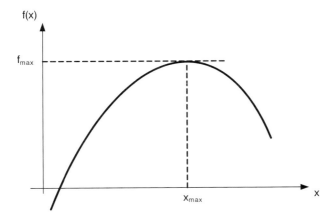

Fig. 5.1. A function of one variable $f(x)$ has a maximum f_{max} at $x = x_{max}$

$$x = \begin{bmatrix} x_1 \\ x_2 \end{bmatrix}. \quad (5.5)$$

This notation is particularly useful when we need to form vectors of partial derivatives of functions with respect to subsets of their arguments. Such situations will be encountered below. Using (5.4) and the vector notation for x_1, x_2, we can write the condition for an extremum (5.3) as

$$\nabla f(x) = \frac{\partial f(x)}{\partial x} = 0, \quad (5.6)$$

where there is a two-dimensional zero vector on the right hand side.

Resolving the distinction between a maximum and a minimum can be done by use of the Hessian matrix

$$Hf(x) = Hf(x_1, x_2) = \begin{bmatrix} \dfrac{\partial^2 f(x_1, x_2)}{\partial x_1^2} & \dfrac{\partial^2 f(x_1, x_2)}{\partial x_1 \partial x_2} \\ \dfrac{\partial^2 f(x_1, x_2)}{\partial x_1 \partial x_2} & \dfrac{\partial^2 f(x_1, x_2)}{\partial x_2^2} \end{bmatrix}. \quad (5.7)$$

If (5.6) holds at $x = x^*$ and $Hf(x)$ given by (5.7) is positive definite then the function $f(x)$ has its local minimum at $x = x^*$, and if Hessian matrix $Hf(x)$ given by (5.7) is negative definite at $x = x^*$ then x^* corresponds to a local maximum.

Level sets (curves for two dimensions, surfaces for three dimensions, etc.) are sets for which the function $f(x_1, x_2)$ has a constant value, i.e., $\{x_1, x_2 : f(x_1, x_2) = C = const\}$. Every level set can be associated with a the constant value C of the function. When constructing algorithms for recursive maximization we aim to design a sequence of points, which "climbs uphill"; each point in the sequence belongs to a level line with a higher C than the previous one. If a point x belongs to a level line (or surface or hypersurface in more than two dimensions) $f(x) = C$, then which direction should we head in to increase C? This can be resolved by using the property that the gradient vector $\nabla f(x)$ is perpendicular to the level set of the function $f(x)$ and points in the direction of increase of C. This property is illustrated in Fig. 5.2, where a 3D plot of an exemplary function is drawn in the upper part and corresponding level sets and gradient vectors are depicted in the lower part. So, in many optimization procedures, the direction for the update of recursions is parallel to the gradient vector, possibly with some scaling factor, for function maximization, and the direction is antiparallel to the gradient vector for function minimization.

In (5.3)–(5.7), we assumed a two-dimensional space of vectors x. However, all of the above can be extended in an obvious way to vectors of higher dimensionality. Namely, for a function $f : R^n \to R$ of $x = [x_1, x_2, ..., x_n]^T$ the gradient vector ∇f is

$$\nabla f(x) = \begin{bmatrix} \frac{\partial f(x)}{\partial x_1} \\ \frac{\partial f(x)}{\partial x_2} \\ \vdots \\ \frac{\partial f(x)}{\partial x_2} \end{bmatrix},$$

and its Hessian matrix is given by

$$Hf(x) = \begin{bmatrix} \frac{\partial^2 f(x)}{\partial x_1^2} & \frac{\partial^2 f(x)}{\partial x_1 \partial x_2} & \cdots & \frac{\partial^2 f(x)}{\partial x_1 \partial xn} \\ \frac{\partial^2 f(x)}{\partial x_2 \partial x_1} & \frac{\partial^2 f(x)}{\partial x_2^2} & \cdots & \frac{\partial^2 f(x)}{\partial x_2 \partial xn} \\ \vdots & \vdots & \ddots & \vdots \\ \frac{\partial^2 f(x)}{\partial x_n \partial x_1} & \frac{\partial^2 f(x)}{\partial x_n \partial x_2} & \cdots & \frac{\partial^2 f(x)}{\partial x_n^2} \end{bmatrix}. \quad (5.8)$$

5.1.1 Convexity and Concavity

Convexity and concavity are notions playing an important role in many fields of mathematical modeling (see Sect. 2.6) and among other things, in both static and dynamic optimization. By checking the convexity or concavity of a function one can distinguish between minima and maxima, as mentioned above, and turn necessary conditions for optimality into sufficient conditions.

A set $X \subset R^n$ is convex if the fact that two points x_A and x_B, satisfy, $x_A \in X$ and $x_B \in X$ implies that the whole segment with ends x_A and x_B belongs to X, which can be expressed as

$$\forall_{p \in [0,1]} \; px_A + (1-p)x_B \in X. \quad (5.9)$$

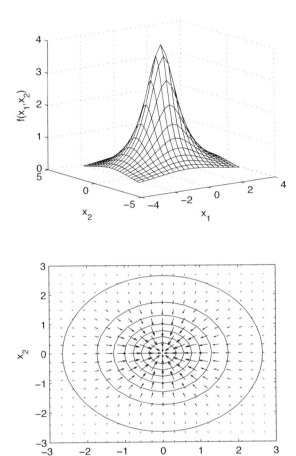

Fig. 5.2. *Top*: 3D plot of an example function. *Bottom*: level curves and gradient vectors

A function f defined over a convex set, $f : X \to R$, is convex if

$$\forall_{p \in [0,1]} \; f[px_A + (1-p)x_B] \leq pf(x_A) + (1-p)f(x_B). \qquad (5.10)$$

A function g defined over a convex set, $g : X \to R$, is concave if it satisfies the inequality converse to (5.10),

$$\forall_{p \in [0,1]} \; g[px_A + (1-p)x_B] \leq pg(x_A) + (1-p)g(x_B), \qquad (5.11)$$

where again x_A and x_B belong to X. If $f(x)$ is convex then $-f(x)$ is concave and vice versa.

If X and Y are two convex sets then their intersection $X \cap Y$ is a convex set. If $f(x)$ is a convex function $f : X \to R$, then the set bounded by the level hypersurface $Z_C = \{x : f(x) \leq C\}$, for every value of the constant C, is convex. If $g(x)$ is a concave function, $g : X \to R$, then the set bounded by the level hypersurface $Z_C = \{x : g(x) \geq C\}$ is again convex. Owing to the property of convexity of intersections of convex sets, the sets defined by

$$X = \{x : f_1(x) \leq C_1,$$
$$f_2(x) \leq C_2,$$
$$\vdots$$
$$f_k(x) \leq C_k\}$$

and

$$X = \{x : g_1(x) \geq C_1,$$
$$g_2(x) \geq C_2,$$
$$\vdots$$
$$g_l(x) \geq C_l\}$$

where $f_1, \ldots, f_k$ are all convex functions and $g_1, \ldots, g_l$ are all concave, are convex.

For smooth functions, convexity and concavity can be verified by use of the second derivatives. Namely, if $f(x)$ has continuous second partial derivatives in some convex set X, then $f(x)$ is convex in X if and only if its Hessian matrix (5.8) is positive semi-definite. If $g(x)$ has continuous second partial derivatives in some convex set X then $f(x)$ is concave if and only if its Hessian matrix (5.8) is negative semi-definite.

The linear function

$$f(x) = a^T x + c,$$

where a is a parameter vector and c is a constant, is both convex and concave. If a function $g(x)$ is convex (or concave) then $g(x) + a^T x + c$ is also convex (or concave, respectively).

5.1.2 Constrained Optimization with Equality Constraints

Very often one needs to optimize a function $f(x)$ with an additional condition that x belongs to some set, for example a the set of points in the plane satisfying an equation $g(x) = 0$ for some $R^2 \to R$ function $g(.)$. This is called a constrained optimization task, and can be stated more formally as

$$\min\ f(x) \tag{5.12}$$

with the constraint

$$g(x) = 0, \tag{5.13}$$

5.1 Static Optimization   129

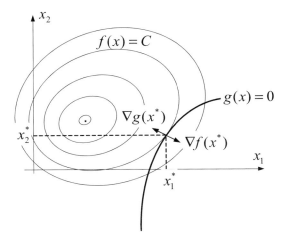

Fig. 5.3. Graphical illustration of the solution to the constrained optimization problem $\min f(x)$ with the constraint $g(x) = 0$, $x = [x_1, x_2]^T$. The constraint curve $g(x) = 0$ is depicted in bold. A family of level curves $f(x) = C = const$ is also shown. At the optimal point $x^* = [x_1^*, x_2^*]^T$, the gradient vectors of functions $f(x)$ and $g(x)$ $\nabla f(x^*)$ and $\nabla g(x^*)$, are parallel or antiparallel

for example. It will help us to illustrate the constrained optimization problem (5.12)–(5.13) graphically, as we have done in Fig. 5.3 for a two-dimensional $x = [x_1\ x_2]^T$. In this figure, the family of level lines of the function $f(x)$ is a family of ovals centered around the same point, and the constraint curve $g(x) = 0$ is depicted in bold. When we try to minimize the function $f(x)$ we aim at moving downhill–as close to the center of the family of ovals as possible. This is limited by the requirement that $g(x) = 0$ and we see that, at the optimal point $x^* = [x_1^*\ x_2^*]^T$ the curves $f(x) = C$ and $g(x) = 0$ are tangential. The tangentiality condition can also be understood meaning that gradient vectors $\nabla f(x^*)$ and $\nabla g(x^*)$ are parallel to each other, and so there must be a scalar number $-\lambda$ such that

$$\nabla f(x^*) = -\lambda \nabla g(x^*). \tag{5.14}$$

We have used a minus sign here to fit to with the usual notation. The number λ is called a Lagrange multiplier, and by associating with the constrained problem (5.12)–(5.13) the Lagrange function

$$L(x, \lambda) = f(x) + \lambda g(x), \tag{5.15}$$

we can express (5.14) at $x = x^*$ as

$$\frac{\partial}{\partial x} L(x, \lambda) = 0. \tag{5.16}$$

We note that computing the coordinates of the minimum x^* in Fig. 5.3 involves solving for three variables, namely the two coordinates of x and the scalar λ. The vector condition (5.16) gives two algebraic equations, and the third equation is the constraint (5.13). Equation (5.13) can be written equivalently as a condition on the gradient of the Langrange function (5.15) with respect to λ,

$$\frac{\partial}{\partial \lambda} L(x, \lambda) = 0. \tag{5.17}$$

One can also derive the optimality conditions (5.16) and (5.17) by algebraic manipulations without invoking a geometric interpretation (Exercise 1).

The situation depicted in Fig. 5.3 involving a two-dimensional argument space can be generalized to spaces of higher dimensionality and to a number of constraint equations of more than one. This leads to the following vector formulation analogous to (5.12)–(5.17).

Lagrange Multiplier Theorem

Consider the constrained problem

$$\min \ f(x), \tag{5.18}$$

with the constraint

$$g(x) = 0, \tag{5.19}$$

where f and g are functions that assign to a vector $x \in R^n$, respectively a scalar and an m-dimensional vector, respectively, i.e., $f : R^n \to R$ and $g : R^n \to R^k$. The necessary conditions for optimality in (5.18)–(5.19) are

$$\frac{\partial}{\partial x} L(x, \lambda) = 0 \tag{5.20}$$

and

$$\frac{\partial}{\partial \lambda} L(x, \lambda) = 0, \tag{5.21}$$

where $L(x, \lambda)$ is the Lagrange function associated with the problem (5.18), (5.19), given by

$$L(x, \lambda) = f(x) + \lambda^T g(x) \tag{5.22}$$

$$= f(x) + \sum_{i=1}^{n} \lambda_i g_i(x) \tag{5.23}$$

with a vector of Lagrange multipliers $\lambda \in R^m$. In (5.18) and (5.19), vector notation has been used, i.e.,

$$x = \begin{bmatrix} x_1 \\ x_2 \\ \vdots \\ x_n \end{bmatrix}, \ g(x) = \begin{bmatrix} g_1(x) \\ g_2(x) \\ \vdots \\ g_m(x) \end{bmatrix}, \ \lambda = \begin{bmatrix} \lambda_1 \\ \lambda_2 \\ \vdots \\ \lambda_m \end{bmatrix}; \tag{5.24}$$

λ^T stands for the row vector resulting from transposition of λ in (5.24), and $\lambda^T g(x)$ is a scalar product given by $\lambda^T g(x) = \sum_{i=1}^{m} \lambda_i g_i(x)$.

5.1.3 Constrained Optimization with Inequality Constraints

Consider the optimization problem

$$\max \ f(x), \tag{5.25}$$

where $x \in R^n$ and $f : R^n \to R$ and with constraints of the form

$$g(x) \geq 0, \tag{5.26}$$

where $g : R^n \to R^k$. The inequality in (5.26) is componentwise; (5.26) is a shortened vector notation for

$$\begin{aligned} g_1(x) &\geq 0, \\ g_2(x) &\geq 0, \\ &\vdots \\ g_m(x) &\geq 0. \end{aligned} \tag{5.27}$$

If $x \in R^n$ fulfills (5.27), then each of the component inequalities in (5.27) can be either active or inactive. Inequality number i is active if $g_i(x) = 0$, and inactive if $g_i(x) > 0$. In a some sense, the problem (5.25), (5.26) is more general than (5.18), (5.19) since each equality constraint $g_i(x) = 0$ can be represented as two opposite-sign inequalities, $g_i(x) \geq 0$, $-g_i(x) \geq 0$. The optimality problem (5.25), (5.26) can be resolved by use of the Kuhn–Tucker theorem.

Kuhn–Tucker Theorem

We define the Lagrange function corresponding to the constrained optimization problem (5.25), (5.26), by

$$L(x, \lambda) = f(x) + \lambda^T g(x) \tag{5.28}$$

$$= f(x) + \sum_{i=1}^{n} \lambda_i g_i(x) \tag{5.29}$$

where λ is a vector of Lagrange multipliers $\lambda \in R^m$. The notation in the above is the same as in (5.22). The Kuhn–Tucker theorem formulates the following necessary conditions for optimality in (5.25), (5.26):

$$\frac{\partial}{\partial x} L(x, \lambda) = 0, \tag{5.30}$$

$$\frac{\partial}{\partial \lambda} L(x, \lambda) \geq 0, \tag{5.31}$$

$$\lambda \geq 0, \tag{5.32}$$

and

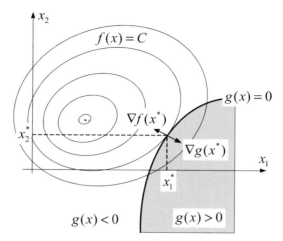

Fig. 5.4. Graphical illustration of the Kuhn–Tucker necessary optimality conditions for max $f(x)$ subject to the constraint $g(x) \geq 0$. The curve $g(x) = 0$ is depicted in bold and the region $\{x : g(x) > 0\}$ is shaded. A family of level curves $f(x) = C = const$ is also shown. At the optimal point $x^* = [x_1^*, x_2^*]^T$ the gradient vectors of the functions $f(x)$ and $g(x)$, $\Delta f(x^*)$ and $\Delta g(x^*)$, must point in opposite directions. In other words the vector equation $\Delta f(x^*) + \lambda \Delta g(x^*) = 0$, where $\lambda > 0$ is the scalar multiplier, must hold.

$$\lambda^T g(x) = 0. \tag{5.33}$$

We shall briefly discuss the idea behind their construction. As previously, let us denote the optimal point by x^*. The condition (5.31) is a repetition of (5.26). The equality (5.33) is called complementarity condition. The components of the vector of Lagrange multipliers which correspond to inequalities that are inactive at x^* are set to zero, i.e., $\lambda_i = 0$ and components corresponding to active inequalities are made strictly greater than zero, i.e., $\lambda_k > 0$. One can see that with this (and only with this) choice, the complementarity condition is satisfied. If one knew which of the constraints (5.27) were active and which were inactive, at the optimal point x^*, then (5.30)-(5.33) would reduce to solving a system of equalities and to checking (5.32) at found solutions. The need to add the condition (5.32) to (5.30) can be explained geometrically, as shown in Fig. 5.4. The gradient vector $\nabla f(x^*)$ must point in the opposite direction to $\nabla g(x^*)$. Otherwise it would be possible to increase $f(x)$ and $g(x)$ simultaneously, which would contradict the optimality of the point x^*. More generally, if we denote by $i1^a > 0, i2^a > 0 \ldots i_K^a$ the indices of active constraints Lagrange multipliers corresponding to active constraints, then

$$\lambda_{i_1^a} \nabla g_{i_1^a}(x^*) + \lambda_{i_2^a} \nabla g_{i_2^a}(x^*) + \ldots + \lambda_{i_K^a} \nabla g_{i_K^a}(x^*) \tag{5.34}$$

is a cone of feasible directions, i.e., the directions along which we can move the argument x without violating the constraints. If moving x along any of the directions in (5.34) were to increase the value of the function $f(x)$, then this would contradict the optimality of x^*.

As stated above, the Kuhn–Tucker conditions are only necessary; they can be satisfied both at local and at global extrema of functions. Situations where these necessary conditions become sufficient can be identified by employing the concepts of convexity and concavity, as presented in the next subsection.

5.1.4 Sufficiency of Optimality Conditions for Constrained Problems

If a convex function attains a local minimum over convex sets then, this is also the global minimum. Analogously, if a concave function attains a local maximum over convex sets, then this is also the global maximum. These properties are used to formulate the sufficiency of optimality conditions.

If, in optimization problem (5.25), (5.26) the scoring function $f(x)$ is strictly concave and all constraint functions $g_1(x), ..., g_m(x)$ are concave then the Kuhn–Tucker conditions (5.30)–(5.33) become both necessary and sufficient.

Consider a problem analogous to (5.25)–(5.26) with maximization replaced by minimization,

$$\min \ f(x) \tag{5.35}$$

subject to constraints

$$g(x) \leq 0. \tag{5.36}$$

Kuhn Tucker necessary optimality conditions (Exercise 4) stated with the use of the Lagrange function (5.28),

$$\frac{\partial}{\partial x} L(x, \lambda) = 0, \tag{5.37}$$

$$\frac{\partial}{\partial \lambda} L(x, \lambda) \leq 0, \tag{5.38}$$

$$\lambda \leq 0 \tag{5.39}$$

and

$$\lambda^T g(x) = 0, \tag{5.40}$$

become both necessary and sufficient if the function $f(x)$ is convex and all component functions $g_1(x), ..., g_m(x)$ in $g: R^n \to R^k$ are convex.

5.1.5 Computing Solutions to Optimization Problems

For many cases of optimization problems analytical solutions leading to very useful results can be obtained by using optimality conditions discussed above. In some situations, however, no analytical expressions for optimal points are available, but proofs can be carried out of the existence and/or uniqueness of optimal points and of recursive methods which converge to optimal points. Some of these will be discussed in the following.

Some functions often appearing in optimization problems are linear and quadratic forms. A function $f : R^n \to R$

$$f(x) = a^T x, \tag{5.41}$$

where a is a constant n-dimensional vector is called a linear form. Its gradient is $\nabla f(x) = a$ and its Hessian matrix is the zero matrix $Hf(x) = 0$.

A function $f : R^n \to R$,

$$f(x) = x^T Q x + a^T x, \tag{5.42}$$

where Q is a symmetric $n \times n$ matrix is called a quadratic form. The common symmetry assumption of $Q = Q^T$ is due to the decomposition of a matrix into symmetric and antisymmetric components $A = (1/2)(A+A^T)+(1/2)(A-A^T)$. Only the symmetric component $(1/2)(A + A^T)$ will contribute to the value of the quadratic form $x^T A x$. The gradient vector of the quadratic form (5.42) is

$$\nabla f(x) = 2Qx + a, \tag{5.43}$$

and the Hessian matrix is given by $Hf(x) = 2Q$.

By comparing the gradient (5.43) with zero we obtain the following (given that the matrix Q is invertible):

$$x^* = -\frac{1}{2} Q^{-1} a, \tag{5.44}$$

which is a unique maximal point, provided that Q is negative definite, and a unique minimal point provided that Q is positive definite.

Simple Linear Regression by Least Squares

As an example of the application of (5.44) let us consider the problem of fitting the parameters a and b of a straight line $y = ax + b$ to some measured data points (x_i, y_i), $i = 1, \ldots, n$. With the notation

$$y = \begin{bmatrix} y_1 \\ y_2 \\ \vdots \\ y_n \end{bmatrix}, \; Y = \begin{bmatrix} x_1 & 1 \\ x_2 & 2 \\ \vdots & \vdots \\ x_n & 1 \end{bmatrix}, \; p = \begin{bmatrix} a \\ b \end{bmatrix}$$

we can represent the sum of the squared errors of the model versus the data, as follows:

$$\sum_{i=1}^{n} (y_i - a x_i - b)^2 = (y - Yp)^T (y - Yp)$$
$$= y^T y - 2 y^T Y p + p^T Y^T Y p.$$

The last expression is a quadratic form $f(p)$ given in (5.42), where $a = Y^T y$, and $Q = Y^T Y$, and so from (5.44), the optimal parameter fit is

$$p^* = \begin{bmatrix} a^* \\ b^* \end{bmatrix} = (Y^T Y)^{-1} Y^T y.$$

Constrained Optimization Problems

Here we analyze some examples, involving linear and quadratic forms, illustrating the Lagrange multiplier and the Kuhn–Tucker constraint optimality conditions. As the first example consider minimization in $x \in R^n$

$$\min x^T Q x \qquad (5.45)$$

subject to the linear constraint

$$a^T x = c. \qquad (5.46)$$

Notation for a, c and Q is the same as above in this section. Lagrangian function for (5.45)-(5.46) is

$$L(x, \lambda) = x^T Q x + \lambda (a^T x - c),$$

where λ is a scalar Lagrange multiplier and using (5.20)-(5.21) we compute

$$x^* = \frac{c}{a^T Q^{-1} a} Q^{-1} a. \qquad (5.47)$$

Knowing that (5.20) and (5.21) are only necessary conditions we recall the remarks from section 5.1.4 to find whether there is maximum or minimum at x^*. If Q is a positive definite matrix then x^* is indeed a minimum of (5.45) subject to (5.46). If Q is negative definite, then a minimum does not exist, function in (5.45) can approach $-\infty$ for some sequences of x, all satisfying (5.46). For negative definite Q, a unique solution to the problem (5.45), (5.46) would exist if minimization were replaced by maximization.

As the second example, let us consider the following:

$$\max x_1^2 + x_2^2 \qquad (5.48)$$

subject to the constraints

$$x_1 \leq 1 \qquad (5.49)$$

and

$$x_2 \leq 2. \qquad (5.50)$$

Here the level curves of the function (5.48) are circles centered at $(x_1, x_2) = (0, 0)$, the sets defined by constraints (5.49) and (5.50) are half planes, and the global maximum does not exist, in the sense that (5.48) can be increased to arbitrarily large values without violating (5.49) and (5.50). Yet the Kuhn–Tucker conditions are satisfied at three points, $(x_1^*, x_2^*) = (1, 0)$, $(x_1^*, x_2^*) = (0, 2)$, and $(x_1^*, x_2^*) = (1, 2)$. The sufficiency conditions discussed in Sect. 5.1.4 are not satisfied, since we are maximizing a convex, not a concave function.

Finally, if we replace maximization in the above by minimization and we change directions of inequalities i.e., if we consider

$$\min x_1^2 + x_2^2 \tag{5.51}$$

subject to the constraints

$$x_1 \geq 1 \tag{5.52}$$

and

$$x_2 \geq 2, \tag{5.53}$$

then the unique solution to the Kuhn–Tucker conditions is $(x_1^*, x_2^*) = (1, 2)$. Here the sufficiency conditions from Sect. 5.1.4 are satisfied.

Both of the above problems (5.48)–(5.53) are easily interpreted by using plots of functions and constraint sets in the plane (x_1, x_2).

5.1.6 Linear Programming

Linear programming is a special optimization problem, of finding the extremal value of a linear form over a set defined by linear inequalities. It can be formulated as follows:

$$\min a^T x \tag{5.54}$$

subject to

$$Bx \leq b. \tag{5.55}$$

In the above expressions (5.54) and (5.55), $x \in R^n$, a is an n-dimensional vector of the parameters of the linear form $a^T x$, B is an $m \times n$-dimensional matrix, b is an m-dimensional vector. The inequalities in (5.55) are understood componentwise. The set in R^n defined by the system of inequalities (5.55) is a (possibly unbounded or degenerate) convex hyperpolyhedron, and the problem defined by (5.54) and (5.55) can be understood as looking for the vertex of the polyhedron located farthest away along the direction defined by the vector a.

The formulation (5.54) and (5.55) is the most general in the sense that any linear programming problem can be transformed to it by introducing suitable definitions. Minimization can be changed to maximization by taking $a = -a_1$. Equality constraints can be represented by pairs of inequality constraints.

There are algorithms and computer software that allow one to solve linear programming problems with very large sizes of the vector x.

5.1.7 Quadratic Programming

The quadratic programming problem is defined as follows:

$$\min(x^T Q x + a^T x) \qquad (5.56)$$

subject to

$$Bx \leq b. \qquad (5.57)$$

In (5.56) Q is a symmetric, positive definite (or positive semidefinite) $n \times n$ matrix. Other parameters, a, B and b have the same meanings as those in (5.54) and (5.55). If the matrix Q is positive definite the problem (5.56)-(5.57) has a unique solution.

The optimization problem (5.56)-(5.57) can be efficiently solved for vectors x of large size by use of appropriate algorithms and related computer software.

5.1.8 Recursive Optimization Algorithms

In general, optimality conditions can be difficult to find solutions for. A solution may not exist owing either to contradictory constraints or the possibility of the value of $f(x)$ diverging to infinity. There may exist multiple solutions, even infinitely many or uncountable sets of solutions. The Kuhn–Tucker conditions (5.30)–(5.33), which involve both equalities and inequalities, are more difficult to find solutions for than systems of algebraic equations occuring in the Lagrange multiplier theorem (5.20)–(5.21); the difficult problem may be identifying the active and inactive constraints.

Even for unconstrained problems, computing optimal points is very often not possible analytically. Therefore numerical algorithms, where the value of the function to be optimized is improved step by step, are very useful and are very often applied. Below, we briefly describe some commonly applied versions of iterative unconstrained optimization algorithms.

Search for Extremum of a Function Without Derivatives

The information about the direction of increase or decrease of a function is contained in its gradient vector. However, sometimes computing the gradient vector of a function is time-expensive or cumbersome. It is therefore worth mentioning algorithms which seek an extremum recursively solely on the basis of values of the function $f(x)$, without derivatives. In the one-dimensional case, searching for an extremum of a function in an interval $(x_{\min}, x_{\max})$ can be accomplished on the basis of its successive division into smaller parts can be applied, such as bisection, or golden-section based on Fibonacci proportions. One algorithm without derivatives designed for multidimensional cases is named Nelder–Mead, moving-simplex or moving-amoeba method [205], Owing to the idea behind its construction, where the aim is to localize the extremal argument of a function inside a simplex and then, recursively, shrink the diameter of the simplex to zero. Despite slow or even problematic convergence, especially in higher numbers of dimensions, it can be very useful and it is included in most software packages for optimization. Let us assume that our aim is to minimize a function $f(x)$ in n dimensions. The algorithm needs the following parameters to be specified: ρ (reflection), χ (expansion), γ (contraction) and σ (shrinkage). The choice of values typically applied is $\rho = 1$, $\chi = 2$, $\gamma = 0.5$, and $\sigma = 0.5$. We shall describe one iteration of the algorithm, which starts from a simplex in R^n with $n+1$ vertices $x_1, x_2, \ldots, x_{n+1}$. Assume that the values of the function are ordered such that $f(x_1) < f(x_2)$, $\ldots, < f(x_{n+1})$. Call vertices $x_1, x_2, \ldots, x_n$ the base of the simplex and the vertex x_{n+1} the peak of the simplex. First, we reflect the peak with respect to the center of the base $\bar{x}$, where

$$\bar{x} = \frac{1}{n} \sum_{i=1}^{n} x_i,$$

using the assumed value of the reflection parameter ρ. The result of this operation is denoted x_R and is given by

$$x_R = \bar{x} + \rho(\bar{x} - x_{n+1}).$$

Now, depending on the relations between $f(x_R)$ and the values of f at the vertices of the simplex, we perform different operations.

Case 1. If $f(x_1) < f(x_R) < f(x_n)$, replace x_{n+1} by x_R and terminate the iteration.

Case 2. If $f(x_R) < f(x_1)$, calculate the expansion point x_E, given by

$$x_E = \bar{x} + \chi(x_R - \bar{x}),$$

evaluate $f(x_E)$ and replace x_{n+1} by x_E if $f(x_E) < f(x_R)$ or by x_R if $f(x_E) > f(x_R)$. Terminate the iteration.

Case 3. If $f(x_n) < f(x_R) < f(x_{n+1})$ compute the outside contraction point x_C, where

$$x_C = \bar{x} + \gamma(x_R - \bar{x}),$$

and evaluate $f(x_C)$. If $f(x_C) < f(x_R)$ replace x_{n+1} by x_C and terminate the iteration. If $f(x_C) > f(x_R)$, perform a shrink operation and terminate the iteration.

Case 4. If $f(x_R) > f(x_{n+1})$, compute the inside contraction point x_{CC}, where

$$x_{CC} = \bar{x} + \gamma(\bar{x} - x_{n+1}),$$

and evaluate $f(x_{CC})$. If $f(x_{CC}) < f(x_{n+1})$, replace x_{n+1} by x_{CC} and terminate the iteration. If $f(x_{CC}) > f(x_{n+1})$, perform a shrink operation and terminate the iteration.

The shrink operation is defined as follows. Replace the vertices $x_1, x_2, \ldots, x_{n+1}$ of the simplex by new vertices $x_1, x'_2, \ldots, x'_{n+1}$, where

$$x'_i = x_1 + \sigma(x_i - x_1), \ i = 2, 3, \ldots, n+1.$$

Counterexamples can be constructed such that the above algorithm will not reach a minimal point despite the convexity of the function $f(x)$. Nevertheless, it is successful in many typical examples and, as stated, can be very useful. At least, it can be tried as a first choice.

Gradient Algorithms

If we want to minimize a function $f(x)$ in n dimensions, then at a given point x_k, the direction of its fastest decrease (steepest descent) is opposite to the gradient vector $\nabla f(x)$. So we can plan a step of a minimization algorithm as follows:

$$x_{k+1} = x_k - \gamma \nabla f(x_k), \tag{5.58}$$

where γ is a suitably defined parameter. The main problem is tuning the step size, adjusted by the parameter γ. If it is too small, convergence to minimum is very slow. Values that are too large will typically cause instabilities in the algorithm. Therefore there are many more or less heuristic modifications of the recursion in (5.58), aimed at producing algorithms that both have a high speed of convergence and are robust to instabilities [89].

Algorithms Using Second Derivatives

Let us start with the one-dimensional case, where a function $f(x)$ is minimized over a scalar x. We assume that the kth iteration of the minimization procedure hits a point x_k in the close vicinity of the optimal argument x^*, and that the function $f(x)$ is twice continuously differentiable, so the following approximate equation holds, since $f'(x^*) = 0$:

$$f(x_k) = f(x^*) + \frac{1}{2} f''(x^*)(x_k - x^*)^2. \tag{5.59}$$

Now we think of $f(x^*)$ and $f''(x^*)$ as constant parameters, which allows us to compute the following expression for $f'(x_k)$,

$$f'(x_k) = f''(x^*)(x_k - x^*), \tag{5.60}$$

and for $f''(x_k)$,

$$f''(x_k) = f''(x^*). \tag{5.61}$$

Using (5.60) and (5.61), we can compute x^* from

$$x^* = x_k - \frac{f'(x_k)}{f''(x_k)}. \tag{5.62}$$

Since (5.59) is only an approximation, we take (5.62) as the new value in the recursion rather than as a final solution, i.e.,

$$x_{k+1} = x_k - \frac{f'(x_k)}{f''(x_k)}. \tag{5.63}$$

The above is called the Newton–Raphson or Gauss–Newton iterative minimum search for the minimum. An analogous derivation applies in the multidimensional case, where $x \in R^n$. If $\nabla f(x_k)$ is the gradient vector of the function $f(x)$ taken at the point of kth iteration, and $Hf(x_k)$ is its Hessian matrix, then the next iteration of minimum search algorithm will is

$$x_{k+1} = x_k - [Hf(x_k)]^{-1} \nabla f(x_k). \tag{5.64}$$

In the close vicinity of the minimum, the convergence of (5.63) or (5.64) is very fast, but if the scheme is started from a random point it can easily end up in instability. Therefore, again scaling is commonly applied to the steps of the algorithm:

$$x_{k+1} = x_k - \gamma [Hf(x_k)]^{-1} \nabla f(x_k), \tag{5.65}$$

where γ is a suitable parameter.

5.2 Dynamic Programming

Dynamic programming [26, 65] is solving optimization problems by organizing the optimizing decisions in the sequential order. The method of dynamic programming has been applied efficiently to large variety of problems. Sometimes formulating a dynamic programming solution to an optimization problems can be tricky, and may need research. We list some properties of discrete dynamic optimization problems. Knowing them should help one to develop a suitable formulation of dynamic programming algorithm. (1) We should be able to decompose the optimization problem into separate decisions and organize the decision-making process into stages. (2) At each stage of decision-making process, we should be able to define a state of the system (a problem) which summarizes the influence of the decisions already made. (3) The scoring index should be expressed such that it can be computed iteratively, stage by stage, and provided optimal value of score for $(i+1)$th stage is known we can find a recursion for ith stage. Developing the recursion in (3) is a basic technique for constructing algorithms for solving discrete dynamic optimization problems. The principle behind the recursive update of the scoring index is the following Bellman's optimality principle [26],

> *An optimal policy has the property that whatever the initial state and initial decision are, the remaining decisions must constitute an optimal policy with regard to the state resulting from the first decision.*

The recursion following from the above principle is called Bellman's equation. Below, we illustrate the principle by going through some examples, both general and more specific.

5.2.1 Dynamic Programming Algorithm for a Discrete-Time System

Consider a discrete-time dynamical system written in the fairly general form

$$x_{k+1} = f_k(x_k, u_k), \qquad (5.66)$$

where $k = 0, 1, \ldots, K$ are discrete time instants that index stages of the optimization process, u_k stands for the decision variables, f_k is a function which gives a model for the discrete-time evolution of the system and x_k is the state of the process. Knowing x_k and $u_k, u_{k+1}, \ldots$ allows us to compute future states $x_{k+1}, x_{k+2}, \ldots$ no matter what previous decisions $u_{k-1}, u_{k-1}, \ldots$ were. We assume that at each time instant k possible decisions are constrained by the requirement that u_k belongs to a set

$$u_k \in U_k(x_k), \qquad (5.67)$$

depending both on the state x_k and on the time instant k. The aim is to find a sequence of decisions u_k, $k = 0, 1, \ldots, K$ such that, given initial state x_0, the scoring index

$$I(x_0) = \sum_{k=1}^{K} s_k(x_k, u_k) \qquad (5.68)$$

is minimized. The above formulation is fairly general, in the sense that we allow the functions f_k, feasibility sets U_k, the components of the scoring function s_k, and the numbers of components of states x_k and decisions u_k, to change in each stage of the decision-making process.

Solution

Define the optimal partial cumulative score as

$$I_k^{opt}(x_k) = \min_{u_k, u_{k+1}, \ldots, u_K} \sum_{i=k}^{K} s_i(x_i, u_i), \qquad (5.69)$$

where we have listed the optimization arguments, but have skipped the constraints (5.67) for brevity. The solution to our dynamic optimization problem is $I_0^{opt}(x_0)$; however, we cannot compute it by direct minimization, as in (5.69), owing to the large number of optimization variables. Knowing the optimality principle, we want to organize the minimization in (5.69) in a recursive style. We start the recursion from the last stage of the decision-making process $k = K$, which leads to

$$I_K^{opt}(x_K) = \min_{u_K \in U_K(x_K)} s_K(x_K, u_K) \qquad (5.70)$$

and

$$u_K^{opt}(x_K) = \arg \min_{u_K \in U_K(x_K)} s_K(x_K, u_K). \tag{5.71}$$

We assume that the above minimization, involving only the last decision u_K, can be performed efficiently. Now, the optimality principle gives a recursion between the optimal partial cumulative scores $I_k^{opt}(x_k)$ and $I_{k+1}^{opt}(x_{k+1})$, i.e., Bellman's equation, in the form

$$\begin{aligned} I_k^{opt}(x_k) &= \min_{u_k \in U_k(x_k)} s_k(x_k, u_k) + I_{k+1}^{opt}(x_{k+1}) \\ &= \min_{u_k \in U_k(x_k)} s_k(x_k, u_k) + I_{k+1}^{opt}[f_k(x_k, u_k)], \end{aligned} \tag{5.72}$$

and allows us to compute the optimal decision

$$u_k^{opt}(x_k) = \arg \min_{u_k \in U_k(x_k)} s_k(x_k, u_k) + I_{k+1}^{opt}[f_k(x_k, u_k)]. \tag{5.73}$$

The optimization in (5.72) again involves only one decision and is assumed to be tractable. Starting from (5.70) and (5.71) and repeating (5.72) and (5.73) recursively, we finally obtain the solution to the whole problem, $I_0^{opt}(x_0)$.

Let us acknowledge one difficulty in the above procedure. In (5.70), we are solving not one optimization problem, but rather a whole family of problems, parametrized by the values of x_K. Similarly, in (5.72), we are solving a family of problems parametrized by the values of x_k. In most practical situations, iterating (5.72) and (5.73) is therefore only possible by tabulating $I_k^{opt}(x_k)$ over grids of points in the state space of x_k. This can become prohibitive if the state x_k is too complex, for example if it is a real vector with many dimensions. This difficulty is called the curse of dimensionality.

The formulation above is fairly general, since we do not make any specific assumptions on variables, functions, and sets that appear in (5.66)–(5.68); they can be real numbers or integers, the sets can be defined by inequalities or by listing their elements, etc. The recursive optimization (5.70)–(5.73) covers all specific cases. However, a restrictive element in our formulation is the fixed, predefined number of steps K. This may be an obstacle if we want to solve, for example, problems of reaching certain points or sets in the plane or in 3D space, problems of traversing graphs, or problems with stopping conditions.

We shall now go through the derivation (5.66)–(5.73) aiming at a modification that would allow us to relax the above limitation. Since our recursive optimization follows backwards from terminal to the initial state, then in order to make the number of steps variable, we may fix the index label of the terminal state and vary the index of the initial state. Such a system of numbering, involving recursions between $I_k^{opt}(x_k)$ and $I_{k-1}^{opt}(x_{k-1})$, can be introduced easily. In view of this consideration, we often formulate dynamical optimization problems like in (5.66)-(5.68) and we understand number of steps K as a free rather than as a fixed parameter.

Nothing was said in (5.66)–(5.73), about admissible ranges of the states x_k, which may be of basic importance when one is programming practical

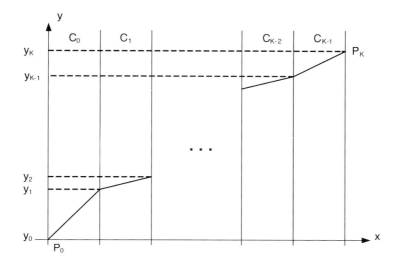

Fig. 5.5. Tracing a minimal-cost path from a point P_0 to a point P_K

solutions to dynamic programming problems. There may exist constraints on state variables in the problem formulation, in the form of inequalities or listing elements of sets, as it was mentioned above. In the context of optimization problems with the variable number of steps, the following recursive definition of admissible sets for states are often useful:

$$X_K = \{x_k : \text{iterations in (5.66) terminate}\}, \qquad (5.74)$$

then

$$X_{K-1} = \{x_{K-1} : \exists_{u_{K-1} \in U(x_{K-1})} \text{ such that } f(x_{K-1}, u(x_{K-1})) \in X_K\} \quad (5.75)$$

and, successively,

$$X_{k-1} = \{x_{k-1} : \exists_{u_{k-1} \in U(x_{k-1})} \text{ such that } f(x_{k-1}, u(x_{k-1})) \in X_k\}. \quad (5.76)$$

Now, at the optimization stage k in recursive optimization (5.70)-(5.73) we include the condition $x_k \in X_k$. Again, we solve the parametric optimization problems (5.72)–(5.73), going backwards, i.e., for K, $K - 1$, ..., with the additional constraint $x_k \in X_k$.

5.2.2 Tracing a Path in a Plane

Let us consider the problem, sketched graphically in Fig. 5.5, of tracing a minimal-cost path, in the plane, starting from a point P_0 and ending at a point P_K. The cost of a fragment of a path of length l is

$$C(l) = c_k l,$$

where c_k is a coefficient, $k = 0, 1, \ldots, K - 1$. The area between P_0 and P_K is divided into vertical strips, each of which has different cost coefficient c_k. The widths of successive vertical strips are denoted by d_k, $k = 0, 1, \ldots, K - 1$. Clearly, if all the cost coefficients were equal, the optimal path from P_0 to P_K would be a straight line. Owing to the unequal costs in different strips, the optimal path is a sequence of straight-line segments and in order to arrange them optimally we can use dynamic programming. We denote the state of the discrete process of decision making at stage k by y_k, equal to the y-th coordinate of the optimal path when it crosses between strips $k - 1$ and k, as shown in Fig. 5.5. The recursion may be

$$y_k = y_{k-1} + u_{k-1}$$

where the value of u_k describes the change between two successive states. The cost of crossing the kth strip can be expressed as $c_k \sqrt{d_k^2 + u_k^2}$, and the scoring index for the optimization problem is then

$$I = \sum_{k=0}^{K-1} c_k \sqrt{d_k^2 + u_k^2}.$$

Since the path must hit the point P_k, we have the following constraint, which is one element set for $k = K - 1$:

$$u_{K-1} \in U_{K-1}(y_{K-1}) = \{y_K - y_{K-1}\}. \tag{5.77}$$

The controls $u_0, u_1, \ldots, u_{K-2}$ are not constrained. As we can see, the above is an instance of the formulation (5.66)–(5.68) and the solution can be obtained recursively as in (5.70)-(5.73). Bellman's equation has the form

$$I_k^{opt}(y_k) = \min_{u_k} \left[c_k \sqrt{d_k^2 + u_k^2} + I_{k+1}^{opt}(y_k + u_k) \right] \tag{5.78}$$

for $k = 1, 2, \ldots, K - 2$, and for $k = K - 1$ we have

$$I_{K-1}^{opt}(y_{K-1}) = c_{K-1} \sqrt{d_{K-1}^2 + (y_K - y_{K-1})^2}.$$

Even in this relatively simple case we are not able to compute an analytical solution. Instead, we approximate the possible range of y_k by a discrete set of for example, $N = 1000$ grid points, and we proceed by updating (5.78) over the grid defined. Rigorously speaking, with this approach we obtain only an approximation to the solution to the formulated problem. We can improve the approximation merely by increasing N? If we want a solution with an accuracy as high as possible, an approach better than dynamic programming would be a variational formulation [183].

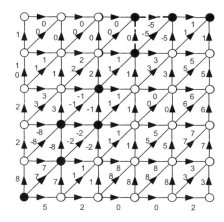

Fig. 5.6. *Left*: the problem of crossing an array of numbers from the bottom left corner to the top right corner, such that the sum of scores is minimized. Feasible moves are →, ↑, ↗. *Right*: equivalent formulation as a graph-traversing problem. The optimal solutions are depicted by bold squares and cicles

5.2.3 Shortest Paths in Arrays and Graphs

In Fig. 5.6, on the left-hand side, we present the problem of programming the optimal crossing through an array of numbers, with the aim of minimizing the score function given by the sum of the numbers in the cells of the array. The path starts at the bottom-left corner and ends in the top-right corner, and the feasible moves (decisions) are →, ↑, and ↗. In this problem, the state of the process at stage k is

$$x_k = [x_k^r \; x_k^c] \tag{5.79}$$

where x_k^r and x_k^c are the indices of rows and columns of the array. We assume that the bottom-left corner of the array corresponds to numbers $x^r = 1$, $x^c = 1$ and the top-right corner to $x^r = R$, $x^c = C$, (in Fig. 5.6, $R = C = 6$). The state transition function is therefore

$$x_{k+1} = f(x_k, u_k) = \begin{cases} [x_k^r + 1 \; x_k^c] & \text{for } u_k = \rightarrow \\ [x_k^r \; x_k^c + 1] & \text{for } u_k = \uparrow \\ [x_k^r + 1 \; x_k^c + 1] & \text{for } u_k = \nearrow \end{cases} \tag{5.80}$$

Denoting the scores in the cells in the array in Fig. 5.6 by

$$s(x^r, x^c),$$

for example, $s(2, 1) = 7$, we can write the scoring index for the problem as

$$I = \sum_{k=1}^{K} s(x_k^r, x_k^c). \tag{5.81}$$

146   5 Optimization

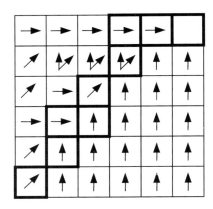

Fig. 5.7. *Left*: matrix of optimal partial cumulative scores for the problem in Fig. 5.6. *Right*: matrix of optimal controls. The optimal path is depicted by bold squares

The number of steps K in the decision making process will depend on the path through the array. Again, we can see the problem (5.79)–(5.81) as an instance of that described in Sect. 5.2.1. Bellman's equation takes the form

$$I_k^{opt}(x_k) = \min_{u_k \in \{\rightarrow,\, \uparrow,\, \nearrow\}} s(x_k^r, x_k^c) + I_{k+1}^{opt}[f(x_k, u_k)] \tag{5.82}$$

and the optimal decision at stage k is

$$u_k^{opt}(x_k) = \arg \min_{u_k \in \{\rightarrow,\, \uparrow,\, \nearrow\}} s(x_k^r, x_k^c) + I_{k+1}^{opt}[f(x_k, u_k)]. \tag{5.83}$$

The admissible sets for controls depend on states since the path cannot cross boundaries of the array, and so they reduce to $u_k \in \{\rightarrow\}$ if $x_k^r = R$ and to $u_k \in \{\uparrow\}$ if $x_k^c = C$. The optimization in (5.82) is very simple, and proceeds by inspection of at most three elements. Also, since the states are discrete and finite, the values of $I_k^{opt}(x_k)$ are easy to tabulate. From (5.82), we see that the optimal partial cumulative scores can be stored in a matrix of a size corresponding to the size of the array of scores $s(x^r, x^c)$. The order of filling in the entries of the matrix of optimal partial cumulative scores must be such that (5.82) is always manageable. In the left part of Fig. 5.7, the array of the values of the optimal partial scores is shown. Also, in the course of computing the optimal partial scores, one can record the optimal decisions, which are shown in the right pat of Fig. 5.7. We can trace the optimal strategy by following arrows in this plot, and we see that there are two different paths of equal score $I_1^{opt}(x_1) = -6$. It is also possible to find an optimal path directly from array of optimal partial scores, without recording optimal decisions, by following the "steepest descent", or, in other words, by repeating (5.82) for the array of optimal partial scores. Both the array of optimal partial cumulative scores and the array of optimal decisions allow us to find optimal path to the final state not only from the initial cell $x^r = 1$, $x^c = 1$ but also from any other cell in the array.

Graphs and Shortest Paths

The problem described above is very closely related to planning shortest paths in graphs. In the right part of Fig. 5.6, a graph is presented, that is equivalent to the array of scores in the left part. Each node corresponds to one cell of the array and each arrow corresponds to a feasible transition between cells. All arrows are directed, and their lengths (weights) correspond to the scores in the cells of the array. The problem of scheduling an optimal path through an array, discussed above, is equivalent to that of designing an optimal path in a graph, with the minimal sum of weights. Clearly, it can be solved by a dynamic programming method analogous to that already described. What makes the solution relatively easy, and can be adequately formulated in terms of graph terminology, is aperiodicity. A directed graph is aperiodic if, after departing from any of nodes, there is no possibility of returning. Clearly the graph in the right part of Fig. 5.6 has this property. For an aperiodic graph there is always a method to assign integer numbers to the nodes such that if, for nodes x and y, number$(x) <$ number(y), then there is no path from y to x. The node with the smallest number will not have any entering vertices (arrows), and the node with the largest number will not have any exiting vertices. Using numbering of nodes, one can easily order the optimizing decisions in an appropriate way and therefore efficiently solve for the shortest path. What if the graph is not aperiodic? In this case it can have cycles; after departing from some node, there might be a possibility to return after traversing some other vertices. For example, if we assume that the possible moves in the array of scores in Fig. 5.6 are now →, ↑, ↗, ←, ↓, ↙ then the corresponding graph can obviously have cycles. Planning paths for graphs with cycles is more difficult than in the aperiodic case. It is also necessary to introduce some conditions on scores (weights), for example, $s(x_k^r, x_k^c) \geq 0$, when minimizing the total score of the path. Without this condition it could happen that a cycle has a score that is negative and one can make the total score go to $-\infty$ by performing loops around this cycle. Actually, we could observe this in the formulation in Fig. 5.6 if we allowed moves →, ↑, ↗, ←, ↓, ↙. An efficient algorithm for solving for the shortest path in general graphs was formulated by Dijkstra [64]; this algorithms can be stated with the use of dynamic programming [65] (Exercise 7).

5.3 Combinatorial Optimization

Combinatorial optimization problems are commonly understood as optimization problems over mathematical objects such as paths, trees, graphs, such that their listing, labeling or enumerating involves using combinatorics. Combinatorial optimization problems can be hard to solve owing to the large solution space, which is difficult to explore. More precisely, combinatorial optimization is a branch of optimization theory dealing with the complexity of optimization and decision problems and the related classification of algorithms for solving optimization problems. Combinatorial optimization has links to branches of computer science and applied mathematics, such as algorithm theory, artificial intelligence, operations research, discrete mathematics, and software engineering. Knowledge and experience in the field of the computational complexity of an instance of an algorithm becomes critically important when the size of the problem increases. Since in bioinformatics the size of the data, i.e., sequences and measurements, is usually very high, exploring the computational complexity of the algorithms is very important. In this section, we overview the classification of optimization or decision problems from the point of computational complexity and give examples of combinatorial optimization problems. Excellent presentations of the present state of the art in combinatorial optimization can be found in the monographs [94, 290, 54].

5.3.1 Examples of Combinatorial Optimization Problems

We start by presenting several examples of combinatorial problems.

Traveling salesman problem. For every pair out of K cities $C_1, \ldots, C_K$, we know the distance or the cost of travel between them, $d(C_i, C_j)$. The problem is to find the shortest (or cheapest) route through the cities $C_1, \ldots, C_K$, such that each of the cities is visited at least once.

Hamiltonian path problem. Given a graph G, verify whether there exists a Hamiltonian cycle for G. A Hamiltonian cycle is a path along the edges of a graph such that every vertex (or node) is visited exactly once.

Shortest-superstring problem. Given a collection of words $w_1, \ldots, w_K$ over an alphabet, find the shortest string that contains all words $w_1, \ldots, w_K$.

Boolean satisfiability problem. Given a Boolean function (expression) $f(b_1, \ldots, b_K)$ over binary variables $b_1, \ldots, b_K$, determine whether we can assign values, zero or one, to each of the variables $b_1, \ldots, b_K$ such that the Boolean formula $f(.)$ is true, in other words, that will set $f(b_1, \ldots, b_K) = 1$.

5.3.2 Time Complexity

We can assign a size to a combinatorial optimization problem. In the problems listed above, the size is given by the number K. The size is proportional to the length of the data string fed to the algorithm for solving the problem.

By the time complexity of a problem or of an algorithm for solving an instance of a problem, we mean the relation between the running time of the algorithm and the size of the problem. More formally, the "running time" can be replaced by the number of steps required by a Turing machine (see Chap. 3) programmed for the execution of the algorithm.

5.3.3 Decision and Optimization Problems

Observe that in the list in the Sect. 5.3.1, there are two types of problems, decision problems (determine whether an object with given properties exists or not) and optimization problems (find the object which optimizes a criterion, i.e., the cheapest, shortest, etc.). However, we can demonstrate that the distinction between decision and optimization problems is not very important from the point of view of their time complexities. For example, let us replace the traveling-salesman optimization problem stated above by the following traveling salesman decision problem: Decide whether there is a route visiting each of the cities $C_1, \ldots, C_K$ at least once and such that its total cost is $\leq \theta$, where θ is a given number. Assuming that we have an algorithm for solving the traveling-salesman decision problem, we can repeat this algorithm several times and use the idea of bisection of an interval to obtain reasonable knowledge about the optimal route. Roughly, the number of repetitions of the decision algorithm necessary will be proportional to $\log_2(\text{size})$. So, having an algorithm of time complexity Time(size) for solving the traveling salesman decision problem we can design an algorithm for solving traveling salesman optimization problem with time complexity $\log_2(\text{size}) \times \text{Time(size)}$. If a combinatorial problem belongs to one of the classes polynomial or exponential, then multiplying it by $\log_2(\text{size})$ does not change the class. Therefore, for the traveling salesman problems, optimization and decision problems belong to the same class. An analogous argument can be applied to other combinatorial problems.

5.3.4 Classes of Problems and Algorithms

The classification of problems is related to their time complexities. The class P includes problems for which there are algorithms with a polynomial time complexity. The classes NP, NP-complete, and NP-hard include problems whose time complexities are most probably higher.

Let us present the classes NP and NP-complete more precisely. On the basis on equivalence between optimization and decision problems demonstrated above, we focus only on decision problems, which have the property that the output of the related algorithm is yes or no. The name "NP" is an abbreviation for "nondeterministic polynomial". Problems that belong to this class have a polynomial-time certificate. A certificate here is an algorithm used to determine whether a decision guess satisfies a condition. For example, in the traveling salesman problem, we may construct (guess) any route through all the cities $C_1, \ldots, C_K$. The existence of a certificate means that, in polynomial time we can find whether the proposed route satisfies "cost $\leq \theta$" or not. NP problems can be solved by a nondeterministic Turing machine. A nondeterministic Turing machine is a Turing machine additionally equipped with a guessing, write-only head. The class NP clearly includes all P-problems, P $\subset$

NP, since they can be not only certificated but also even solved in polynomial time. Among the problems in the class NP there is a subclass, called NP-complete. Problems in the class NP-complete have the property that any problem in the class NP can be reduced to a problem in the class NP-complete in polynomial time. The first result concerning NP-completeness was Cook's theorem, stating that the Boolean satisfiability problem was NP-complete. Following Cook's theorem, many other problems have been proven to belong to the class NP-complete. A large collection of NP-complete problems is given in the book [94]. All problems listed in Sect. 5.3.1 are known to be NP-complete.

One more class of combinatorial problems is the class named NP-hard. The class NP-hard contains all problems H such that every decision problem in the class NP can be reduced to H in polynomial time. The difference between classes NP-hard and NP-complete is that for NP-hard problems we do not demand that they must belong to NP. In other words, these problems may not have certificates.

5.3.5 Suboptimal Algorithms

An important field is the development of combinatorial algorithms for NP-complete problems, called suboptimal, near-optimal, or approximate. These algorithms are of significantly lower complexity; most often they work in polynomial time. Despite the fact that they do not guarantee that the solution will be obtained but only that one will get close to it, the results they provide can be acceptable and useful in many practical applications. Some examples, also mentioned later in the book, are polynomial algorithms for suboptimal solutions of the shortest-superstring problem [274] and polynomial algorithms for approximate solutions of the Hamiltonian path problem [272].

5.3.6 Unsolved Problems

Combinatorial optimization and decision problems have been studied extensively. The research in this area has two main directions. The first involves improvements in performances of algorithms. If the best known algorithm for solving a specific problem has a time complexity $C \exp(K)$, where K is the size of the problem, then developing an improvement leading to time complexity $(C/2) \exp(K)$ may be a useful and publishable result. Developing an algorithm, which improves the time complexity from $O(K^2)$ to $O(K \ln K)$ is a substantial advance, which can result in the appearance of new methods and new applications in related areas.

The second involves proving results concerning classification of problems. As already stated the classification for large number of problems, according to the above rules, has been established. Establishing time complexity classification of many problems, for example of the linear programming problem or the problem of factorization of an integer, involved many years of research.

Combinatorial optimization and the theory of algorithms contain many unsolved problems. First, the class of many NP problems is unknown; they have not been neither proven to belong to the NP-complete class, nor has a polynomial time algorithm been found. Moreover, the famous hypothesis P = NP is still unsolved. The common belief is that classes P and NP-complete are disjoint. But nobody has proven that any problem from the NP-complete class cannot be solved in polynomial time.

5.4 Exercises

1. Derive the necessary optimality conditions (5.16) and (5.17) by algebraic manipulations, using (5.1), but without calling on the geometric interpretation in Fig. 5.3.
2. Solve the following constrained optimization problems
   a) $\min_{x,y} x^2 + xy + y^2$
      subject to the constraint $x + y = 1$;
   b) $\min_{x,y} x + y$
      subject to the constraint $x^2 + y^2 = 1$;
   c) $\max_{x,y} x + y$
      subject to the constraint $x^2 + y^2 = 1$;
   d) $\max_{x,y,z} x^2 + y^2 + z^2$
      subject to the constraints $x^2 + y^2 = 1$ and $y^2 + z^2 = 1$;
   e) $\min_{x,y,z} x^2 + y^2 + z^2$
      subject to the constraints $x^2 + y^2 = 1$ and $y^2 + z^2 = 1$;
   f) $\min_{x,y,z} x^2 + y^2 + z^2$
      subject to the constraints $x^2 + y^2 = 1$ and $y^2 + z^2 \geq 1$;
   g) $\min_{x,y,z} x^2 + y^2 + z^2$
      subject to the constraints $x^2 + y^2 = 1$ and $y^2 + z^2 \leq 1$.
3. How does the formulation of the Lagrange multiplier theorem change if we replace minimization by maximization in (5.18)?
4. Derive the Kuhn–Tucker conditions corresponding to the minimization problem (5.35), (5.36).
5. Derive alternative formulations of the Kuhn–Tucker conditions (5.30)–(5.33) for the cases where (a) the maximization in (5.25) is replaced by minimization, and (b) the inequality sign "$\geq$" in (5.26) is replaced by the opposite sign "$\leq$".
6. Derive Lagrange-multiplier optimality conditions for the problem
$$\min_x x^T Q x$$
subject to the constraint
$$x^T R x = C,$$
where Q and R are symmetric positive definite matrices. (This result applies to material on singular value decomposition Chaps. 4 and 11.)

7. We collect observations $x_1, x_2, ..., x_K$ and $y_1, y_2, ..., y_K$ in an experiment, which is modeled by quadratic relation

$$y = ax^2 + bx + c.$$

Describe the use of the least-squares method for estimating parameters a, b and c.

8. Write a computer program for performing the steepest-descent iterations (5.58). Experimenting with different functions and with different values of the step size γ, try to observe different kinds of behaviors, from stability to unstable oscillations.

9. Repeat Exercise 8 for the Gauss–Newton iterations (5.65).

10. Write a computer program that will solve the optimization task presented in Sect. 5.2.2. Check that when all cost coefficients are equal, the optimal path becomes a straight line. This verifies the correctness of the program.

11. Design an algorithm for assigning the correct numbers to nodes in an aperiodic graph.

12. Write a computer program to find the shortest path in an aperiodic graph.

13. Develop an algorithm to find the shortest path in a graph without the aperiodicity condition, or study the solution in the literature [65, 64].

14. Using the optimality principle derive iterative solution to the following optimization problem: minimize

$$I = \sum_{k=0}^{K} x_k^2 + u_k^2$$

over scalar, real controls $u_0, u_1, u_2, ..., u_K$ subject to

$$x_{k+1} = x_k + u_k,$$

with x_0 given .

15. Assume that the dynamic optimization problem in Sect. 5.2.1 is modified, such that the scoring index (5.68) is replaced by the following one:

$$I = \sum_{k=1}^{K} s_k(x_k, x_{k+1}, u_k, u_{k+1}).$$

Derive a dynamic programming algorithm to solve this modified problem.

Part II

Applications

6
Sequence Alignment

Sequence alignment involves establishing correspondences between bases or codons of DNA or RNA strings or between amino acids forming linear sequences in proteins. Aligning DNA, RNA or amino acid sequences is of basic importance in genomics, proteomics and transcriptomics and can be used for a variety of research purposes. It can find similarity between two DNA sequences resulting from the existence of a recent common ancestor, which these two sequences originate from. By measuring or computing distances between the aligned sequences, one draw inferences about the evolutionary processes they have gone through. This inference about the evolutionary process may involve estimating the time that has passed from the common ancestor to the present, but may also involve stating hypotheses or reconstructing a single evolutionary event in the past or a sequence of them. Aligning two sequences can allow one to detect their overlap or to notice that one sequence is a part of the other or that the two sequences share a subsequence. Instead of two sequences, one can also align many sequences or match a sequence against a DNA, RNA, or protein database. Multiple alignments of RNA or amino acid sequences in proteins allow one to infer their secondary and tertiary structures as well as active or functionally important sites in proteins.

There is a wide range of literature on sequence alignment. Depending on which variant of sequence alignment is being performed, the mathematical approach can vary. An approach using the dynamic programming method [76, 204, 261, 281] is most useful for pairwise comparisons or and comparing multiple sequences. For tasks involving looking for (exact or approximate) occurrences of a given sequence in a large database, direct application of dynamic programming would be impossible, owing to the excessive computational time. Instead, two-step or multistep approaches are applied, where the first step uses hash tables or finite automata to pick out candidate sequences, genes or proteins that share enough similarity with a template sequence, without aligning them; in the next step, pairwise or block alignments are carried out for the sequences selected in the first step, [68, 140, 141, 179, 288].

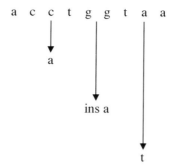

Fig. 6.1. Diagram showing three evolutionary events that may happen to the sequence $acctggtaa$ (substitution, insertion and another substitution in this case)

When constructing an alignment for two or more sequences, one needs to have in mind a model of the evolutionary process that led to the observed differences. The simplest events in the evolutionary process of replication of a DNA sequence are single base substitutions, insertions and deletions (indels). For example, the sequence

$$s = acctggtaaa \tag{6.1}$$

after undergoing substitution of its second base c to a and of its third base a to t, and insertion of a base c into the nucleotide pair gg, which can be represented as shown in Fig. 6.1 will lead to a sequence s_1

$$s_1 = acatgcgtata. \tag{6.2}$$

Since we know the evolutionary history behind sequences s and s_1, we also know true correspondences between bases in s and s_1. The resulting alignment between s and s_1 is commonly represented as follows:

$$\begin{array}{l} s = \;\; a\,c\,c\,t\,g - g\,t\,a\,a\,a \\ \;\; :\,:\;\;\;:\,:\;\;\;\;\;\;:\,:\,:\;\;\;: \\ s_1 = a\,c\,a\,t\,g\,c\;\;g\,t\,a\,t\,a \end{array} \tag{6.3}$$

where the colon symbols indicate matches between bases. The symbol "$-$" called a gap, added to the alphabet of four bases, allows us to represent insertions and deletions (indels). When we are aligning DNA sequences obtained in real experiments, their evolutionary history is not available and so the base substitutions and indels can only be hypothesized. Also, the direction of the events is unknown. For example, when comparing s and s_1 without knowledge of the history of mutations, the correspondence $c - $ " $-$ " could equally likely result both from an insertion of c in gg, or from a deletion of c in gcg. Similarly, the correspondence $c - a$ could result from the substitution $c \rightarrow a$ in s

or from a substitution $a \to c$ in s_1. Also, there is no information about the time order of the three mutational events.

In the example above, the positions of both the beginning and end of the two sequences coincided. However, correspondences between the bases of two strings can have other configurations, for example

$$c\,g\,c\,a\,t\,t\,g\,c\,a\,c\,a\,c\,t\,t\,a\,c\,c\,t\,g-t\,a\,a\,a\,a$$
$$\vdots\qquad\qquad\vdots\,\vdots\quad\vdots\,\vdots\quad\vdots\,\vdots\,\vdots\quad\vdots$$
$$g\,c\,c\,t\,t\,c\,g\,g\,a\,a\,a\,c\,a\,t\,g\,c\;t\,a\,a\,t\,a$$

or

$$g\,t\,a\,c\,c\,a\,t\,a-a\,c-t\,t\,g\,t\,a\,t\,a\,a$$
$$\vdots\,\vdots\,\vdots\,\vdots\,\vdots\,\vdots\qquad\qquad\vdots\,\vdots\,\vdots\,\vdots\,\vdots\,\vdots$$
$$c\,t\,g\,g\,t\,c\,a\,a\,c\,c\,a\,t\,a\,c\;t\,g\,c\;g\,c\,g\,t\,a\,t\,a\,a$$

If it is known that the aligned sequences must coincide at their beginning and end, then the natural term is "global alignment". In the case of an alignment where all fragments of the sequences can slide over each other and matches may involve only parts of sequences, the term "local alignment" is used.

6.1 Number of Possible Alignments

Assume that an alignment problem for two sequences s_1 and s_2 of lengths n and m, is to be solved. We form a rectangular matrix with rows corresponding to the characters in the first string s_1 and columns corresponding to the characters in the second string s_2, such that the order of characters is to the right and down. The problem of searching for an alignment between s_1 and s_2 can be formulated as tracing out a path through this matrix, starting from the upper left corner and terminating at the lower right corner. The feasible moves are horizontally right, vertically down and diagonally down and right. Moving diagonally down and to the right means adding correspondences between consecutive letters of s_1 and s_2, moving down vertically means inserting gaps in s_1, and moving horizontally to the right means inserting gaps in s_2. An example of representing a possible alignment of $s_1 = ttcgga$ and $s_2 = acgtgagagt$ as a path through a matrix is presented in Fig. 6.2. We start from $row = 0$ and $column = 0$ and visiting matrix entry $row = i$ and $column = j$ corresponds to the situation where i letters of the string s_1 and j letters of string s_2 have been already aligned.

The number of paths through this matrix, stating from $row = 0$ and $column = 0$ and ending at $row = n$ and $column = m$ is

$$q(n,m) = \sum_{k=0}^{\min(n,m)} \frac{(n+m-k)!}{(n-k)!(m-k)!k!}. \tag{6.4}$$

The above equality can be derived by decomposing the possible paths into classes with $k = 0$, $k = 1$, ..., $k = \min(n,m)$ right-down diagonal moves, $n-k$

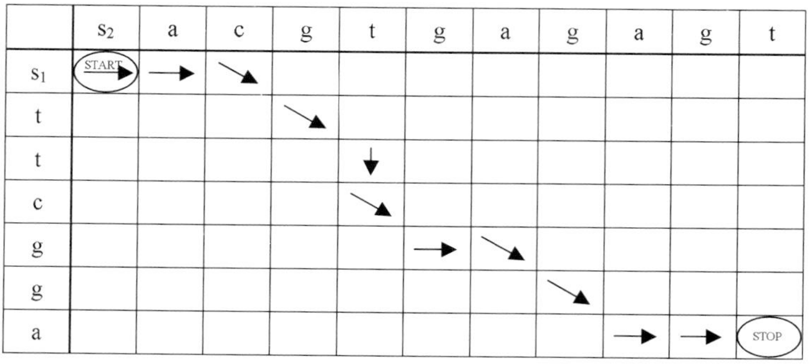

	s_2	a	c	g	t	g	a	g	a	g	t
s_1	START	→	↘								
t				↘							
t					↓						
c					↘						
g						→	↘				
g								↘			
a										→	→ STOP

```
s₁ :  -  -  t  c  g  -  g  a  -  -
s₂ :  a  c  g  t  -  g  a  g  a  g  t
```

Fig. 6.2. Example of representation of an alignment of $s_1 = ttcgga$ and $s_2 = acgtgagagt$ as tracing out a path through a matrix with rows tagged by the symbols in s_1 and columns tagged by the symbols in s_2. The alignment corresponding to the path crossing the matrix shown here is depicted at the *bottom*

down moves and $m-k$ right moves. We denote a right-down diagonal move by the symbol "↘", right move by "→" and a down move by "↓". The number of possible paths in each category is the number of possible strings of length $n+m-k$ with k symbols "↘", $n-k$ symbols "↓", and $m-k$ symbols "→", which is equal to

$$\frac{(n+m-k)!}{(n-k)!(m-k)!k!}$$

and implies (6.4). By algebraic transformations, we can represent the above as

$$\frac{(n+m-k)!}{(n-k)!(m-k)!k!} = \frac{\binom{m+n-k}{m-k}}{\binom{m}{m-k}} \binom{m}{k}\binom{n}{k},$$

and since

$$\binom{m+n-k}{m-k} \geq \binom{m}{m-k},$$

we obtain the following lower bound on $q(n,m)$:

$$q(n,m) \geq \sum_{k=0}^{\min(n,m)} \binom{m}{k}\binom{n}{k} = \binom{n+m}{n}. \qquad (6.5)$$

The above equality is called the Vandermode or Cauchy identity [104]. For large n and m, the number of alignments grows combinatorially and it is not possible to find an alignment by going through all possibilities.

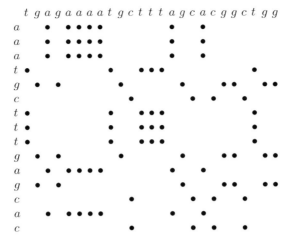

Fig. 6.3. Dot matrix comparison of sequences $s_1 = tgagaaaatgctttagcacggctgg$ and $s_2 = aaatgctttgagcac$. The dots show possible correspondences between the characters of the strings s_1 and s_2

6.2 Dot Matrices

The dot matrix is a simple and very useful concept for aligning two DNA sequences. Assume that the DNA sequences to be aligned are

$$s_1 = tgagaaaatgctttagcacggctgg$$

and

$$s_2 = aaatgctttgagcac.$$

Form a rectangular $n \times m$ matrix with rows corresponding to the characters in the first string s_1 and columns corresponding to the characters in the second string s_2, such that the order of characters is to the right and down. Place a dot in each matrix entry, where a base from s_1 matches a base from s_2. The result, shown in Fig. 6.3 is called a dot matrix.

The dots show possible correspondences between the characters of the strings s_1 and s_2. There are many dots related to accidental matches between letters of the two strings. We can eliminate some of these by removing dots unlikely to represent a nonrandom correspondence between characters of the strings s_1 and s_2 with the use of some intuitive criterion. If we introduce the requirement that, in order that a dot is not removed, there must be at least k neighboring matches along the right-down diagonal direction, then this will result in some of the random accidental matches being filtered out. If k is too small, many accidental matches will remain in the dot matrix plot. On the other hand, if it is too large, some of the true correspondences between strings

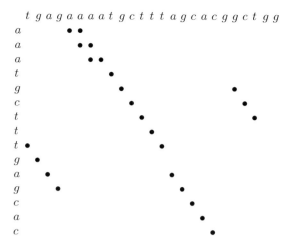

Fig. 6.4. Filtered dot matrix comparison of sequences $s_1 =$ *tgagaaaatgct-ttagcacggctgg* and $s_2 =$ *aaatgctttgagcac*. The dots are now arranged in diagonal paths, which more clearly show the possible correspondences between the characters of the strings s_1 and s_2.

may be unintentionally omitted. If we take $k = 3$ we obtain the filtered dot matrix shown in Fig. 6.4, which is much easier to interpret than the original one.

From the filtered dot matrix, can we construct the following alignment between s_1 and s_2

$$s_1 = \qquad\qquad a\,a\,a\,t\,g\,c\,t\,t\,t\,g\ \ a\,g\,c\,a\,c$$
$$\vdots\,\vdots\,\vdots\,\vdots\,\vdots\,\vdots\,\vdots\,\vdots\,\vdots\,\vdots\ \ \vdots\,\vdots\,\vdots\,\vdots\,\vdots$$
$$s_2 = t\,g\,a\,g\,a\,a\,a\,a\,t\,g\,c\,t\,t\,t\,-\,a\,g\,c\,a\,c\,g\,g\,c\,t\,g\,g$$

Using dot matrices is rather intuitive, since the alignment is performed by following long lines of dots in the plot. Nevertheless, there is a scoring system behind it. For example, we may assign a score of 1 for every single match between letters of strings, and we should not introduce indels unless it results in a large enough number of new scores. We should also penalize correspondence between mismatching symbols.

6.3 Scoring Correspondences and Mismatches

Dot matrix plots provide a simple and easy way to review reasonable alignments between two short sequences. However, in order to (1) solve large alignment problems (thousands of kilobases) or multiple alignments and (2) be

6.3 Scoring Correspondences and Mismatches

precise in statements concerning comparing possible matches, one needs to develop more sophisticated methods and algorithms. One aim of this section is to describe how one can start to develop scoring systems more suitable than just counting matches and mismatches.

The idea of scoring by incrementing a performance index for every match between letters of s_1 and s_2 and penalizing gaps and mismatches can be formalized by defining a scoring function

$$S = n_m w_m + n_s w_s + n_g w_g \qquad (6.6)$$

where n_m denotes the number of matched characters in the alignment, n_s stands for the number of mismatches and n_g stands for the number of gaps. The weights (scores) w_m, w_s and w_g are usually chosen such that matches are rewarded and mismatches and gaps are penalized, i.e., $w_m > 0$, $w_s < 0$, $w_g < 0$. The above scoring index seems intuitively justified, but rather heuristic. A simple argument [281] allows one to interpret a scoring index (6.6) as a log-likelihood of the alignment. Assume that the mutational process, which transforms the sequence s given by (6.1) to sequence s_1 given by (6.2), $s \to s_1$, has the following probabilities: p_0, a base does not change (a match occurs), p_s, a base is substituted by another one (a mismatch occurs); and p_g, an indel occurs. In the alignment (6.3) there are $n_m = 8$ matches, $n_s = 2$ mismatches, and $n_g = 1$ gaps, which under the assumption of independence of the three processes, leads to the following likelihood of $s \to s_1$ (see Sect. 2.3.5):

$$l = (p_0)^{n_m} (p_s)^{n_s} (p_g)^{n_g}.$$

The log-likelihood becomes

$$L = n_m \ln p_0 + n_s \ln p_s + n_g \ln p_g, \qquad (6.7)$$

which has the same structure as (6.6) with $w_m = \ln p_0$, $w_s = \ln p_s$, and $n_g = \ln p_g$.

Developing further the idea of using the maximum likelihood for scoring alignments, we can modify the index expressed by (6.7) by assuming more realistic models describing substitutions, insertions, and deletions in DNA, RNA, or amino acid sequences. The most general formulation, valid both for strings of bases and for strings of amino acids, involves introducing a function defining a score or penalty for correspondences between possible symbols of aligned strings and for correspondences between gaps and letters of strings. Let Ξ be the alphabet of all possible letters in aligned strings. We denote the entries of the score function by $d(\xi, \eta)$ and by $d(\xi, -)$, where $\xi, \eta \in \Xi$. These entries define scores for aligning ξ versus η and for aligning ξ versus a gap $-$. In all computations we assume the symmetry $d(\xi, \eta) = d(\eta, \xi)$ and $d(\xi, -) = d(-, \xi)$.

The sequence alignment problem with a score function $d(\xi, \eta)$ leads to two issues. (1) There is a need to decide (i.e., estimate) the exact values of scores, which should be done in conjunction with researching and modeling the random process of substitutions, insertions and deletions of bases or amino acids. (2) The formulation of a performance or scoring index leads to stating the alignment problem as a discrete optimization problem, which can be solved by a dynamic programming method. In the forthcoming sections of this chapter we will discuss problems (1) and (2) in more detail. As will become evident, these two problems are not disjoint; they are, rather, substantially interrelated and reasonable approaches are needed to obtain satisfactory solutions based on available data.

6.4 Developing Scoring Functions

The conclusion from Sect. 6.3 is that in order to compute the likelihoods of alignments we need estimates of the probabilities of substitutions of bases in DNA or RNA, or of amino acids in proteins. The probabilities must be estimated, on the basis of empirical data and here we describe some appropriate methods and approaches. We start from estimating probabilities for nucleotide substitution, and then we cover amino acid substitutions.

6.4.1 Estimating Probabilities of Nucleotide Substitution

We think of the process of replicating a single DNA base, from generation to generation, over a long evolutionary time. We assume that this process can be adequately modeled as a Markov chain (see Chap. 2) where states correspond to nucleotides A, C, G, and T and state the transitions correspond to nucleotide substitutions. The equation for evolution of the distribution of nucleotides (i.e., states) at a given site of the DNA sequence has the form of

$$\pi(k+1) = \pi(k) \begin{bmatrix} p_{AA} & p_{AC} & p_{AG} & p_{AT} \\ p_{CA} & p_{CC} & p_{CG} & p_{CT} \\ p_{GA} & p_{GC} & p_{GG} & p_{GT} \\ p_{TA} & p_{TC} & p_{TG} & p_{TT} \end{bmatrix}. \tag{6.8}$$

where the row vector $\pi(k)$ of state probabilities at discrete time k has four components:

$$\pi(k) = \begin{bmatrix} \pi_A(k), \pi_C(k), \pi_G(k), \pi_T(k) \end{bmatrix}$$

and the matrix of transition probabilities has entries p_{ij} defined as conditional probabilities of state j at time $k+1$ given state i at time k. According to classification of mutations in Chap. 8, we classify substitutions within a class of nucleotides (purine → purine or pyrimidine → pyrimidine) as transitions and between classes (purine → pyrimidine or pyrimidine → purine) as transversions, i.e.,

$$\left.\begin{array}{l} A \to G,\ G \to A \\ C \to T,\ T \to C \end{array}\right\} \text{transitions}$$

$$\left.\begin{array}{l} A \to C,\ T \to A \\ C \to G,\ G \to C \end{array}\right\} \text{transversions.}$$

Consider the problem of maximum likelihood estimation of the state transition probabilities $p_{AA}, ... p_{TT}$, i.e., the state transition probabilities matrix P, on the basis of a sequence of states observed over a long time. Since $p_{AA}, ..., p_{TT}$ are conditional probabilities, then the expressions for the maximum likelihood estimates are, for example,

$$\hat{p}_{AA}^{ml} = \frac{\#\text{ all transitions from } A \text{ to } A}{\#\text{ all occurrences of } A} \tag{6.9}$$

or

$$\hat{p}_{CT}^{ml} = \frac{\#\text{ all transitions from } C \text{ to } T}{\#\text{ all occurrences of } C}, \tag{6.10}$$

where "#" stands for "number of". The problem with using the estimates (6.9) and (6.10) is that the necessary data on the numbers of transitions are never available. Practical estimates of transition probabilities are not based on time sequence data, as required in (6.9), (6.10), but rather on comparisons of homologous nucleotide sequences in different but related organisms. This implies that (1) the direction of state transitions are not known, and (2) one cannot exclude the possibility that an observed difference between states results from more than one substitution, for example, $A \to C$ could also result from $A \to G \to C$, and so forth.

6.4.2 Parametric Models of Nucleotide Substitution

There are several models for nucleotide substitution. Selecting any particular model imposes further assumptions on the mechanisms of transitions and transversions, which leads to more specific parameters of the state transition matrix of the Markov chain. These models also usually ensure ergodicity and reversibility of Markov chain corresponding to nucleotide substitutions. A collection of Markov chain models, along with an overview of methods of fitting their parameters to DNA data is provided in Chap. 13 of [76]. Here we shall pick out three models, often applied, the simplest, one parameter Jukes-Cantor model [135] and the more flexible Felsenstein [80] and HKY (Hasegawa, Kishino, and Yano) [115] models. In the following, we define P as

$$P = \begin{bmatrix} p_{AA} & p_{AC} & p_{AG} & p_{AT} \\ p_{CA} & p_{CC} & p_{CG} & p_{CT} \\ p_{GA} & p_{GC} & p_{GG} & p_{GT} \\ p_{TA} & p_{TC} & p_{TG} & p_{TT} \end{bmatrix}$$

where the entries are described further by more specific expressions, and Q always denotes the transition intensity matrix of the continuous-time version of the Markov model.

Jukes–Cantor Model

This model has the following transition probability matrix, for the discrete case:

$$P = \begin{bmatrix} 1-3\alpha & \alpha & \alpha & \alpha \\ \alpha & 1-3\alpha & \alpha & \alpha \\ \alpha & \alpha & 1-3\alpha & \alpha \\ \alpha & \alpha & \alpha & 1-3\alpha \end{bmatrix}.$$

Since the one-step substitution probability α usually is extremely small, then it is convenient to pass to a continuous-time Markov process with a state transition intensity matrix

$$Q = \begin{bmatrix} -3\alpha & \alpha & \alpha & \alpha \\ \alpha & -3\alpha & \alpha & \alpha \\ \alpha & \alpha & -3\alpha & \alpha \\ \alpha & \alpha & \alpha & -3\alpha \end{bmatrix}. \tag{6.11}$$

The Jukes–Cantor Markov process with the state transition intensity (6.11) is ergodic and reversible. All substitutions (state transitions) are equally probable, and the stationary distribution is uniform. The model relies on only one parameter, and clearly cannot be sufficiently flexible in many situations.

Felsenstein Model

The Felsenstein model has the following discrete state transition probabilities:

$$P = \begin{bmatrix} 1-u\varphi_{CTG} & u\varphi_C & u\varphi_G & u\varphi_T \\ u\varphi_A & 1-u\varphi_{AGT} & u\varphi_G & u\varphi_T \\ u\varphi_A & u\varphi_C & 1-u\varphi_{ACT} & u\varphi_T \\ u\varphi_A & u\varphi_C & u\varphi_G & 1-u\varphi_{ACG} \end{bmatrix} \tag{6.12}$$

where

$$\varphi_{CTG} = \varphi_C + \varphi_G + \varphi_T,$$
$$\varphi_{AGT} = \varphi_A + \varphi_G + \varphi_T,$$
$$\varphi_{ACT} = \varphi_A + \varphi_C + \varphi_T,$$
$$\varphi_{ACG} = \varphi_A + \varphi_C + \varphi_G. \tag{6.13}$$

This model is reversible and the stationary distribution is $(\varphi_A \ \varphi_C \ \varphi_G \ \varphi_T)$, which can be easily seen from the fact that all reversibility conditions are satisfied, i.e., $\varphi_A p_{AC} = \varphi_C p_{CA}$, and so forth. This is achieved by assuming that each state transition (substitution) probability (e.g., p_{AC}) is proportional to the frequency of the substituting nucleotide, φ_C, with a proportionality coefficient (intensity coefficient) u. State transition intensity matrix for the continuous-time counterpart of (6.12) is

$$Q = \begin{bmatrix} -u\varphi_{CTG} & u\varphi_C & u\varphi_G & u\varphi_T \\ u\varphi_A & -u\varphi_{AGT} & u\varphi_G & u\varphi_T \\ u\varphi_A & u\varphi_C & -u\varphi_{ACT} & u\varphi_T \\ u\varphi_A & u\varphi_C & u\varphi_G & -u\varphi_{ACG} \end{bmatrix}. \quad (6.14)$$

HKY Model

This model is the most flexible and it is possible to fit most data with it. In the HKY model, the probability of substitution (state transition) is again proportional to the frequency of the substituting nucleotide. However, a generalization is made, compared with the Felsenstein model, in that the proportionality constants are different between transitions (u) and transversions (v). This leads to state transition probabilities in the discrete case

$$P = \begin{bmatrix} 1-\psi_{CTG} & v\varphi_C & u\varphi_G & v\varphi_T \\ v\varphi_A & 1-\psi_{AGT} & v\varphi_G & u\varphi_T \\ u\varphi_A & v\varphi_C & 1-\psi_{ACT} & v\varphi_T \\ v\varphi_A & u\varphi_C & v\varphi_G & 1-\psi_{ACG} \end{bmatrix}, \quad (6.15)$$

where

$$\psi_{CTG} = u\varphi_G + v(\varphi_C + \varphi_T),$$
$$\psi_{AGT} = u\varphi_T + v(\varphi_A + \varphi_G),$$
$$\psi_{ACT} = u\varphi_A + v(\varphi_C + \varphi_T),$$
$$\psi_{ACG} = u\varphi_C + v(\varphi_A + \varphi_A).$$

The state transition intensity matrix Q in the continuous-time version of the model has the following form:

$$Q = \begin{bmatrix} -\psi_{CTG} & v\varphi_C & u\varphi_G & v\varphi_T \\ v\varphi_A & -\psi_{AGT} & v\varphi_G & u\varphi_T \\ u\varphi_A & v\varphi_C & -\psi_{ACT} & v\varphi_T \\ v\varphi_A & u\varphi_C & v\varphi_G & -\psi_{ACG} \end{bmatrix}. \quad (6.16)$$

The HKY model is reversible and the stationary distribution is again (φ_A, $varphi_C$, $varphi_G$, φ_T). The reversibility conditions are readily verified.

6.4.3 Computing Transition Probabilities

We consider only the time-continuous case as it is more natural for estimations concerning the evolutionary time t. The discrete-time case can be computed using the same technique. In the three Markov models, the Jukes–Cantor (6.11), Felsenstein (6.14) and HKY (6.16) models, computing state transition matrices at time t,

$$P(t) = \exp(Qt) \quad (6.17)$$

can be done analytically, by expressing intensity matrices Q using the Jordan canonical representation [25]

$$Q = U_Q D_Q V_Q^T. \tag{6.18}$$

In the above, U_Q and V_Q^T are transformation matrices built of the column and row eigenvectors of Q, respectively, which satisfy

$$U_Q = (V_Q^T)^{-1},$$

and D_Q is a diagonal matrix containing the eigenvalues of Q. We introduce the notation

$$U_Q = \begin{bmatrix} u_{Q1} & u_{Q2} & u_{Q3} & u_{Q4} \end{bmatrix}$$

for the 4×4 matrix composed of the column eigenvectors,

$$V_Q^T = \begin{bmatrix} v_{Q1}^T \\ v_{Q2}^T \\ v_{Q3}^T \\ v_{Q4}^T \end{bmatrix}$$

for the 4×4 matrix composed of the row eigenvectors, and

$$D_Q = diag(q_1, q_2, q_3, q_4)$$

for the diagonal matrix of the eigenvalues of Q. The matrices $P(t)$ and Q share the same transformation matrices U_Q and V_Q^T, and

$$P(t) = U_Q \exp(D_Q t) V_Q^T, \tag{6.19}$$

where

$$\exp(D_Q t) = diag\left[\exp(q_1 t), \exp(q_2 t), \exp(q_3 t), \exp(q_4 t)\right].$$

Using (6.19) we can obtain the transition probabilities $p_{ij}(t)$, $i, j \in \{A, C, G, T\}$ as follows:

$$p_{ij}(t) = \mathbf{1}_i^T P(t) \mathbf{1}_j = u_Q^{Ti} P(t) v_Q^j, \tag{6.20}$$

where $\mathbf{1}_i^T$ and $\mathbf{1}_j$ denote a row vector where all elements are 0 except the ith element, which is set equal to 0, and a column vector where all elements are 0 except the jth element, which is set equal to 1, and u_Q^{Ti} and v_Q^j denote the ith row of matrix U_Q and the jth column of matrix V_Q^T, respectively.

As mentioned earlier, the eigenvalues and eigenvectors of the matrices Q in (6.11), (6.14) and (6.16) can be computed analytically, which leads to the following expressions.

For the Jukes–Cantor model (6.11) the eigenvalues of the intensity matrix are

$$D_Q = diag(0, -4\alpha, -4\alpha, -4\alpha), \tag{6.21}$$

the column eigenvectors are

$$U_Q = \begin{bmatrix} 1 & -1 & -1 & -1 \\ 1 & 1 & 0 & 0 \\ 1 & 0 & 1 & 0 \\ 1 & 0 & 0 & 1 \end{bmatrix}, \tag{6.22}$$

and the row eigenvectors are

$$V_Q^T = \begin{bmatrix} 0.25 & 0.25 & 0.25 & 0.25 \\ -0.25 & 0.75 & -0.25 & -0.25 \\ -0.25 & -0.25 & 0.75 & -0.25 \\ -0.25 & -0.25 & -0.25 & 0.75 \end{bmatrix}. \tag{6.23}$$

For the Felsenstein model (6.14) the eigenvalues of the intensity matrix are

$$D_Q = diag(0, -u, -u, -u), \tag{6.24}$$

the column eigenvectors are

$$U_Q = \begin{bmatrix} 1 & -\varphi_C/\varphi_A & -\varphi_G/\varphi_A & -\varphi_T/\varphi_A \\ 1 & 1 & 0 & 0 \\ 1 & 0 & 1 & 0 \\ 1 & 0 & 0 & 1 \end{bmatrix}, \tag{6.25}$$

and the row eigenvectors are

$$V_Q^T = \begin{bmatrix} \varphi_A & \varphi_C & \varphi_G & \varphi_T \\ -\varphi_A & \varphi_{AGT} & -\varphi_G & -\varphi_T \\ -\varphi_A & -\varphi_C & \varphi_{ACT} & -\varphi_T \\ -\varphi_A & -\varphi_C & -\varphi_G & \varphi_{ACG} \end{bmatrix}. \tag{6.26}$$

In (6.26) we have used the notation defined in (6.13).

For the HKY model (6.16) the eigenvalues of the intensity matrix are

$$D_Q = diag[0, -v, -v(\varphi_A + \varphi_G) - u(\varphi_C + \varphi_T), -v(\varphi_C + \varphi_T) - u(\varphi_A + \varphi_G)], \tag{6.27}$$

the column eigenvectors are

$$U_Q = \begin{bmatrix} 1 & -\dfrac{\varphi_C + \varphi_T}{\varphi_A + \varphi_G} & 0 & -\dfrac{\varphi_G}{\varphi_A} \\ 1 & 1 & -\dfrac{\varphi_T}{\varphi_C} & 0 \\ 1 & -\dfrac{\varphi_C + \varphi_T}{\varphi_A + \varphi_G} & 0 & 1 \\ 1 & 1 & 1 & 0 \end{bmatrix}, \tag{6.28}$$

and the row eigenvectors are

$$V_Q^T = \begin{bmatrix} \varphi_A & \varphi_C & \varphi_G & \varphi_T \\ -\varphi_A & \dfrac{\varphi_C(\varphi_A + \varphi_G)}{\varphi_C + \varphi_T} & -\varphi_G & \dfrac{\varphi_T(\varphi_A + \varphi_G)}{\varphi_C + \varphi_T} \\ 0 & -\dfrac{\varphi_C}{\varphi_C + \varphi_T} & 0 & \dfrac{\varphi_C}{\varphi_C + \varphi_T} \\ -\dfrac{\varphi_A}{\varphi_A + \varphi_G} & 0 & \dfrac{\varphi_A}{\varphi_A + \varphi_G} & 0 \end{bmatrix}. \tag{6.29}$$

Using the decompositions (6.21)-(6.29) listed above and (6.20), we can compute analytically all state transition probabilities. Some examples are listed below.

For the Jukes–Cantor model (6.11), the transition probabilities are

$$p_{ij}(t) = 0.25 - 0.25\exp(-\alpha t), \ i \neq j, \qquad (6.30)$$

and

$$p_{ii}(t) = 0.25 + 0.75\exp(-\alpha t), \qquad (6.31)$$

where $i, j \in \{A, C, G, T\}$.

For the Felsenstein model (6.14), the transition probabilities are

$$p_{ij}(t) = \varphi_j[1 - \exp(-ut)], \ i \neq j, \qquad (6.32)$$

and

$$p_{ii}(t) = \varphi_i + (1 - \varphi_i)\exp(-ut), \qquad (6.33)$$

where again $i, j \in \{A, C, G, T\}$.

For the HKY model, we can derive, for example for the transversion $A \to C$,

$$p_{AC}(t) = \varphi_C[1 - \exp(-vt)], \qquad (6.34)$$

for the transition $G \to A$,

$$p_{GA}(t) = \varphi_A + \frac{\varphi_A(\varphi_C + \varphi_T)}{\varphi_A + \varphi_G}\exp(-vt) \\ - \frac{\varphi_A}{\varphi_A + \varphi_G}\exp[-v(\varphi_C + \varphi_T)t - u(\varphi_A + \varphi_G)t], \qquad (6.35)$$

and for the base conservation $A \to A$,

$$p_{AA}(t) = \varphi_A + \frac{\varphi_A(\varphi_C + \varphi_T)}{\varphi_A + \varphi_G}\exp(-vt) \\ + \frac{\varphi_G}{\varphi_A + \varphi_G}\exp[-v(\varphi_C + \varphi_T)t - u(\varphi_A + \varphi_G)t]. \qquad (6.36)$$

6.4.4 Fitting Nucleotide Substitution Models to Data

Suppose that, on the basis of a comparison of two aligned, equal length DNA sequences, such as,

$$seq1 : AGGCTTAACTGATCGCTACCAAGTAGGCACGAG$$
$$seq2 : AGGCTTCACTGATCGCTACCAAGTAGGCACGAG, $$
$$(6.37)$$

we wish to estimate the parameters of a model of evolutionary nucleotide substitution. The lengths of both sequences is denoted by K and we think of the comparison between *seq*1 and *seq*2 as K independent repetitions of pairwise comparison experiments, where each experiment has 10 possible outcomes. These outcomes are 6 unordered pairs ij, where $i < j$ (by "<" we mean alphabetical order), and 4 pairs ii. The indices belong to the set of symbols of nucleotides, i.e., $i, j \in \{A, C, G, T\}$. The pairs are unordered because we do not observe transitions between states. What is recorded in our experimental data is only correspondence, for example, $A - C$, which can result from either $A \rightarrow C$ or $A \leftarrow C$.

The probability distribution corresponding to the data in (6.37) is the multinomial distribution (2.33) with 10 possible outcomes ("types" of pairs), and the log-likelihood function (2.41) is

$$L = \sum_{i<j} k_{ij} \ln q_{ij} + \sum_i k_{ii} \ln q_{ii}, \tag{6.38}$$

where we have dropped the constant term. By q_{ij} we denote the probabilities of each of the outcomes of the experiment. The numbers k_{ij}, analogously to (2.41), are the multiplicities of the observed outcomes. Note that

$$K = \sum_i k_{ii} + \sum_i \sum_{j<i} k_{ij}. \tag{6.39}$$

A method for computing the probabilities q_{ij} follows from the assumption that there exists a Markov chain or process behind the differences observed between *seq*1 and *seq*2. However, there is no one-to-one correspondence between the transition probabilities $p_{AA}, p_{AC}, p_{AG},...p_{TT}$ in (6.8) and the probabilities q_{ij} of the occurrences of unordered pairs in (6.38). Appropriate computations must be carried out in order to take this into account. Below, we discuss several aspects of fitting Markov models to data in (6.37).

One scenario for fitting a Markov chain (process) model to the data in (6.37) is the following. (1) Use the Markov model to compute the probabilities q_{ij}. (2) Maximize the likelihood (6.38) over the parameters of the Markov model.

An alternative approach is the following. (1) Estimate q_{ij} using the maximum likelihood method. (2) Derive relation between $p_{AA}, p_{AC}, p_{AG},...p_{TT}$ in (6.8) and q_{ij} in (6.38). (3) Obtain estimates of $p_{AA}, p_{AC}, p_{AG},...p_{TT}$ by inverting the relation obtained in step (1). This method may be simpler than the first one since it decomposes the problem into two steps, each of which may be easier to perform. The maximum likelihood estimates for q_{ij} in (6.38) are

$$\hat{q}_{ii} = \frac{k_{ii}}{K} = \frac{k_{ii}}{\sum_i k_{ii} + \sum_i \sum_{j<i} k_{ij}} \tag{6.40}$$

and

$$\hat{q}_{ij} = \frac{k_{ij}}{K} = \frac{k_{ij}}{\sum_i k_{ii} + \sum_i \sum_{j<i} k_{ij}}. \tag{6.41}$$

Stationary Distribution

We assume stationarity of the Markov chain, with a stationary probability distribution

$$\pi^{ST} = \begin{bmatrix} \pi_A & \pi_C & \pi_G & \pi_T \end{bmatrix}. \qquad (6.42)$$

The probabilities π_A, π_C, π_G and π_T can be estimated by nucleotide counts in (6.37), i.e.,

$$\hat{\pi}_i = \frac{\text{\# occurrences of nucleotide } i \text{ in } seq1 \text{ and } seq2}{2K}$$

$$= \frac{2k_{ii} + \sum_{j \neq i} k_{ij}}{2(\sum_i k_{ii} + \sum_i \sum_{j<i} k_{ij})}. \qquad (6.43)$$

One-Step Markov Chain

This paragraph serves mainly to explain some facts, which will be used later. Let us start from the rather unrealistic assumption that observed differences between $seq1$ and $seq2$ result from the transitions probabilities in one step of a Markov chain model (6.8). Our aim is to estimate the probabilities p_{AA}, $p_{AC}, p_{AG},...,p_{TT}$, under the above assumption. Using the one-step model (6.8) with the stationary distribution (6.42), we derive the probability of recording an $i - i$ pair in the observed data (6.37),

$$q_{ii} = \pi_i p_{ii} \qquad (6.44)$$

and the probability of recording an unordered $i - j$ pair,

$$q_{ij} = \pi_i p_{ij} + \pi_j p_{ji}, \ i < j, \qquad (6.45)$$

where, again $i, j \in \{A, C, G, T\}$. We replace q_{ii}, q_{ij}, and π_i in the above by their estimates $\hat{q}_{ii}$, $\hat{q}_{ij}$, and $\hat{\pi}_i$, obtained from (6.40), (6.41), and (6.43) and try to invert (6.44) and (6.45). This is impossible because there are three unknowns p_{ii}, p_{ij}, and p_{ji} and two equations. It is also not surprising, since by restricting ourselves to recording unordered pairs we lose information about the direction of the process. We can sensibly estimate the transition probabilities under the additional assumption of reversibility of the Markov chain (6.8), which is related to the additional condition

$$\pi_i p_{ij} = \pi_j p_{ji}. \qquad (6.46)$$

Using the detailed balance condition (6.46) as the missing third equation we can now reasonably estimate p_{ii} and p_{ij} as

$$\hat{p}_{ii} = \frac{\hat{q}_{ii}}{\hat{\pi}_i} = \frac{2k_{ii}}{2k_{ii} + \sum_{j \neq i} k_{ij}}, \qquad (6.47)$$

and

$$\hat{p}_{ij} = \frac{\hat{q}_{ij}}{2\hat{\pi}_i} = \frac{k_{ij}}{2k_{ii} + \sum_{j \neq i} k_{ij}}, \ j \neq i. \qquad (6.48)$$

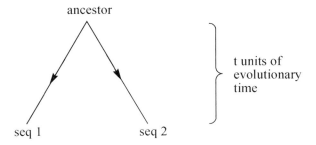

Fig. 6.5. Sequences $seq1$, $seq2$, and their most recent common ancestor

Markov Process over Evolutionary Time t

It is more sound to assume that the divergence between sequences $seq1$ and $seq2$ is related to the fact that each of the nucleotides in $seq1$ and $seq2$ has undergone the evolutionary scenario depicted in Fig. 6.5 starting from the most recent common ancestor of both sequences. This scenario involves many iterations of Markov transitions or, better, can be explained by a continuous-time Markov process model with ancestor of $seq1$ and $seq2$ occurring t units of evolutionary time ago. Compared with the situation depicted in the previous paragraph, here we must additionally include the possibility that an observed pair ij cumulates several substitutional events.

The substitution process at each base is a time-continuous Markov process described by a substitution intensity matrix Q. The substitution probabilities for nucleotides distant in evolutionary time by t units are given by the transition probability matrix

$$P(t) = \exp(Qt). \tag{6.49}$$

Using the entries, $p_{ij}(t)$ of the matrix $P(t)$ in (6.49) and the stationary distribution (6.42), we derive expressions for the probabilities q_{ij} in (6.38), by summing over all possible states of the ancestral nucleotide, as follows

$$q_{ii} = \sum_{n \in \{A,C,G,T\}} \pi_n p_{ni}^2(t) \tag{6.50}$$

and

$$q_{ij} = \sum_{n \in \{A,C,G,T\}} 2\pi_n p_{ni}(t) p_{nj}(t), \ i < j. \tag{6.51}$$

Again, we assume reversibility, which by substituting $\pi_n p_{ni}(t) = \pi_i p_{in}(t)$ or $\pi_n p_{nj}(t) = \pi_j p_{jn}(t)$ in (6.50) and (6.51) and using the Chapman–Kolmogorov equation (see Chap. 2), leads to

$$q_{ii} = \pi_i p_{ii}(2t) \tag{6.52}$$

and
$$\frac{1}{2} q_{ij} = \pi_i p_{ij}(2t) = \pi_j p_{ji}(2t). \tag{6.53}$$

Alternatively, (6.52) and (6.53) can be derived by reversing one of the arrows in Fig. 6.5, which, by reversibility, is feasible. We can estimate the transition probabilities by

$$\hat{p}_{ii}(2t) = \frac{\hat{q}_{ii}}{\hat{\pi}_i} \tag{6.54}$$

and

$$\hat{p}_{ij}(2t) = \frac{\hat{q}_{ij}}{2\hat{\pi}_i}. \tag{6.55}$$

We can see that these expressions are analogous to (6.47) and (6.48).

Estimates of Parameters in Parametric Models of Substitution

By combining the derived equations (6.52) and (6.53) with the parametric expressions derived in Sect. 6.4.3, we can easily compute estimates of the parameters of models described in Sect. 6.4.3. Symmetries in the model parameters suggest summing over some of the indexes in q_{ij}. Thus, for the Jukes–Cantor model we obtain

$$\widehat{\alpha t} = -\frac{1}{8} \ln\left(1 - \frac{4}{3}\hat{d}\right) \tag{6.56}$$

where

$$\hat{d} = \frac{\text{\# pairs } ij, j<i}{K} = \frac{\sum_{j<i} k_{ij}}{K}, \tag{6.57}$$

and for the Felsenstein model we obtain

$$\hat{\varphi}_i = \hat{\pi}_i = \frac{2\sum_i k_{ii} + \sum_{j\neq i} k_{ij}}{2K}, \tag{6.58}$$

$$\widehat{ut} = -\frac{1}{2} \ln(1 - \hat{d}), \tag{6.59}$$

where

$$\hat{d} = \frac{\sum_{j<i} k_{ij}}{2\sum_{j<i} \hat{\varphi}_i \hat{\varphi}_j}. \tag{6.60}$$

Estimating the parameters of the HKY model is left as an exercise for the reason.

Observe that the intensities of the substitution processes are estimated as composite parameters αt and ut. In (6.56), if the frequency of observed differences becomes close to $\hat{d} = 3/4$ the estimate $\widehat{\alpha t}$ diverges to infinity. This is related to the fact that we would still observe, on average, that about 1/4 of conserved bases were conserved in a comparison of randomly chosen sequences.

6.4.5 Breaking the Loop of Dependencies

The task of aligning DNA and amino acid sequences has a loop structure. For computing reliable alignments we need a good scoring function, which we can derive if we know substitution probabilities, and for estimating the substitution probabilities we need reliable alignments. It is necessary to break this chain of dependence at some point to start a study of the problem. Practically, research in this area uses an iterative approach, where one aligns sequences with the use of an ad hoc scoring function, uses the alignment obtained for estimating probabilities, and aligns sequences again with an improved scoring function. By a few iterations of this process, one can obtain reasonable estimates. Also there is feedback from the biological side, since biologists can score alignments using their knowledge and experience, which, by trial and error, can be "translated" to modification of the weights of scoring functions.

The convergence of this iterative scheme depends to a substantial extent on how sensitive the alignments are to the choice of scoring functions. As one might expect, nucleotide sequences are much less sensitive to the choice of scoring than the amino acid sequences. Therefore, when studying alignments of amino acid sequences, a lot of care must be devoted to developing reliable alignments.

6.4.6 Scaling Substitution Probabilities

We are aligning sequences which are distinct from each other by some number of units of evolutionary time. Assume that this time t is known and that we have a reliable estimate of the substitution intensity matrix Q. Under these hypotheses, we can compute $P(t) = \exp(Qt)$ and use the entries of the state transition probability matrix $P(t)$ to score correspondences in the alignment problem.

However, since the value of t for the evolutionary distance between the sequences being compared is typically not available, a more practical alternative is to scale the substitution probabilities by the expected total amount of changes observed between the sequences. This parameter is easy to estimate. For example, the in Jukes-Cantor model for $\alpha t = 0$ we expect 75% of nucleotides to be conserved between sequences, for $\alpha t = 1$ it is 53%, for $\alpha t = 2$ it is 35%, and for $\alpha t \to \infty$ it becomes 25%.

We can measure the number of nucleotide changes between the sequences being compared and use it to scale the substitution probabilities appropriately.

6.4.7 Amino Acid Substitution Matrices

The process of substitution in amino acid sequences is, in the terms of its basic principle, analogous to that discussed above for nucleotides. Again we assume that substitution can be modeled by a Markov chain with an appropriate matrix of state transition probabilities. However, the alignment and estimation problems become more complicated for the following reasons. (1) There are 20 different amino acids and the transition probability matrix has 20×20 entries, which is almost two orders of magnitude more than in the case of nucleotides. (2) Alignments of amino acid sequences are more sensitive to the choice of the values of coefficients in the scoring functions.

There are two widely used methods for estimation of amino acid substitution matrices, PAM (Percent Accepted Mutation) [61] and BLOSUM (Block Substitution Matrices) [122]. Both can be constructed on the basis of the theory presented above, and we shall briefly present them. Both methods rely on data obtained from aligned, ungapped blocks of conserved amino acid sequences. By "conserved" we mean that each sequence in a block must have a certain amount (e.g., 85%) of similarity to at least one other sequence. Such blocks can be found when one compares sequences of amino acids from proteins in related organisms. When the conservation requirement is introduced, it is believed that alignments become more reliable, i.e., the differences between sequences correspond to substitution events in the past.

PAM Substitution Matrix

PAM substitution matrices were derived in [61], on the basis of 71 blocks of aligned, ungapped amino acid sequences. These blocks of conserved sequences were found by imposing the requirement that all sequences must share at least 85% of similarity, as mentioned above. Maximum parsimony phylogenies were constructed for these blocks (see Chap. 7) leading to estimations of the evolutionary histories of all sequences in the blocks. From phylogenies of all blocks, statistics of aligned pairs of amino acids were constructed.

Let us introduce a notation, analogous to that already employed, to explain the results in [61] in quantitative terms. Namely,

$$k_{ij} \text{ is the number of aligned pairs } ij, \qquad (6.61)$$

where $i, j \in \{1, \ldots, 20\}$ index the amino acids. We assume that these indices are ordered by some criterion, and when listing k_{ij} in (6.61) we can always take $i < j$, since, as previously, we do not observe the directionality of changes. The definition in (6.61) is explained in Fig. 6.6 where an example of a maximum-parsimony phylogeny for 5 sequences is presented in the upper part, and the resulting counts for k_{ij} are shown in the table in the lower part. As can be seen, the counts in (6.61) result from all pairwise comparisons along the branches of the maximum-parsimony tree. The tree in Fig. 6.6 has six branches, $AACD \rightarrow BACD$, $AACD \rightarrow AACD$, $AACD \rightarrow AACD$, $AACD \rightarrow AACA$, $ABCD \rightarrow AACD$, and $ABCD \rightarrow ABCA$. Since the sequence length is 4, then there are $6 \times 4 = 24$ correspondences between characters A, B, C and D. By counting how many correspondences yield $A \rightarrow A$ we obtain $k_{AA} = 9$ as depicted in the table in the lower plot in Fig. 6.6.

It often happens that there are many maximum-parsimony trees (see Chap. 7, in which case we average over all of them. Also we sum the statistics over

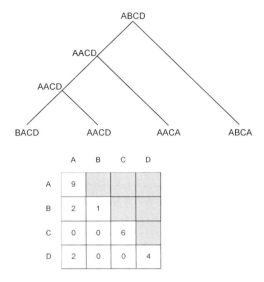

Fig. 6.6. *Top*: Illustration of computing the counts k_{ij} using a maximum parsimony phylogeny. *Bottom*: There are six possible sequence comparisons for sequences of length 4, which gives a total of $6 \times 4 = 24$ comparisons in the table

different blocks. After summing over blocks and averaging over the most parsimonious trees, we obtain composite numbers of pairs in the amino acid sequences, which we denote as follows

$\bar{k}_{ij}$ is the number of aligned pairs ij after summing over blocks and averaging over parsimony trees.

For estimating the transition probabilities we now use the same approach as that described in Sect. 6.4.4. This means that we assume reversibility, so we can derive expressions analogous to (6.47) and (6.48), (6.54) and (6.55), namely

$$\hat{p}_{ii}(\tau) = \frac{\hat{q}_{ii}}{\hat{\pi}_i} = \frac{2\bar{k}_{ii}}{2\bar{k}_{ii} + \sum_{j \neq i} \bar{k}_{ij}} \quad (6.62)$$

and

$$\hat{p}_{ij}(\tau) = \frac{\hat{q}_{ij}}{2\hat{\pi}_i} = \frac{\bar{k}_{ij}}{2\bar{k}_{ii} + \sum_{j \neq i} \bar{k}_{ij}}, \; j \neq i, \quad (6.63)$$

where $i, j \in \{1, \ldots, 20\}$. In (6.62) and (6.63), we have introduced the symbol τ to denote the "average" evolutionary time between the amino acid sequences. This time is, however, not known and as said in Sect. 6.4.6 the practical approach is to rescale probabilities in (6.62) and (6.63) in the terms of the expected total amount of change between sequences. Since the expected ratio of the number of differences to sequence length, for two sequences related by substitution scenario described by $P(\tau)$ is (see Exercise 10)

$$\frac{E(\# \text{ of differences})}{\text{sequence length}} = \sum_i \sum_{j \neq i} \pi_i p_{ij}(\tau), \qquad (6.64)$$

we compute the scaling constant

$$c = \sum_i \sum_{j \neq i} \hat{\pi}_i \hat{p}_{ij}(\tau),$$

and we rescale the probabilities in (6.63) and (6.62) such that the expected ratio in (6.64) becomes 0.01. This leads to the estimates

$$\hat{p}_{ij}^1 = \frac{0.01}{c} \hat{p}_{ij}(\tau) \qquad (6.65)$$

and

$$\hat{p}_{ii}^1 = 1 - \sum_{j \neq i} \hat{p}_{ij}^1. \qquad (6.66)$$

Amino acid sequences whose evolutionary distance is described by a Markov transition with probabilities (6.65) and (6.66) (i.e., sequences where we expect 1% of differences between symbols) are commonly said to be 1 PAM distant.

The PAM matrix with entries given in (6.63) and (6.62) is called PAM1. It is best for scoring alignments of amino acid sequences where we expect, on average, 1% of differences between symbols. If we want to align more distant amino acid sequences, then the PAM matrix is raised to the power x, and the result is called PAMx (e.g., PAM160, PAM250, etc.). Also, for fast practical use of PAM matrices, the logarithms to base 2 of the values of their transition probabilities are taken and rounded to the nearest integer.

BLOSUM Matrices

BLOSUM matrices were introduced in [122] as an alternative to PAM matrices. Their construction was based on a larger collection of over 2000 ungapped, aligned amino acid sequence blocks. The rationale for modifying the PAM matrices was that instead of extrapolating by raising PAM1 to high powers, it might be sound to base the estimation of transition probabilities on blocks with a more diverse set of distances. The distance, described as the expected proportion of conserved amino acids between sequences is used to index the specific BLOSUM matrix. For example, BLOSUM 85 is a BLOSUM matrix with transition probabilities such that 85% of amino acids are, on average, conserved between sequences. This nomenclature makes the use of these matrices more convenient.

A second change is that the construction of maximum-parsimony phylogenies is omitted. Instead, the analysis is based on numbers k_{ij} observed in all pairwise comparisons of sequences in aligned blocks.

Finally, the third change is the replacement of the separate estimations of the transition probabilities p_{ij} of $i \to j$ and p_{ji} of $j \to i$ by estimation of one parameter, namely the odds ratio

$$\hat{e}_{ij} = \begin{cases} 2\hat{\pi}_i \hat{\pi}_j / \hat{q}_{ij} & \text{if } i \neq j \\ \hat{\pi}_i^2 / \hat{q}_{ii} & \text{otherwise} \end{cases}. \tag{6.67}$$

The estimate $\hat{e}_{ij}$ in (6.67) can be readily computed by using, for $\hat{\pi}_i$, $\hat{\pi}_j$, and $\hat{q}_{ii}$, which appear on the right hand side, the expressions derived in (6.40), (6.41) and (6.43). The expression (6.67) serves the purpose of estimating only one probability of observing an unordered pair ij, instead of deriving separate expressions for transition probabilities p_{ij} of $i \to j$ and p_{ji} of $j \to i$. This is justified by the structure of the data, which records numbers of unordered pairs rather than (unobservable) transitions.

From (6.67) it follows that BLOSUM matrices are symmetric, the values of the odds ratios can be fully defined by filling in the diagonal and either upper or lower part of the matrix. Similarly to the case of PAM matrices, logarithms to base 2 of the odds are taken and rounded to the nearest integer for speeding up the computations in sequence alignment.

The authors of [122] showed several examples, where the use of BLOSUM matrices indeed led to results that were more sound biologically than those obtained by using PAM matrices. This concerned particularly the more evolutionarily distant proteins.

6.4.8 Gaps

In the above procedure for estimating the probabilities of changes in DNA and amino acid sequences, we have assumed ungapped alignments, due to the necessity of formalizing the analysis as a Markov chain. However, the method of biological sequence alignments discussed in the next section includes searching for insertions and deletions in DNA. Allowing for insertions and deletions in aligned sequences is also a strong element of the algorithms described there, compared to other methods of sequence analysis, which are based only on ungapped strings.

In order to equip alignment procedures with the possibility of predicting insertions and deletions, we need to develop a scoring system for penalizing occurrences of gaps in alignments. Again we need to experiment with alignments with temporary penalties for gaps and then modify the scores with accumulating experience. Scoring systems for gaps typically include different scores for gap initiation and gap extension. Well-balanced penalty system for gaps is very important. If gaps are overpenalized, they appear to rarely and instead many false nucleotide or amino acid alignments are produced. In contrast, if gaps are underpenalized, they appear too frequently and, as a result, they replace many true alignments of sequence symbols, i.e., nucleotides or amino acids.

6.5 Sequence Alignment by Dynamic Programming

Several of the previous sections of this chapter were devoted to descriptions of the scoring functions for sequence alignments. Now, we assume that scoring systems for alignments are available and we state the formal problem of defining the alignment, which maximizes a defined scoring function. This problem was solved by using the technique of dynamic programming (Sect. 5.2) by Needleman and Wunsch [204] and later developed further by Smith and Waterman [261]. Here we describe these methods.

6.5.1 The Needleman–Wunsch Alignment Algorithm

The algorithm for aligning two sequences developed by Needleman and Wunsch [204] can be formulated as a version of the dynamic programming method to find the best path through an array, presented in Section 5.2. As previously, we denote the two sequences to be aligned by s_1 and s_2. We assume that their lengths are, n and m respectively. We also denote the ith letter of s_1 by $s_1(i)$, $i = 1, ..., n$, and the jth letter of s_2 by $s_2(j)$, $j = 1, ..., m$. These letters belong to the alphabet Ξ. We use a general symbolic notation for an alphabet in order to cover both nucleotide and amino acid sequences. Again, a gap symbol "−" is added to allow for scoring insertions and deletions.

We assume that the scoring functions $d(\xi, \eta)$ and $d(\xi, -)$, $\xi, \eta \in \Xi$ are available from previous research or from references. Solving the alignment problem involves tracing a path through an $n \times m$ matrix with rows and columns related to s_1 and s_2 or, equivalently, making a sequence of decisions $u_1, u_2, \ldots, u_L$. Each of the decisions can be one of

$$u_k = \begin{cases} \searrow & \text{diagonally right and down,} \\ \rightarrow & \text{right,} \\ \downarrow & \text{down.} \end{cases}$$

From the sequence of decisions there follows the sequence of states $x_1, x_2...x_L$. States are two element vectors composed of matrix coordinates (indices)

$$x_k = [x_k^r \ x_k^c],$$

where x_k^r is the row index and x_k^c the column index of the kth state. The state transition rule $\Phi(x_k, u_k)$ defines the next state x_{k+1} given the present state x_k and the present decision u_k:

$$x_{k+1} = \Phi(x_k, u_k) = \begin{cases} [x_k^r + 1 \ x_k^c + 1] & \text{if } u_k = \searrow \\ [x_k^r \ x_k^c + 1] & \text{if } u_k = \rightarrow \\ [x_k^r + 1 \ x_k^c] & \text{if } u_k = \downarrow \end{cases} \quad (6.68)$$

6.5 Sequence Alignment by Dynamic Programming

A constraint has to be added to make the above precise. If a row index of a state equals n, i.e., $x_k^r = n$, then it cannot be incremented, i.e., $x_{k+1}^r = n$ must hold. A similar constraint applies to the column index: if $x_k^c = m$, then again $x_{k+1}^c = m$. The constraint (on the decision) to be added is $x_k^r = n \Rightarrow u_k = \rightarrow$ and $x_k^c = m \Rightarrow u_k = \downarrow$. The cumulative scoring index for the alignment given by $u_0, u_1, \ldots, u_L$ and $x_1, x_2, \ldots, x_L$ is

$$I = \sum_{k=1}^{L} f(x_{k+1}, u_k), \qquad (6.69)$$

where the function $f(x_k, u_k)$ is determined by the correspondences between the letters of strings s_1 and s_2. The numbers of letters are defined by coordinates x_k^r, x_k^c, which leads to

$$f(x_{k+1}, u_k) = \begin{cases} d(s_1(x_{k+1}^r), s_2(x_{k+1}^c)) & \text{if } u_k = \searrow \\ d(-, s_2(x_{k+1}^c)) & \text{if } u_k = \rightarrow \\ d(s_1(x_{k+1}^r), -) & \text{if } u_k = \downarrow \end{cases}.$$

The alignment problem can now be formulated as maximization of the cumulative score

$$S = \max_{u_1, u_2, \ldots, u_L} \sum_{k=1}^{L} f(x_{k+1}, u_k).$$

Introducing the lth cumulative partial score and defining

$$S(x_l) = \max_{u_l, u_{l+1}, \ldots, u_L} \sum_{k=1}^{L} f(x_{k+1}, u_k),$$

we can formulate the following Bellman equation for the cumulative partial scores $S(x_l)$

$$S(x_l) = \max_{u_l \in \{\searrow, \rightarrow, \downarrow\}} f(\Phi(x_l, u_l), u_l) + S(\Phi(x_l, u_l)) \qquad (6.70)$$

where x_{l+1} is related to x_l by the transition rule (6.68). The boundary conditions are $x_0 = [0\ 0]$ and $x_L = [n\ m]$. Defining $S(x_L) = 0$, we can now use the Bellman equation recursively, starting from $x_L = [n\ m]$ and successively passing to lower values of indices $[i\ j]$. When the dynamic programming problem (6.70) is being solved, the values of the optimal cumulative partial scores are saved in a rectangular matrix corresponding to the possible states.

Using the above dynamic programming solution we have aligned the strings

$$s_1 = ttcgga \qquad (6.71)$$

and

$$s_2 = acgtgagagt, \qquad (6.72)$$

	s_2	a	c	g	t	g	a	g	a	g	t
s_1											
t		-1	-1	-1	3	-1	-1	-1	-1	-1	3
t		-1	-1	-1	3	-1	-1	-1	-1	-1	3
c		-1	3	-1	-1	-1	-1	-1	-1	-1	-1
g		-1	-1	3	-1	3	-1	3	-1	3	-1
g		-1	-1	3	-1	3	-1	3	-1	3	-1
a		3	-1	-1	-1	-1	3	-1	3	-1	-1

Fig. 6.7. Needleman–Wunsch scoring matrix for the problem of aligning strings $s_1 = ttcgga$ and $s_2 = acgtgagagt$

	s_2	a	c	g	t	g	a	g	a	g	t
s_1	5	6	7	7	3	0	0	-2	-2	-2	-2
t	6	6	7	8	4	1	1	-1	-1	-1	-5
t	6	7	4	5	5	2	2	0	0	-4	-4
c	2	3	4	5	6	3	3	1	1	-3	-3
g	-2	-1	0	1	2	3	4	1	2	-2	-2
g	-6	-5	-4	-3	-2	-1	0	1	-2	-1	-1
a	-11	-10	-9	-8	-7	-6	-5	-4	-2	-1	0

Fig. 6.8. Matrix of optimal partial scores

previously used in the example illustrated in Fig. 6.2, assuming a score for a match $d(\xi,\xi) = 3$, $\xi = a,c,t,g$, and penalties for mismatches and gaps $d(\xi,\eta) = -1$, $\xi \neq \eta$, $d(\xi,-) = -1$. This leads to the problem of traversing, from upper left corner to the bottom right corner, of the matrix of scores shown in Fig. 6.7. The solution to this dynamic programming problem is shown in Fig. 6.8, where the matrix of optimal partial cumulative scores is given and in Fig. 6.9, where the array of optimal decisions is shown. Tracing the array of optimal decisions we can see that there are 7 solutions to the alignment problem, all with the same $score = 5$. The alignments implied by these solutions are listed in Fig. 6.10.

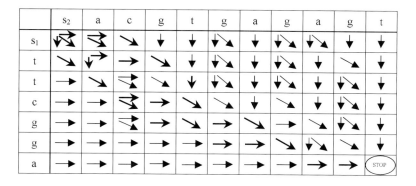

Fig. 6.9. Matrix of optimal decisions

$$
\begin{array}{lllllllllll}
s_1 : & - & - & t & t & c & g & - & g & a & - & - \\
s_2 : & a & c & g & t & - & g & a & g & a & g & t
\end{array}
$$

$$
\begin{array}{lllllllllll}
s_1 : & - & t & - & t & c & g & - & g & a & - & - \\
s_2 : & a & c & g & t & - & g & a & g & a & g & t
\end{array}
$$

$$
\begin{array}{lllllllllll}
s_1 : & t & - & - & t & c & g & - & g & a & - & - \\
s_2 : & a & c & g & t & - & g & a & g & a & g & t
\end{array}
$$

$$
\begin{array}{lllllllllll}
s_1 : & t & t & c & - & - & g & - & g & a & - & - \\
s_2 : & a & - & c & g & t & g & a & g & a & g & t
\end{array}
$$

$$
\begin{array}{lllllllllll}
s_1 : & t & t & c & g & - & g & - & - & a & - & - \\
s_2 : & a & - & c & g & t & g & a & g & a & g & t
\end{array}
$$

$$
\begin{array}{lllllllllll}
s_1 : & t & t & c & - & - & g & - & g & a & - & - \\
s_2 : & - & a & c & g & t & g & a & g & a & g & t
\end{array}
$$

$$
\begin{array}{lllllllllll}
s_1 : & t & t & c & g & - & g & - & - & a & - & - \\
s_2 : & - & a & c & g & t & g & a & g & a & g & t
\end{array}
$$

Fig. 6.10. Optimal alignments. All correspond to $score = 5$

6.5.2 The Smith–Waterman Algorithm

It may seem quite controversial to align sequences of different length, as we did for (6.71) and (6.72), with the requirement that their starting and ending points coincide. More precisely, we are not constraining the solution based on coincidence, but rather we are penalizing trailing gaps. Consequently, the solutions presented in Fig. 6.8 seem of unequal quality, despite the fact that they all share the same value of the $score = 5$. This example emphasizes

s_2		a	c	g	t	g	a	g	a	g	t
s_1		9	5	5	2	2	2	1	2	3	0
t	4	10	6	6	3	3	3	1	2	3	0
t	10	11	7	7	4	4	4	2	2	0	0
c	6	7	8	7	8	5	5	2	3	0	0
g	2	3	4	5→6	5	6	2	3	0	0
g	3	0	0	1	2	3	2	3	0	0	0
a	0	0	0	0	0	0	0	0	0	0	0

Fig. 6.11. Matrix of optimal partial cumulative scores for Smith-Waterman algorithm applied to the problem of aligning strings $s_1 = ttcgga$ and $s_2 = acgtgagagt$. The largest score is 11. The optimal path is marked by arrows

the importance of proper scoring and proper formulation of the optimization problem to obtaining acceptable results.

Smith and Waterman [261] modified the Needleman and Wunsch method by allowing matches and mismatches between sequences to be scored locally. To achieve this, they introduced two rules in the dynamic programming iterations related to traversing the score matrix:

1. If an optimal cumulative score becomes negative, it is reset to zero.
2. The starting point of the alignment occurs at the largest score in the optimal cumulative score matrix.

When we apply the above rules to the strings in (6.71) and (6.72) using again the scores for matches $d(\xi, \xi) = 3$, $\xi = a, c, t, g$, and penalties for mismatches and gaps $d(\xi, \eta) = -1$, $\xi \neq \eta$, $d(\xi, -) = -1$, we obtain the matrix of optimal partial cumulative scores shown in Fig. 6.11. The largest score is 11. The optimal path is marked by arrows. Summing up, application of the Smith-Waterman algorithm leads to the unique local alignment

$$s_1 = t\ t\ \mathbf{c}\ \mathbf{g} - \mathbf{g}\ \mathbf{a} - - - -$$
$$\vdots\ \vdots\ \vdots\ \vdots$$
$$s_2 = -\ a\ \mathbf{c}\ \mathbf{g}\ t\ \mathbf{g}\ \mathbf{a}\ g\ a\ g\ t$$

which has a score of 11.

6.6 Aligning Sequences Against Databases

Given a sequence of nucleotides or amino acids, it is of basic interest to look for similar sequences in the existing large databases of DNA, RNA, or amino acid sequences in proteins. Considering the large number of sequences to be analyzed it is infeasible to perform all possible pairwise alignments. Some efficient approaches to identifying sequences in a large database sharing similarities with a given sequence, are based on the idea of hashing [68, 288], described in Chap. 3. A related family of algorithms, called FASTA, proceed along the following steps. First, hash tables are looked up to establish how many subsequences of given length (typically $11-15$ nucleotides for DNA and RNA and $2-3$ amino acids for proteins) a database sequence shares with a target sequence. In the next step only the database sequences with the highest scores are selected. Finally, the distances between the selected sequences and the target sequence are recomputed on the basis of the Smith-Waterman alignments.

The idea of computing co-occurrences of subsequences between a target sequence and databases was further developed by taking into account that not all co-occurrences are of equal importance, which is especially important for amino acid sequences. A statistical theory has been developed, [76, 140, 141] for assessing the significance of co-occurrences of words in molecular sequences. On the basis of this theory, an appropriate scoring system was elaborated leading to efficient algorithms for aligning a sequence against a database. There are several different variants of these algorithms, they are known generally as BLAST [326].

6.7 Methods of Multiple Alignment

One can imagine aligning more than two sequences for the purpose of estimation of parameters of a substitution process, such as the three sequences depicted below, obtained by adding a new sequence $seq3$ to the two in (6.37):

$seq1 : AGGCTTAACTGATCGCTACCAAGTAGGCACGAG$
$seq2 : AGGCTTCACTGATCGCTACCAAGTAGGCACGAG$
$seq3 : AGGCTTCACTGATCGCTACCAAGCAGGCACGAG.$

(6.73)

Aligning blocks of multiple sequences, instead of pairs, has two important aspects: (i) it allows us to incorporate more data when estimating substitution frequencies, and (ii) intuitively it can lead to more reliable alignments in the sense that aligning two (or more) different symbols is related to substitution events in the past, not to putting together random items by mistake.

The optimal alignment of two sequences can be formulated as searching for a path through a plane. In the case of three sequences, as in (6.73) we can state the problem of multiple alignment as searching through a three-dimensional space, and develop a dynamic programming algorithm analogous to those described above. However, problems often solved involve multiple alignments of numerous sequences, which would lead to a need for searching through a high-dimensional space, which is numerically intractable. Two possibilities emerge. (1) We analyze all possible pairs, assume that the evolutionary divergence is similar for all pairs of sequences and ignore other evolutionary aspects of the data. (2) We try to analyze the evolutionary history of the sequences and incorporate this data into the process of estimating parameters. Again, possibility (2) leads to a loop, since evolutionary inference needs good models of substitution. And, again, this loop must be broken by some method. Both of the approaches (1) and (2) are applied in different various to alignment problems.

The practical approaches to simultaneous alignment of sequences are based on the paradigm that data on substitution probabilities and on the ancestry tree of the sequences should be incorporated into the alignment algorithm. Reliable heuristic algorithm CLUSTAL W [273], with its associated internet server, is based on the following steps. First, all pairs of sequences are "temporarily" aligned separately. On the basis of these alignments, a distance matrix is computed. Using the distance matrix obtained, a neighbor-joining tree is built, and the final alignment is obtained by progressively aligning sequences according to the branching order in the tree.

6.8 Exercises

1. Derive an expression for the number of different paths through the matrix associated with s_1 and s_2 given in (6.4).
2. Prove the Vandermode-Chu identity (6.5).
3. Write a computer program for drawing dot matrices of two DNA sequences. Try to use it for aligning sequences.
4. The Kimura Models 1 and 2 [76], [148] are described by the following (discrete) state transition matrices: for Model 1,

$$P = \begin{bmatrix} 1-\alpha-2\beta & \beta & \alpha & \beta \\ \beta & 1-\alpha-2\beta & \beta & \alpha \\ \alpha & \beta & 1-\alpha-2\beta & \beta \\ \beta & \alpha & \beta & 1-\alpha-2\beta \end{bmatrix}, \quad (6.74)$$

and for Model 2,

$$P = \begin{bmatrix} 1-\alpha-\beta-\gamma & \alpha & \beta & \gamma \\ \alpha & 1-\alpha-2\beta & \gamma & \beta \\ \beta & \gamma & 1-\alpha-2\beta & \alpha \\ \gamma & \beta & \alpha & 1-\alpha-2\beta \end{bmatrix}. \quad (6.75)$$

Compute the Jordan canonical decompositions for these matrices, and derive expressions for the transition probabilities in the continuous-time version of the Markov chain. An easy method for finding decompositions of matrices is to use symbolic-computation software such as Mathematica, Maple, or Matlab.

5. Derive estimates of the parameters for the Kimura models (6.74) and (6.75).
6. Develop a method for computing estimates of the parameters of the HKY model, based on a comparison between the two DNA sequences (6.37).
7. Derive an expression for the variance of the estimator in (6.56), [76].
8. Write a computer program for simulating the time evolution of DNA or amino acid sequences, where each base (or each amino acid) undergoes random substitution described by the substitution models described in this chapter.
9. Using the program developed in Exercise 8, (a) generate some random sequence data and (b) using the data from (a), estimate the model parameters. Compare with the true parameters of the model.
10. Derive (6.64). Use the program developed in Exercise 8 to verify, by averaging over multiple random simulations, that this equation is true.
11. Compute the expected proportion of changes between amino acid sequences (a) distant by PAM160 and (b) distant by PAM250.
12. Introduce a mechanism for insertions and deletions into the program developed in Exercise 8.
13. Write a computer program for aligning two sequences by use of the Needleman–Wunsch algorithm. Apply this program to data obtained from Exercise 12.
14. Write a computer program for aligning two sequences with the use of Smith–Waterman algorithm. Apply this program to data obtained from Exercise 12.
15. Download from the Internet an amino acid sequence of a protein, for example the human growth hormone. Use BLAST to find several variants (homologs) of this sequence, coming from organisms other than human. Align the set of sequences by using CLUSTAL W Internet server.

7
Molecular Phylogenetics

Molecular phylogenetics is the study of evolutionary relationships among organisms, genes, or proteins, using a combination of molecular biology and statistical techniques. Molecular data are powerful: DNA and protein sequences generally evolve in a more regular manner than morphological and physiological characters. Also, sequence data are amenable to quantitative treatment.

Phylogenetic relationships are usually depicted in the form of binary trees. The structure of the tree illustrates possible ancestor-descendant relationships between unknown variants of a sequence existing in the past, which are ancestral with respect to the contemporary (extant) variants (the external nodes). Inference about the past branchings in the tree (the internal nodes) can be carried out if a principle or model of evolution of the sequence is assumed. Different approaches based on many different models, may yield inconsistent results. Frequently, the discrepancies can be interpreted in biological terms. Therefore it is common practice that several different tree-making principles are applied to a given data set.

In this chapter we present the most commonly used tree reconstruction algorithms, based on the distance between sequences and on the maximum likelihood and the maximum parsimony principles. We also mention some other developments and methods in tree-building algorithms. In addition, we review some elements of coalescence theory, which deals with the genetic mechanisms and forces influencing evolutionary trees.

There are many Internet resources for inferring phylogenetic trees. One of the best known is PHYLIP [330], which provides procedures for all tree-building methods. It also contains links to many other Internet sites related to phylogenetic trees.

7.1 Trees: Vocabulary and Methods

As stated above, phylogenetic relationships can be represented in the form of trees, usually positioned upside-down. Observations, usually in the form of sequences, are available only at the bottom of the tree (i.e., at tree's crown). The task of molecular phylogenetics is to find the structure (topology) of the tree, and the branch lengths, representing the structure of the relatedness of the extant sequences and the time depth of these relationships. We shall start with a vocabulary of basic terms and remarks concerning the basic approaches to tree-making.

7.1.1 The Vocabulary of Trees

Phylogenetic (binary) tree. A tree is a graph composed of nodes and branches, in which any two nodes are connected by a unique path. A binary tree is a tree with directed branches, such that each of the nodes has no more than two descendants. A phylogenetic tree is a tree whose nodes and branches have interpretations as species or molecular sequences and relations between them.

Nodes. Nodes in phylogenetic trees are called taxonomic units. (TUs) Usually, taxonomic units are represented by sequences (DNA or RNA nucleotides or amino acids). They can also correspond to species or individuals in populations and be represented by parameters describing individuals, such as lengths, angles, or colors.

Branches. Branches in phylogenetic trees indicate descent/ancestry relationships among the TUs.

Terminal (external) nodes. The terminal nodes are also called the external nodes, leaves, or tips of the tree. For phylogenetic trees, the names of the terminal nodes are extant taxonomic units or operational taxonomic units. (OTUs)

Internal nodes. The internal nodes are nodes, which are not terminal. They are also called ancestral TUs.

Root. The root is a node from which a unique path leads to any other node, in the direction of evolutionary time. The root is the common ancestor of all TU's under study.

Rooted/unrooted tree. In Fig. 7.1, we present an example of a rooted versus an unrooted tree for the same set of extant nodes A, B, C, D, E. In a rooted tree, the direction of the evolutionary path (or time) is always specified. In an unrooted tree, the extant nodes are uniquely determined but there are many possible evolutionary paths, depending on the location of the root.

Topology. The topology is the branching pattern of a tree. The number of possible topologies is generally enormous. If the number of extant TUs is n, the number of different labeled unrooted trees is

$$N^{\text{unrooted}}(n) = \frac{(2n-5)!}{2^{n-3}(n-3)!}, \tag{7.1}$$

and the number of different labeled rooted trees is

$$N^{\text{rooted}}(n) = \frac{(2n-3)!}{2^{n-2}(n-2)!}. \tag{7.2}$$

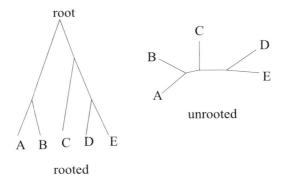

Fig. 7.1. Rooted (*left*) versus unrooted tree (*right*)

The above expressions can be derived by using an iterative procedure of adding branches to existing trees (see Exercise 2).

Branch length. The lengths of the branches determine the metrics of a tree. In phylogenetic trees, lengths of branches are measured in units of evolutionary time. The passage of evolutionary time results in the accumulation of evolutionary changes. Therefore, when we are inferring the metrics of a phylogenetic tree, the numbers of evolutionary changes between species are estimators of the branch lengths.

7.2 Overview of Tree-Building Methodologies

Phylogenetic trees may be built on the basis of very different approaches, which might be loosely divided into data-oriented and model-oriented methods. Examples of the data-oriented methods are the *distance methods*: the tree is constructed by joining sequences with a small distance between them. Another example is the *maximum parsimony method*: the tree that explains the observed data using the smallest number of substitutions is accepted. No model of evolution is assumed in the distance and maximum parsimony methods. This might be the reason why the data-oriented approaches are more appealing to biologists and are usually considered "model-free".

The model-based approaches include, among other, the *maximum likelihood method* and methods based on the *coalescent*. In maximum likelihood methods, a probabilistic model of evolution is assumed and its fit to the sequence data maximizes the likelihood of all possible trees. Calculating the likelihoods is computationally intensive, but the method can be extended in several directions, including evolution under selective pressure, which may be helpful in the identification of active sites in proteins. The coalescent is the ancestral tree of genetic variants in the Wright-Fisher model, the most commonly used model of mutation and drift in population genetics. The usefulness of the coalescent is mostly limited to genetic data and to the time depth over which it can be expected that the assumptions of the Wright-Fisher model hold. However, whenever applicable, the coalescent has a lot of explanatory power.

Another family of methods is that of Bayesian approaches, which recently have gained a lot in popularity. The philosophical underpinning of this methodology is the principle that it is advantageous to supplement the information directly provided by the data with additional "prior" information (preconception) supplied by previous experience, a higher-order theory, or "common sense" . The technical tool used in this approach is usually the posterior probability theorem (known also as the Second Bayes' Theorem)

$$\Pr[\text{parameters}|\text{data}] = \frac{\Pr[\text{data}|\text{parameters}]\Pr[\text{parameters}]}{\int_{\text{all parameters}} \Pr[\text{data}|\text{parameters}]\Pr[\text{parameters}]}$$
$$= \frac{\text{Likelihood}[\text{parameters};\text{data}]\Pr[\text{parameters}]}{\int_{\text{all parameters}} \text{Likelihood}[\text{parameters};\text{data}]\Pr[\text{parameters}]}.$$
(7.3)

The probability $\Pr[\text{parameters}|\text{data}]$ is known as the *posterior probability* and the preconceptions are expressed by the *prior probability* $\Pr[\text{parameters}]$. The data influence the outcome through the *likelihood,* $\text{Likelihood}[\text{parameters};\text{data}] = \Pr[\text{data}|\text{parameters}]$. Note that the parameters are treated as random variables. In the classical estimation theory, estimates of parameters (treated as numbers) are usually the values that maximize the likelihood. In the Bayesian approach, the posterior probability (or probability distribution) of the parameters serves to define regions within which the "true" parameters reside. The outcome is a fusion of data and preconceptions. This fact can be considered both a strength and a weakness of the Bayesian approach.

Technically, the main difficulty is the computation of the denominator, which almost never can be expressed in an elementary manner. Moreover, in many important applications such as tree-making, most parameter values make the data very unlikely and finding the values that make nontrivial contributions to the integral is very difficult. The practical solution is to use Monte Carlo methods (see Chap. 2), with various biased sampling schemes. The literature on the topic is extensive (see [82] and references therein). Bayesian methods will not be covered in detail in this chapter.

7.3 Distance-Based Trees

The underlying principle of tree-building is that the distance between the members of any pair of extant (i.e., currently existing) species (nodes or sequences) is informative about, or even proportional to the time separating these sequences from their respective common ancestor. If this is the case, then information extracted from all the pairwise distances should be sufficient to estimate the topology of the tree and the lengths of branches in the tree corresponding to the data set. Note that in the ideal case, in which the flow of time in all branches is uniform (which means, biologically, that the species or sequences evolving along these branches do so at about the same rate), all subtrees of order 3 (with three leaves) satisfy the property of ultrametricity: for any three extant nodes, x, y and z, two distances between nodes are equal and the third is less than these two, for example, $d(x,y) < d(x,z) = d(z,y)$. From a practical viewpoint, it is more interesting if, given a set of nodes with pairwise distances satisfying ultrametricity, it is possible to build a tree with branches the lengths of which are consistent with these distances. The answer is in the affirmative, as will be demonstrated in the following subsections. However, what if ultrametricity is satisfied only approximately or if it is not satisfied at all? In this latter case, we may expect difficulties, and new approaches will be needed, as explained later.

The material in this section is based on [76, 82, 197, 246].

7.3.1 Tree-Derived Distance

The usual axioms of distance are:

- nonnegativity, $d(x,y) \geq 0$;
- nondegeneracy, $d(x,y) = 0 \iff x = y$;
- symmetry, $d(x,y) = d(y,x)$;
- the triangle property, $d(x,y) \leq d(x,z) + d(z,y)$.

We call a function $d(s_i, s_j)$ defined over a set of species $s_1, s_2, \ldots s_n$ a tree derived distance if there is a tree such that $d(s_i, s_j)$ is the distance between s_i and s_j given by sum of the lengths of the branches along the path between s_i and s_j. An example of a tree with four OTUs and the corresponding tree-derived distance is shown in Fig. 7.2. Given a tree, computing tree-derived distances is straightforward. In the following subsections, we shall overview algorithms concerned with the inverse problem, i.e., given the distances, to recover a tree.

7.3.2 Ultrametric Distances and Molecular-Clock Trees

Let us supplement the axioms of distance with one more condition:

- *Ultrametricity:* for any three nodes s_i, s_j, and s_k, two distances between them are equal and the third is less than these two, for example, $d(s_i, s_j) < d(s_i, s_k) = d(s_j, s_k)$.

The ultrametricity condition is related to a type of trees which we call molecular clock trees. Here we demonstrate the relation of the molecular-clock hypothesis to ultrametricity. The molecular-clock hypothesis states that the extant species $s_1, s_2, \ldots s_n$ have evolved at the same rate of evolutionary change. That means that all species $s_1, s_2, \ldots s_n$ are at the same evolutionary

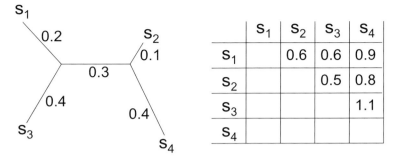

Fig. 7.2. *Left:* tree with extant species s_1, s_2, s_3, s_4. *Right:* table of tree-derived distances

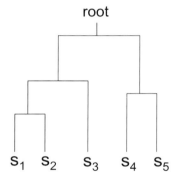

Fig. 7.3. An example of a molecular-clock tree, for OTUs s_1 - s_5. The distance in this tree is measured along the vertical direction only.

distance from their common ancestor. This property is also true for any subset of the species s_1, s_2, ... s_n. A molecular-clock tree is one which satisfies the molecular-clock hypothesis. It always has a unique root, the most recent common ancestor of the species $s_1, s_2, \ldots, s_n$. An example of a molecular-clock tree, for OTUs s_1 - s_5, is shown in Fig. 7.3. The distance in this tree is measured along the vertical direction only.

The following theorem states the equivalence between ultrametric and molecular-clock trees.

Theorem. In a molecular-clock tree, all triplets are ultrametric. Conversely, if all triplets are ultrametric, the distances uniquely define a rooted-molecular clock tree. The proof can be carried out by induction, i.e., by successively adding new species to an existing tree [76].

7.3.3 Unweighted Pair Group Method with Arithmetic Mean (UPGMA) Algorithm

We assume that a set of extant species $s_1, s_2, \ldots, s_n$ is given and that for each pair of species the distance is defined. Also, we assume that the distances are ultrametric. The UPGMA algorithm [197] allows perfect reconstruction of trees on the basis of ultrametric distances. It proceeds iteratively; in each iteration two nodes are merged. Merging nodes leads to formation of clusters. The distance $d(C_i, C_j)$ between two clusters C_i and C_j, where $C_i \cap C_j = \emptyset$, is defined as the average of the distances between their species:

$$d(C_i, C_j) = \frac{1}{\#C_i \#C_j} \sum_{x \in C_i} \sum_{y \in C_j} d(x, y). \tag{7.4}$$

In the above, we have used letters x and y for species from the set s_1, s_2, ... s_n and we have defined $\#C_i$ and $\#C_j$ to mean the numbers of species in clusters C_i and C_j. There is a recursive method for using the definition (7.4). The effect of merging clusters on the distances is given in the following rule. Let C_m be the union of clusters C_i and C_j. The distance between cluster C_m and any of the remaining clusters C_l is then given by the following equation

$$d(C_m, C_l) = \frac{d(C_i, C_l)\#C_i + d(C_j, C_l)\#C_j}{\#C_i + \#C_j}. \tag{7.5}$$

On the basis of (7.5) we can construct the UPGMA algorithm for reconstruction of the tree for the species $s_1, s_2, \ldots, s_n$.

Initialization. Define n clusters $C_1, C_2, \ldots, C_n$, each containing one species, $s_1, s_2, \ldots s_n$. Compute the distance matrix.

Iterative Step. Merge two clusters with the smallest distance $d(C_i, C_j)$. Set $n := n - 1$. Update distance matrix using (7.5).

The iterative step is repeated until there is only one cluster, containing all species.

7.3.4 Neighbor-Joining Trees

One useful property of ultrametric trees, which is shared by "close to ultrametric" distance matrices, is that the closest nodes have to be neighbors. For distance matrices generated by strongly nonultrametric trees, this does not have to be the case. An example is the tree depicted in Fig. 7.4, in which nodes s_1 and s_2, which are the closest, are not neighbors. Note that the tree in Fig. 7.4 is not ultrametric. Distances in the tree in Fig. 7.2 also have the same property. Joining the nodes with the shortest distances will not lead to recovering the true tree topology.

194   7 Molecular Phylogenetics

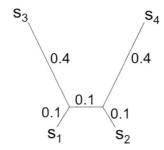

Fig. 7.4. A tree with four extant species s_1 - s_4 in which nodes s_1 and s_2 are the closest to each other, yet they are not neighbors

The principle used to estimate trees from strongly nonultrametric data is to introduce the *neighbor-joining distance* $\delta(x, y)$, which does not necessarily satisfy the usual distance axioms, but which instead is minimized when x and y are neighbors:

$$\delta(x,y) = (N-4)d(x,y) - \sum_{z \neq x,y} |d(x,y) + d(y,z)| \tag{7.6}$$

More precisely, we have the following theorem. Suppose that S is a set of species and d is a tree-derived distance on S obtained from an unrooted tree (not necessarily ultrametric). If x and y are such that $\delta(x, y)$ is a minimum, then x and y are neighbors. The intuitive support for using $\delta(x, y)$ is that the formula (7.6) is "corrected" for longer edges.

The algorithm that takes advantage of this works by joining the nodes separated by the minimum neighbor-joining distance $\delta(x, y)$ in the matrix. Then it recalculates the distances $d(x, y)$ so that the distances from all other nodes to x and to y are replaced by the distance to the joined nodes. The recalculation of distances is based on the following equation:

$$d(s,z) = \frac{1}{2}[d(x,s) + d(y,s) - d(x,y)]. \tag{7.7}$$

Using (7.7) we can compute the distance between any node s and the node z resulting from joining the nodes x and y. After that it calculates the new distances $\delta(x, y)$ and finds their minimum. It joins the two nodes separated by the minimum distance $\delta(x, y)$, and so forth.

7.4 Maximum Likelihood (Felsenstein) Trees

Maximum likelihood trees employ probabilistic models of evolution of sequences along the branches of the tree [80, 82]. These models are usually

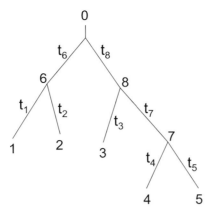

Fig. 7.5. An example of a tree with five external nodes 1-5, a root labeled as node 0, and eight branches with lengths t_1-t_8.

finite-state time-continuous Markov chains. The number of states depends on the type of sequence data. For example, when we are considering DNA sequences, we use 4 states labeled A, C, G, and T, and Markov chains describing probabilities of state transitions (see Chap. 6). When we are considering amino acid sequences, the number of states is equal to 20. The aim is to find a tree with the highest probability of reproducing the data on the OTUs. The parameters which we manipulate are the topology of the tree and the lengths of the branches given the topology. In technical terms, we look for parameter values which maximize the likelihood L given the data.

To understand the construction of the likelihood function, let us consider the evolution of a given (kth) site of the sequence representing node. Consider two adjacent nodes i and j separated by an evolutionary time t, flowing in the direction from i to j. The conditional transition probability $P_{s_i(k)s_j(k)}(t)$ that the site k at node j is equal to $s_j(k)$, given that it was equal to $s_i(k)$ at node i, can be computed from the expression for the entries of the transition matrix $P(t)$

$$P(t) = \exp(Qt) \tag{7.8}$$

where Q is the transition intensity matrix of the process. The reader may compare this equation to (2.122) in the chapter on probability and statistics. In addition, several models of nucleotide substitution are presented in Chap. 6.

Let us now focus on an example of a tree with 5 external nodes and the topology depicted in Fig. 7.5. We shall build the likelihood function for this example in several steps. The generalization to arbitrary trees is evident.

7.4.1 Hypotheses and Steps:

1. We assume the independence of evolution at different sites. Therefore, it is possible to compute the probability of evolution from sequence s_i to sequence s_j in time t as a the product of site transition probabilities:

$$P_{s_i s_j}(t) = \prod_{k=1}^{K} P_{s_i(k) s_j(k)}(t). \tag{7.9}$$

2. If the sequences in all nodes are known, the likelihood is the product of probabilities of change along each branch in the tree and the prior probability of the initial state:

$$L = \pi_{s_0} P_{s_0 s_6}(t_6) P_{s_6 s_1}(t_1) P_{s_6 s_2}(t_2)$$
$$\times P_{s_0 s_8}(t_8) P_{s_8 s_3}(t_3) P_{s_8 s_7}(t_7) P_{s_7 s_4}(t_4) P_{s_7 s_5}(t_5). \tag{7.10}$$

3. Since only extant (OTU) sequences are known, the likelihood has to be summed over all possible internal sequences:

$$L = \sum_{s_0} \sum_{s_6} \sum_{s_7} \sum_{s_8} [\pi_{s_0} P_{s_0 s_6}(t_6) P_{s_6 s_1}(t_1) P_{s_6 s_2}(t_2)$$
$$\times P_{s_0 s_8}(t_8) P_{s_8 s_3}(t_3) P_{s_8 s_7}(t_7) P_{s_7 s_4}(t_4) P_{s_7 s_5}(t_5)]. \tag{7.11}$$

4. This would be a computationally intensive procedure. Fortunately, we can deal with it recursively. Let us suppose that $L_{s_k}^{(k)}$ is the likelihood based on data at or below node k of the tree, given that node k is known to have amino acid s_k at the specific site under consideration. Suppose further that two descendant nodes of node k are labeled i and j. We then have the following recurrence:

$$L_{s_k}^{(k)} = \left(\sum_{s_i} P_{s_k s_i}(t_i) L_{s_i}^{(i)} \right) \left(\sum_{s_j} P_{s_k s_j}(t_j) L_{s_j}^{(j)} \right). \tag{7.12}$$

The above formula (7.12) allows us to compute the likelihood by starting from the leaves and proceeding up to the root. For example, for the tree in Fig. 7.5, it is possible to compute first $L_{s_6}^{(6)}$ and $L_{s_7}^{(7)}$ on the basis of the external nodes, then $L_{s_8}^{(8)}$ on the basis of an external node and $L_{s_7}^{(7)}$, and finally $L_{s_0}^{(0)} = L$ on the basis of $L_{s_6}^{(6)}$ and $L_{s_8}^{(8)}$. Eventually, we obtain the following computationally economical expression:

$$L = \sum_{s_0} \pi_{s_0} \left(\sum_{s_6} P_{s_0 s_6}(t_6) P_{s_6 s_1}(t_1) P_{s_6 s_2}(t_2) \right)$$
$$\times \left(\sum_{s_8} P_{s_0 s_8}(t_8) P_{s_8 s_3}(t_3) P_{s_8 s_7}(t_7) \left(\sum_{s_7} P_{s_7 s_4}(t_4) P_{s_7 s_5}(t_5) \right) \right),$$
$$\tag{7.13}$$

where the hierarchy of parentheses indicates the order of summation.

Note that, in the setting presented above, we are not assuming ultrametricity of the maximum likelihood tree, since we have not imposed ultrametric constraints on the lengths of its branches $t_1, t_2, \ldots, t_8$.

7.4.2 The Pulley Principle

Usually, we assume a reversible Markov chain model for state transitions $s_i(k) \to s_j(k)$; for example, this could be one of the substitution models described in Chap. 6, Jukes-Cantor, Felsenstein, or HKY model. Reversibility of the Markov chain leads to a property of the likelihood function L in (7.10) called the pulley principle. The pulley principle states that we can move the root of the tree to any of the nodes without changing the likelihood L. For example, let us demonstrate that we can move the root of the tree in Fig. 7.5 from node 0 to node 6. From reversibility, we have

$$\pi_{s_0} P_{s_0 s_6}(t_6) = \pi_{s_6} P_{s_6 s_0}(t_6). \tag{7.14}$$

Substituting (7.14) in (7.11), we obtain the following equivalent expression for the likelihood L:

$$L = \sum_{s_6} \sum_{s_0} \sum_{s_7} \sum_{s_8} [\pi_{s_6} P_{s_6 s_0}(t_6) P_{s_6 s_1}(t_1) P_{s_6 s_2}(t_2) \\ \times P_{s_0 s_8}(t_8) P_{s_8 s_3}(t_3) P_{s_8 s_7}(t_7) P_{s_7 s_4}(t_4) P_{s_7 s_5}(t_5)], \tag{7.15}$$

where, along with replacing $\pi_{s_0} P_{s_0 s_6}(t_6)$ by $\pi_{s_6} P_{s_6 s_0}(t_6)$, we have also changed the order of summation. It is easy to observe that the likelihood function in (7.15) corresponds to a tree with its root at node 6, presented in Fig. 7.6 on the left. By repeating steps like (7.14) and (7.15), we can move the root to any other node, for example, to node 8, as shown in Fig. 7.6 on the right. In this figure, the evolutionary time flows along the branches, starting from the root node. This convention is different from an equally common convention according to which the time flows vertically downwards, and the horizontal distances are added only for graphical convenience.

The above arguments demonstrate that the method of computation of tree likelihood presented above is suitable for unrooted trees. This does not, however, mean generally that the maximum likelihood approach must be confined only to unrooted trees. We can, for example, add an ultrametricity condition on the lengths of branches to adjust the maximum likelihood methodology to rooted trees.

7.4.3 Estimating Branch Lengths

In the above we have shown a method for computing the likelihood of a tree with given topology and given lengths of branches. Estimation of branch

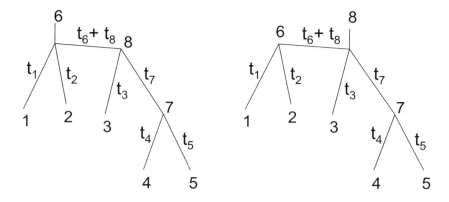

Fig. 7.6. The pulley principle. The root can be placed either at node 6 or at node 8 without changing the value of the likelihood function

lengths, given the topology, can be accomplished by maximizing (7.13) with respect to t_1, t_2, ..., t_8. One can obtain a set of equations by computing derivatives of (7.13) with respect to t_1, t_2,..., t_8, but there is no method to solve it analytically. It is possible to devise iterations that converge to the optimal t_1, t_2, ..., t_8 [80]. These iterations also have an interpretation as EM (Expectation-Maximization) recursions.

7.4.4 Estimating the Tree Topology

Finally, we should search among different topologies for the tree with the maximum likelihood. Owing to the large number of possible topologies of trees, one cannot traverse all of them and pick out the one with the highest probability. Instead, heuristic methods for searching for the maximum likelihood are applied, and it has been verified computationally that they yield acceptable outcomes. It is also sound to apply Markov chain Monte Carlo (MCMC) methods, presented in Chap. 2, for maximization of the likelihood over topologies. One can construct a Markov chain such that different trees correspond to its states. Then, by using the Metropolis–Hastings algorithm, we visit (sample) trees with frequencies corresponding to their probabilities. This allows us to limit the search space to topologies with sufficiently high likelihoods.

7.5 Maximum-Parsimony Trees

In the maximum parsimony method, [82, 88, 174], the tree is sought which requires the smallest number of evolutionary changes to explain the difference observed among the OTUs under study. Since the method is based on counting evolutionary events (substitutions), without differentiating between different types of substitution, its applications have been focused basically on trees representing nucleotide substitution. For amino acids, the hypothesis that all substitutions are equally weighted seems too simplistic.

Since the numbers of changes at different sites add up to the total number of changes it makes sense to single out sites for which all possible tree topologies result in the same minimum number of changes. These are called the non-informative sites. The sites for which topologies matter are called informative. A criterion for a site to be informative is that there are at least two different nucleotides and both are seen at least twice at the site under study. It can be verified by using the maximum-parsimony rule presented in the next subsection, that it this criterion is not satisfied, the minimal number of evolutionary events will be the same for all topologies.

7.5.1 Minimal Number of Evolutionary Events for a Given Tree

Assume the topology of a tree, such as the one with five extant nucleotides presented in the upper part of Fig. 7.7. Given the tree topology, we aim at assigning the nucleotide states of the interior tree nodes such that the number of evolutionary events (substitutions) is minimal. The basic element of the maximum-parsimony algorithms is the rule for updating the possible sets of states at tree nodes, which leads to computing the minimal number of nucleotide substitutions necessary to explain the configuration. This rule is as follows: the set at an interior node is the intersection of its two immediate descendants if the intersection is not empty; otherwise it is the union of the descendant sets. This is illustrated in Fig. 7.7, in the upper part, where the sets of states following from applying the rule are depicted. In the lower part, three different maximum-parsimony solutions are shown, each requiring three evolutionary changes.

7.5.2 Searching for the Optimal Tree Topology

Using the above idea for inferring maximum-parsimony phylogenies for sequences including many nucleotides leads to the following algorithm:

(i) Assume the topology of the tree.

(ii) Given the tree, compute for each nucleotide the minimal number of evolutionary events. Sum over all nucleotides in the sequence.

(iii) Modify the topology of the tree and return to step (i).

Steps (i)-(iii) are repeated until the tree with as few evolutionary events as possible is reached. Heuristic approaches for modification of the tree topology in step (iii) have been developed, such that the index, given by the total number of evolutionary events decreases in successive steps. The method of maximum parsimony is known for its nonuniqueness. For given data, a number of trees with the same minimum number of evolutionary events may be computed. Therefore the maximum parsimony method is often combined with

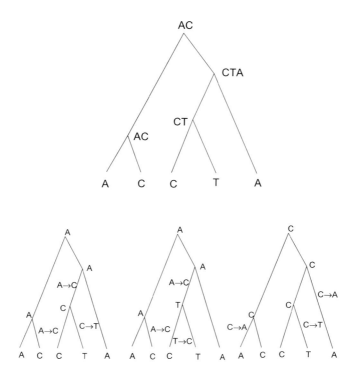

Fig. 7.7. *Top:* illustration of the rule for computing the maximum-parsimony evolutionary scenario. The set of nucleotides at an interior node is the intersection of its two immediate descendants if the intersection is not empty; otherwise, it is the union of the descendant sets. *Bottom:* three different solutions with the minimal number (three) of substitutions necessary to explain the observed data

methods of searching for a consensus tree, which represents a family of trees by one containing their common features.

7.6 Miscellaneous Topics in Phylogenetic Tree Models

The methods for tree reconstruction presented above are the most basic. In this section, we present a sample of possible extensions in this area. Our review is not exhaustive. We focus on strong relations between tree building and other areas of bioinformatics.

7.6.1 The Nonparametric Bootstrap Method

Before a tree is inferred for a sample of DNA or amino acid sequences, these sequences must first be aligned, as described in the previous chapter. The quality of the results of the multiple alignment procedure will clearly influence the reliability of the inferred tree. The nonparametric bootstrap method invented by Felsenstein [81], is one of the most popular methods for evaluating the data support for trees. Its aim is to test how possible errors in sequence alignment before the tree is constructed or the existence of polymorphisms in sequences from a given species might deform the resulting tree. Felsenstein's bootstrap method is limited to trees based on sequence information, but not exclusively to maximum likelihood trees.

The idea is to split the multiple sequence alignment (MSA) into individual columns (sites) and resample a new MSA by sampling columns of the original MSA with return. For each resampled MSA, a new "bootstrap tree" is computed. This is repeated as many times as necessary. Various statistics based on the bootstrap trees can be collected, for example the fraction of trees including a given branch (the "bootstrap support" for the branch). As stated before, bootstrapping allows one to investigate the influence of misalignments on tree structure and branch length.

7.6.2 Variable Substitution Rates, the Felsenstein-Churchill Algorithm and Related Methods

It is known that substitution rates for both nucleotides and amino acids differ between different sites along a sequence. Therefore methods and algorithms have been developed for extending the tree reconstruction methodologies to sequences with variable substitution rates. In [83], an approach was presented using a hidden Markov model for simultaneous estimation of the phylogeny and site-specific substitution rates. The method allows unknown, unequal evolutionary rates at different sites, as well as correlations between rates at neighboring sites. It uses a Markov process to assign rates to sites. A related approach, called the PAML method [294], uses Felsenstein's algorithm to reconstruct phylogeny and allows for each site to evolve at a different rate. It returns rates of evolution for each site.

7.6.3 The Evolutionary Trace Method and Functional Sites in Proteins

As we mention in Chap. 9, one can infer functionally important sites of proteins by analysis of phylogenetic trees of amino acid sequences. A related method, called the evolutionary trace methods, has been developed and presented in a series of papers [170, 171]. Functionally important residues can be related to active sites of a protein, responsible for interactions of the protein with other proteins or molecules, as well as, for example, residues whose interactions are responsible for determining the shape of the protein molecule. The evolutionary trace method looks for conserved residues in branches of the ancestry tree of a group of homologous proteins and maps functionally important residues onto the surface or interior of the protein.

7.7 Coalescence Theory

Coalescence theory is a theory dealing with the mechanisms and forces in the process of formation of evolutionary trees. The force that makes branches of evolutionary trees join is called genetic drift. The basic coalescence theory models interactions between two genetic forces, drift and mutation. The principles of this approach are presented below. Coalescence theory belongs to a branch of biology, population genetics, [75, 114], which involves the description and study of the genetic structures of populations and the mechanisms of their changes.

7.7.1 Neutral Evolution: Interaction of Genetic Drift and Mutation

Species of DNA sequences evolve in a process of replication, which leads to the passing of DNA sequences from one generation to the next. A model of this process is shown in the left panel of Fig. 7.8. We assume discrete, nonoverlapping generations, G_1, G_2, …. The number of individuals in a population is assumed constant and equal to $2N$. The notation $2N$ is motivated by our thinking of DNA sequences at autosomal loci, each present as two copies located on homologous chromosomes. In the model in Fig. 7.8, DNA sequences, represented by circles, replicated and passed from one generation to the next, are sampled with replacement. Therefore some pass more than one copy and others are left out. This mechanism, called genetic drift, shrinks genetic diversity and eventually leads to the fixation of only one allele in the whole population. This is seen in Fig. 7.8 where two alleles are depicted by two colors, black and white. In the course of evolution, with probability one, one of the alleles becomes extinct and one becomes fixed. The same is true for more than two alleles. The second force in the model is mutation, represented graphically in the right panel of Fig. 7.8. Mutations are errors in DNA replication. Mutation introduces new alleles into the population and increases the genetic diversity. The two genetic forces described act in opposite ways, and their interaction results in the observed distributions of quantities that describe the genetic structure of the population, such as the numbers of pairwise differences, the numbers of segregating sites, and the frequencies of alleles.

The most efficient way to analyze and model the joint effects of genetic drift and mutation is through the use of a coalescence approach [150] in which one considers the past of an n-sample of sequences existing at the present. The possible events that may happen in the past are coalescences, leading to common ancestors of sequences, and mutations along the branches of the resulting ancestral tree. The use of coalescence theory allows the efficient formulation of appropriate models and provides a good basis for approaching model analysis problems, hypothesis testing, or parameter estimation.

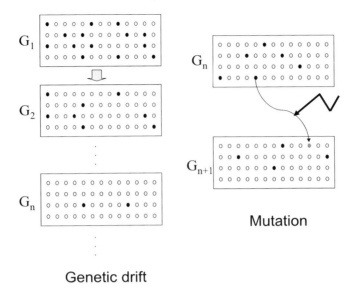

Fig. 7.8. *Left:* the illustration of genetic drift. Different alleles are represented by circles of different colors. Replication is understood as sampling with replacement from a finite population. After a sufficiently large number of generations, one of the alleles becomes fixed and the other one becomes extinct. *Right:* mutation, a genetic force which introduces new alleles

7.7.2 Modeling Genetic Drift

In order to develop the mathematical model known as the Fisher-Wright process, for genetic drift, we start by randomly picking up two individuals (DNA sequences), and we ask in which generation in the past their most recent common ancestor (MRCA) occurred. This leads to a geometric distribution of the numbers of generations separating these two individuals from their MRCA:

$$\pi_2(k) = \frac{1}{2N}\left(1 - \frac{1}{2N}\right)^{k-1}. \tag{7.16}$$

In the above, $\pi_2(k)$ is the probability that the MRCA of the two sequences occurred k generations ago, and $2N$ is the population size.

Another geometric distribution related to genetic drift is

$$\pi_m(k) = \frac{\binom{m}{2}}{2N}\left(1 - \frac{\binom{m}{2}}{2N}\right)^{k-1}, \tag{7.17}$$

where $\pi_m(k)$ denotes the probability that the first coalescence event for a set of m individuals occurred k generations ago. The formula in (7.17) follows from the fact that any of the $\binom{m}{2}$ pairs can coalesce.

204   7 Molecular Phylogenetics

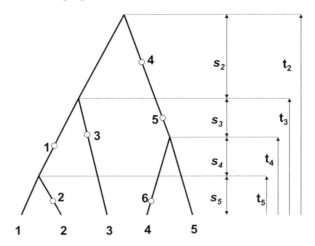

Fig. 7.9. Ancestral tree of five DNA sequences, with notation for lengths of branches and with circles representing events of mutation

The above distributions may be transformed by (i) passing from a discrete to a continuous time scale, for example the mutational time scale used in the next subsections, and (ii) assuming different demographic scenarios, for example time-varying or structured.

7.7.3 Modeling Mutation

Mutation is assumed to follow a Poisson process with an intensity μ, measured per locus (i.e., per site) per generation. The most common mutation models are

1. The infinite-sites model, where it is assumed that each mutation takes place at a DNA site that has never mutated before.
2. The infinite-alleles model, where each mutation produces an allele never before present in a population.
3. The recurrent-mutation model, where multiple changes of a nucleotide at a site are possible.
4. The stepwise-mutation model, where mutation acts bidirectionally, increasing or reducing the number of repeats of a fixed DNA motif.

In Chap. 6 and Sect. 7.6, we focused on recurrent mutations, described by a Markov chain model. Here we apply the infinite-sites mutation model, which is easier to use in coalescence theory.

7.7.4 Coalescence Under Different Demographic Scenarios

An example of an ancestral tree for $n = 5$ DNA sequences, labeled by numbers $1, 2, \ldots, 5$, is given in Fig. 7.9. This figure also introduces some notation.

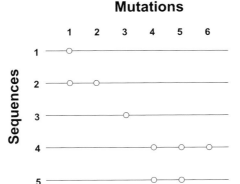

Fig. 7.10. DNA sequences 1,2,...,5 with mutations 1,2,...,6. A sample of such a structure of mutations will be observed if the sequences were evolving as shown in Fig. 7.9

The topology of the tree is given by the configuration of branches and nodes. The nodes are common ancestors of sequences in the sample. The root of the tree is the most recent common ancestor of all sequences in the sample. Mutations that occurred in the course of evolution of the DNA sequences are marked by open circles. There are 6 mutations, labeled with numbers $1, 2, ..., 6$. The tree is also characterized by times involved in the coalescence process. Random variables given by coalescence times for the sample of size n, are denoted by $T_n, T_{n-1}, \ldots, T_2$, and their realizations by corresponding lower-case letters $t_n, t_{n-1}, \ldots, t_2$. Times between coalescence events are denoted by $S_n, S_{n-1}, \ldots, S_2$, and their realizations by $s_n, s_{n-1}, \ldots, s_2$, respectively. As seen in Fig. 7.9, the coalescence times $T_n, T_{n-1}, \ldots, T_2$ are measured backwards, from the present to the past.

The tree depicted in Fig. 7.9 gives a model of the evolution that led to the DNA sequences $1, 2, ..., 5$, but was not directly observed. Many ancestral trees lead to the same DNA sequence data. Data used for inference about population evolution and related parameters look rather like those shown in Fig. 7.10, where the structure of mutations in sequences consistent with the tree in Fig. 7.9 is presented. The labels for samples and mutations are the same as in Fig. 7.9. The infinite-sites model of mutations has been assumed, and so all mutations that happened in the history until the MRCA are seen in the sample.

Homogeneous Population of Constant Size

In the case of a homogeneous population of constant size, the times between coalescence events $S_n, S_{n-1}, \ldots, S_2$, are exponentially distributed, independent random variables. The basic parameters are mutation intensity μ and the effective size of the population N. The pdf of $S_n, S_{n-1}, \ldots, S_2$ depends on the product parameter $\theta = 4\mu N$, and has the following form [93, 283]:

$$p(s_2, \ldots, s_n) = \prod_{k=2}^{n} \frac{\binom{k}{2}}{\theta} \exp\left(-\frac{\binom{k}{2}}{\theta} s_k\right), \qquad (7.18)$$

where $\binom{k}{2}$ is the binomial symbol. The mutational time scale $t = 2\mu\tau$ is used to measure times $S_n, S_{n-1}, \ldots, S_2$ (τ is the time in numbers of generations). On the mutational time scale, the intensity of the mutation process becomes $1/2$. The exponents $\binom{k}{2}/\theta$ are the intensities of the coalescence process, which change after each coalescence event.

Population with Time-Varying Size

The mutational time scale is used analogously to the way it is used in the previous paragraph. If the effective size of the population $N(t)$ changes with time, then the product parameter is also a function of time $\theta(t) = 4\mu N(t)$. The times between coalescence events, $S_n, S_{n-1}, \ldots, S_2$, are no longer independent. It is more convenient to write an expression for the distribution in terms of the coalescence times $T_n, T_{n-1}, \ldots, T_2$. The joint probability density function becomes [109, 161]

$$p(t_2, \ldots, t_n) = \prod_{k=1}^{n} \frac{\binom{k}{2}}{\theta(t_k)} \exp\left(-\int_{t_{k+1}}^{t_k} \frac{\binom{k}{2} \, d\sigma}{\theta(\sigma)}\right), \qquad (7.19)$$

where $t_2 \geq t_3 \geq \ldots \geq t_n$, $t_{n+1} = 0$.

Geographic Structure

We consider M subpopulations. We assume that their effective sizes $N_1, N_2, \ldots, N_M$ are constant. The product parameters are $\theta_m = 4\mu N_m$, $m = 1, 2, \ldots, M$. A new type of event can happen, migrations between subpopulations. The intensity of the migration process from subpopulation j to subpopulation i, per sequence per generation, is denoted by $\mathbf{m}_{ji}$; the ratios of the migration and mutation intensities are denoted by $m_{ji} = \mathbf{m}_{ji}/\mu$.

An expression for the probability density function for the metrics of the ancestral tree can be written conditionally on the sequence of events that happened in the past. It takes the following form [11, 24]:

$$p(u) = \prod_{k=1}^{T} \left(\delta_k m_{wk,vk} + (1 - \delta_k)\frac{\binom{n_{wk}}{2}}{\theta_j}\right)$$

$$\times \exp\left(-u_k \sum_{j=1}^{s} \left[\frac{\binom{n_{kj}}{2}}{\theta_j} + n_{kj} \sum_{m \neq j}^{s} m_{jm}\right]\right). \qquad (7.20)$$

In the above equation, T is the number of events that have happened in the past; $u = [u_1, \ldots, u_T]$ is the vector of times (the mutational time scale is again used) between events; n_{kj} denotes the number of lineages in subpopulation j during the time interval k, s is the number of nonempty $(n_{kj} > 0)$ subpopulations, during the time interval k, δ_k is an indicator variable that records the type of event: equal to 1 when the event at the bottom of interval k is a migration, and 0 if it is a coalescence, and wk, vk is a pair of indices standing for "from population w to population v at time u_k", and wk - "coalescence in population w at time u_k".

7.7.5 Statistical Inference on Demographic Hypotheses and Parameters

The main tool used for statistical inference about demographic hypotheses and parameters is the computation of likelihoods. If we denote by D the data (the set of DNA sequences) and by G the genealogy (which includes both the topology of the ancestral tree and the coalescence times in it), then the likelihood of the sample $P(D)$ can be written as

$$P(D) = \int_{\{G\}} P(D|G) dP(G) \quad (7.21)$$

where $P(D|G)$ is the conditional probability of the data given the genealogy, $P(G)$ denotes the probability of the genealogy, and $\{G\}$ denotes the set of all possible genealogies. When computing $P(G)$, the hypothesis of independence of metrics (coalescence times) and topology is used. All topologies of trees (with ordered branches) are equally probable. The distributions of the metrics (branch lengths) of trees are determined by coalescence process which depends on the demographic hypotheses and population parameters as described in the previous sections. The conditional probability $P(D|G)$ can be computed as a product of Poisson probabilities following from the model of the mutation process.

7.7.6 Markov Chain Monte Carlo (MCMC) Methods

Generally, it is not possible to perform the integration in (7.21) directly, owing to the large number of genealogies. Instead, Monte Carlo techniques are employed. The most straightforward Monte Carlo approach is as follows. (1) Generate a random ancestral tree with a number of leaves equal to the number of DNA sequences analyzed. (2) Introduce random mutations according to a Poisson process. (3) Compute an approximate value of (7.21) by repeating (1) and (2) and summing over the conditional probabilities of the data, given the generated genealogies. This approach is, however, highly inefficient, especially for larger data sets, owing to the fact that of the very large number of ancestral trees, most are very improbable or impossible, given the data. In the case of the infinite-sites model of mutations, the above random simulation procedure will typically lead to DNA sequences with a mutation structure inconsistent with the data and, therefore, of zero probability. Feasible mutation patterns will be encountered very rarely. In the case of the recurrent-mutation model, the situation is similar. Probabilities are greater than zero, but typically very small, so they do not contribute substantially to the sum approximating (7.21).

A solution to the above problem is to confine the area of sampling of the genealogies to those with a high enough posterior probability. In the case of the infinite-sites model, methods for defining all trees consistent with the data, for various hypotheses concerning population evolution, have been devised in [108, 110] (constant population size), [109] (time-varying population size), and [11] (geographic structure with possible changes in the sizes of subpopulations). In the case of recurrent mutations, the first step of the numerical procedure is a maximum likelihood tree reconstruction: a tree close to the most probable (with maximum likelihood) is found by a partly heuristic algorithm [80]. Then the likelihood of the DNA sample is computed by introducing random changes into the tree topology and summing over the generated trees. An algorithm for constant population size was given in [160], the case of a time-variable population size was analyzed in [161], and the case of a geographically structured population was analyzed in [24].

In order to take account of the probabilities of trees properly and to avoid generating improbable trees, a Metropolis-Hastings sampling scheme (see Chap. 2) can be used, where a Markov chain is defined with states corresponding to possible ancestral trees. Appropriate transition rules enforce reversibility of the Markov chain defined, and the desired values of its stationary probabilities.

Computer software is available on the Internet for the algorithms described in the papers referred to above. For example, given the data set shown in Fig. 7.10, the likelihood curve for the parameter θ, for a constant-population-size model, can be computed using the program Genetree [11]; the result is shown in Fig. 7.11. The maximum likelihood estimate of the parameter θ obtained from Fig. 7.11 is $\hat{\theta} = 3.73$.

7.7.7 Approximate Approaches

Despite the use of large computation power and high-efficiency algorithms, maximum likelihood techniques can be difficult or impossible to apply to large DNA samples. Also, methods which are strictly numerical give little insight into the relations between model parameters and the outcomes of computations. Therefore techniques based on approximations and simplifications are an important and always promising area of research.

Under the assumption that the population is homogeneous and remains at a constant size in the course of its evolution, estimates of the product parameter $\theta = 4\mu N$ have been proposed by several authors under various hypotheses

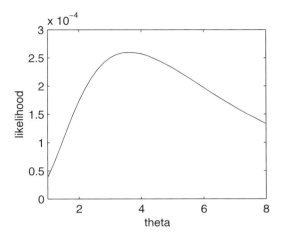

Fig. 7.11. Likelihood curve for parameter θ (theta) under constant population size hypothesis, for sample shown in Fig. 7.10, obtained with the use of the program Genetree

about the mutation model [75, 79, 93, 283]. If the Poisson process is conditioned on branch lengths $S_n, S_{n-1}, \ldots, S_2$, the expression for the probability generating function (pgf) for the number of all mutations N_S (segregating sites) in a sample of n sequences, under the assumption of an infinite-sites mutation model, assumes the following form (a convolution of independent geometric distributions) [93, 283]

$$P_{N_S}(z) = E(z^{N_S}) = \prod_{k=2}^{n} \frac{1}{1 + \theta/(k-1) - z\theta/(k-1)}. \qquad (7.22)$$

From (7.22), the expectation $E(N_S)$ is

$$E(N_S) = \theta \sum_{k=2}^{n} \frac{1}{k-1}, \qquad (7.23)$$

and so a simple moment estimate of the product parameter θ (called the Watterson estimate) is

$$\hat{\theta}_W = (\text{observed } N_S) / \sum_{k=2}^{n} \frac{1}{k-1}. \qquad (7.24)$$

Another estimate is based on the number of pairwise differences D_P. We define $D_P(i,j)$ as the number of differences seen when comparing a pair of sequences i and j, and D_P as the average number of pairwise differences in the sample. For example, in Fig. 7.10 we have $D_P(1,2) = 1$, $D_P(2,4) = 5$, and $D_P = 3.0$. The distribution of D_P (a geometric distribution) is a special case of (7.22), for $n = 2$, and the expected value for D_P is $E(D_P) = \theta$, which gives Tajima's estimate

$$\hat{\theta}_T = \text{observed } D_P \qquad (7.25)$$

For data in Fig. 7.10, we have $\hat{\theta}_W = 2.88$ and $\hat{\theta}_T = 3.0$, which do not differ drastically from the maximum likelihood estimate $\hat{\theta} = 3.73$ obtained from Fig. 7.11. However, it can be demonstrated [79] that both $\hat{\theta}_W$ and $\hat{\theta}_T$ have significantly larger variances than the maximum likelihood estimate. A good and simple estimate of θ was obtained in [93], on the basis of linear-quadratic techniques. It was shown there that the proposed estimate, for large n, becomes equivalent to the maximum likelihood estimate.

For the case of a time-varying population size, several approximate approaches to estimating $\theta(t)$ have also been proposed in the literature. Assuming an infinite-sites mutation model, simple estimates of the time function $\theta(t)$ have been obtained on the basis of the statistics of pairwise the differences D_P. The coalescence intensity function for pairs is a special case of (7.19), with $n = 2$ and $t = t_2$:

$$p(t) = \frac{1}{\theta(t)} \exp\left(-\int_0^t \frac{d\sigma}{\theta(\sigma)}\right) \qquad (7.26)$$

Combining (7.26) with the Poisson distribution, one obtains the following expression for the pgf of the number of pairwise differences $P_{D_P}(z)$:

$$P_{D_P}(z) = \int_0^\infty \exp[(z-1)t] p(t) dt. \qquad (7.27)$$

The exponential term under the integral is the pgf of the Poisson distribution. In [243] a method for fitting a parametric scenario of a stepwise change of the effective population size at time t_s before now, based on (7.26) and (7.27), was developed. This method when applied for data on the worldwide pairwise differences between samples of mitochondrial DNA [49], yielded an estimate of the history of the effective size of the human population in the form of a step function $\theta_{\text{present}} = 410.69$, $\theta_{\text{ancestral}} = 2.44$, and $t_s = 7.18$. A nonparametric method for inferring $\theta(t)$, again based on (7.26) and (7.27), was described in [229]. This method uses the observation that the estimation of $\theta(t)$ can be understood as the two-step inverse problem defined by the relations (7.26) and (7.27). Computing $p(t)$ from $P_{D_P}(z)$ can be formulated, by virtue of (7.27), as a numerical inversion of the Laplace transform [27], and the inverse relation for computing $\theta(t)$ from $p(t)$ follows from the definition of the hazard function in survival theory [57],

$$\theta(t) = \frac{\int_t^\infty p(\sigma) d\sigma}{p(t)}. \qquad (7.28)$$

In Fig. 7.12, data on pairwise differences between sequences of mitochondrial DNA from [49] are presented, together with the resulting estimates of $\theta(t)$ given in [243] and [229]. Both estimates predict a sharp increase in the size

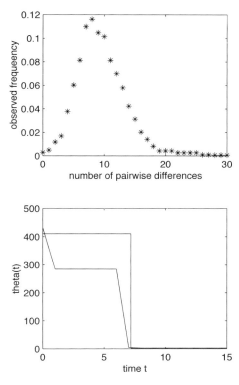

Fig. 7.12. *Top:* data on pairwise differences between sequences of mitochondrial DNA, from [49]. *Bottom:* resulting estimates of $\theta(t)$, from [243] and [229]

of the human population at approximately 7 units of mutational time ago, which may correspond to known archaeological findings [243].

Detection of population expansion under the assumption of a stepwise mutation model was researched in [147, 149]. The imbalance index was proposed, whose value indicates the past population growth or its equilibrium state. This index is a ratio of two estimates of θ: one based on genetic variance and the other one based on average homozygosity. It was demonstrated, with the use of this imbalance index, that the human population shows signatures of a bottleneck followed by expansion. In [230] and references therein, a problem of inferring the history of population size, on the basis of data on single nucleotide polymorphisms (SNPs) was studied.

An approach to demographic inference, based on the idea of using (7.19) as if the coalescence times $t_n, t_{n-1}, \ldots, t_2$ were known, was described in [79] for a constant population size and in [233] for a variable population size. With the assumption that $t_n, t_{n-1}, \ldots, t_2$ are given, maximum likelihood estimates (parametric or nonparametric) of $\theta(t)$ are easily obtained by maximizing $p(t_n, t_{n-1}, \ldots, t_2)$. Estimates of population parameters of this type are used to compute lower bounds on the variances of estimates and to research sources of bias in estimation. They also can be applied practically when combined with simple methods for inference about coalescence times $t_n, t_{n-1}, \ldots, t_2$, such as the UPGMA method. In [233], an application of this method to inferring patterns of growth in populations of HIV viruses was presented.

7.8 Exercises

1. Design a computer program for drawing binary trees, rooted and unrooted.
   a) Design a computer program for drawing all rooted binary trees with n OTUs.
   b) Design a computer program for drawing all unrooted binary trees with n OTUs.
2. By using the principle of mathematical induction, derive (7.2) and (7.1).
3. Derive an algorithm, for 4 extant species, for verifying whether their distance matrix is tree-derived or not.
4. Derive (7.5), which describes the effect of merging clusters on distances.
5. Using trees in Figs. 7.2 and 7.4, verify that the neighbor-joining distance (7.6) indeed allows one to pick out neighboring nodes on the basis of the array $\delta(x, y)$.
6. Design a computer program for random simulation of the states of the nodes A, C, G, T, in a tree, given the tree topology and metrics and given a substitution model. Use one of the substitution models presented in Chap. 6, namely the Jukes-Cantor, Felsenstein or HKY model. This program may be useful for testing the programs developed in Exercises 7 and 8.
7. Design a computer program for computing the likelihood of the tree in Fig. 7.5, under different models of the nucleotide substitution processes, i.e., the Jukes-Cantor, Felsenstein and HKY models.
8. Study the problem of generalizing the program developed in Exercise 7 in such a way that it will work for any tree.
9. On the basis of the solution to Exercise 7, develop an algorithm for estimating lengths of tree branches.

8
Genomics

Genomics is the branch of the biological sciences that deals with the structure and information encoded in the genomes of organisms, that is in the complete DNA sequences of organisms. DNA, (deoxyribonucleic acid) an organic polymer present in the cells of all living organisms has the ability to perform two basic functions: (i) replicating itself, and (ii) storing information on the linear composition of the amino acids in proteins, which are basic elements in the makeup and activity of living creatures. Genomics is a central topic in the biological and biomedical sciences and often an excellent starting point to a study of them, since all aspects of the construction and activity of organisms are reflected in the contents of their DNA. The genomes of organisms are the longest units of data carrying information, and from the information-theoretic viewpoint, the analysis of genome sequences is the greatest challenge. Although RNA and protein polymers seem to exhibit a much more variable functional spatial structure than does DNA, their linear content is always a copy of a short fragment of the genome.

The experimental tools for the study of genome structure and function include sequencing techniques that allow reading of the DNA sequence, and numerous approaches that allow one to research the relation of the DNA sequence to its function. Experimental techniques must be combined with informatic and mathematical-modeling tools in order to (i) organize and store the experimental results, for example, construct and manage genomic databases; (ii) search for structure and order, and perform comparisons; and (iii) come up with new hypotheses based on the data.

We start this chapter with a review of the basic facts regarding the structural and functional aspects of the genome. Genomes differ significantly between species, in the size, topology, types and function of the DNA sequences [118, 232]. In the main presentation we focus on the nucleic genomes of eukaryotes, that is, organisms whose cells are divided into separate compartments by membranes. In eukaryotes (animals, plants and fungi) the genetic material is located in a separate compartment–the nucleus. Eukaryotic organisms are most often multicellular, with the cells specialized for different functions.

214   8 Genomics

Simpler organisms, i.e., prokaryotes, do not have cell nuclei, they lack many cell organelles, and they are most often unicellular.

In order to discuss the functional aspects of genome sequences we introduce the central dogma of molecular biology [58] in this chapter along with some basic information on nucleic acids and proteins. On the basis of these ideas, we then discuss some mathematical and computational approaches that allow one to inquire about structure and function of DNA.

8.1 The DNA Molecule and the Central Dogma of Molecular Biology

The polymer DNA is built of smaller components–nucleotides. These are composed of sugars (deoxyriboses), phosphate groups, and four types of organic nitrogen bases, namely adenine (A), cytosine (C), guanine (G), and thymine (T). Chemical formulas and symbolic representations of the components of DNA are shown in Figs. 8.1 and 8.2. The deoxyribose molecule is shown in Fig. 8.1. It contains, in total, five carbon atoms, labeled according to convention, by numbers $1'$–$5'$. Four of the five carbon atoms belong to the planar ring, while the fifth one, marked by $5'$, sticks out in the direction perpendicular to the surface of the ring. This $5'$ carbon atom is used to define the spatial orientation of the deoxyribose molecule. The $5'$ direction means the direction given by the vector $3' \rightarrow 5'$. The direction opposite to $5'$ is called $3'$. Figure 8.2 shows the four organic nitrogen bases present in DNA. Adenine and guanine are two-ring chemical compounds and are called purines; cytosine and thymine are one-ring compounds and are called pyrimidines.

In 1953 Watson and Crick formulated their model of DNA [282], which explains how the DNA polymer is made up spatially from its components. Watson and Crick discovered the structure of DNA using X-ray diffraction diagrams of crystallized DNA obtained experimentally by Franklin, Gosling and Wilkins, [92, 291], and the knowledge existing at that time about the chemical components of DNA and their possible bindings. This discovery was honored by a Nobel Prize in 1962.

Watson and Crick observed that adenine can pair with thymine and cytosine can pair with guanine, by hydrogen bonds, leading to two approximately planar complexes of the shapes that are similar to each other. This feature is shown in Fig. 8.3 and explained further in the caption. The symbolic representations of A, C, T, and G, in Figs. 8.2 and 8.3 are chosen to reflect the complementarity between bases. Combining all of the facts Watson and Crick proposed that the structure of DNA was two helical chains of alternating deoxyriboses and phosphate groups, stabilized by a ladder-like sequence of complementary pairs of bases AT and CG. The alternating chain of deoxyriboses and phosphate groups is called the backbone of DNA.

A schematic diagram of the composition of the DNA polymer according to the Watson–Crick model is shown in Fig. 8.4. This diagram uses symbolic

8.1 The DNA Molecule and the Central Dogma of Molecular Biology 215

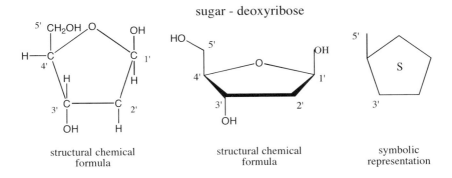

Fig. 8.1. The sugar deoxyribose. The five carbon atoms present in deoxyribose are numbered from $1'$ to $5'$ according to the common convention. The chemical formula for deoxyribose is $C_5H_9O_4$. The chemical formula gives counts of the atoms included in the molecule. The structure of the atomic bonds in the molecule is represented by a structural chemical formula. There are several conventions for showing structures of chemical bonds in a planar representations. *Left*: a structural chemical representation which marks all atoms and shows approximately the relative positions of all atoms and bonds. *Middle*: a structural representation where carbon and hydrogen atoms are not shown. Their positions are implied by the structure of the compound. *Right*: symbolic representation used in descriptions of the structure of DNA leter in this boook

representations of the components of DNA, i.e., the deoxyriboses and nitrogen bases, introduced in figures 8.1–8.2. An additional symbol representing a phosphate group contains a circle with capital letter P inside. The repetitive element in DNA polymer is the complex of a deoxyribose, a phosphate group, and a nitrogen base. The two chains in the DNA polymer are antiparallel, as seen in Fig. 8.4. Along the vertical direction from bottom to up the left chain has the orientation $3' \rightarrow 5'$, and the right chain has the opposite orientation $5' \rightarrow 3'$. The schematic representation in Fig. 8.4 can be complemented by more realistic models including the exact positions of atoms and the bonds between them, as shown in Fig. 12.1. Such a representation, can be drawn by using publicly available programs and tools for molecular graphics. The plots in Fig. 12.1 were produced with the help of the programs 3DNA [302] and Ras Mol [332].

The information carried by DNA is encoded in the order of nucleotides in the DNA polymer, and is commonly represented by listing the nucleotides in the sequence of their appearance. For the left chain of the fragment of DNA strand in Fig. 8.4, the textual representation would be $5'-GACTG-3'$ and for the right chain, complementary to the left one, the representation would be $3'-CTGAC-5'$. The whole polymer molecule can be represented as

216   8 Genomics

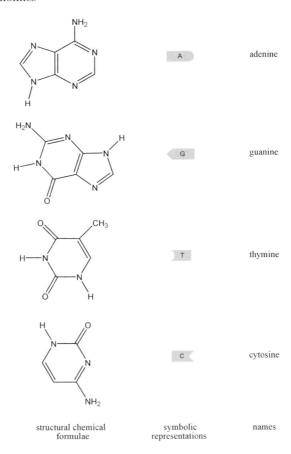

Fig. 8.2. The four nitrogen bases of DNA, adenine, guanine, thymine and cytosine. The chemical formulae of the bases are $C_5H_5N_5$ (adenine), $C_5H_5N_5O$ (guanine), $C_5H_5N_2O_2$ (thymine), and $C_4H_5N_3O$ (cytosine). *Left*: structural chemical representations of bases (with carbon and hydrogen atoms not shown). *Middle*: symbols used later in describing the structure of DNA. *Right*: names of bases

$$5' - GACTG - 3'$$
$$3' - CTGAC - 5'$$

where the upper strand corresponds to the left strand in Fig. 8.4 and the lower one to the right strand. Commonly, symbols 5' and 3' are dropped and by convention, nucleotides in DNA strands are written in the direction they are replicated and transcribed, i.e., from the 5' to the 3' end. In both replication (creating a new copy of DNA) and transcription (creating an RNA sequence complementary to a DNA fragment) protein enzymes called polymerases slide along the template DNA strand from the 3' to the 5' direction (see Fig. 8.5),

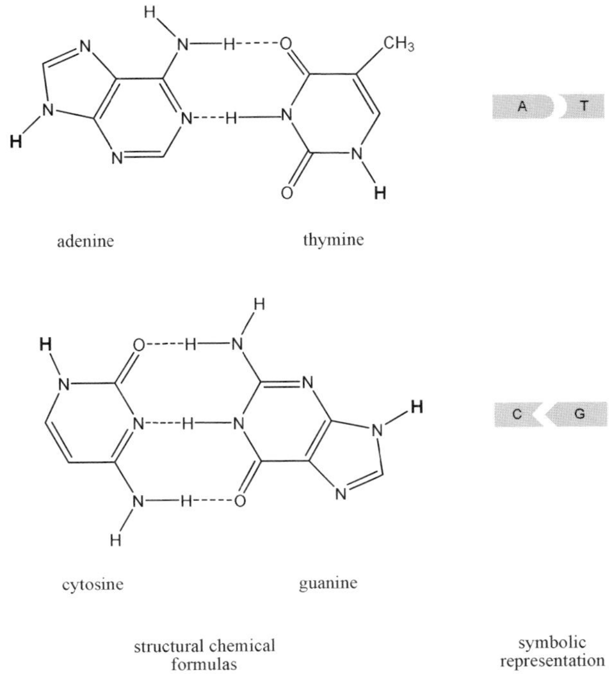

8.1 The DNA Molecule and the Central Dogma of Molecular Biology

Fig. 8.3. The complexes AT and CG, formed by hydrogen bonds. The hydrogen bonds are shown by dashed lines. The complexes AT and CG are approximately planar. The complementarity of A and T and of C and G results in the fact that all complexes AT, TA, CG, and GC are very similar in shape and the distances between the hydrogen atoms bonded to DNA backbones are similar.

which leads to creation of a complementary strand. Order of nucleotides in the complementary strand follows from pairing rules A–T, C–G.

The compounds in the DNA backbone (deoxyriboses and phosphate groups) are kept together by quite strong phosphodiester bonds, so its structure is rather durable. The hydrogen bonds between complementary pairs of bases are weaker and it is relatively easy to separate the two strands of DNA. Double-strand DNA can be split into two separate strands by appropriate enzymes and separated strands can serve as templates in the process of building of complementary structures, leading to the replication of the DNA or to transcription of DNA to RNA. The processes of replication and transcription are nonsymmetric. In replication, two synthesis processes, along both of the separated DNA chains, proceed in parallel. However, since the direction of replication is $5'$–$3'$, the replication along the $3'$ template strand proceeds continuously, while the process in other direction, along the $5'$ strand, takes place

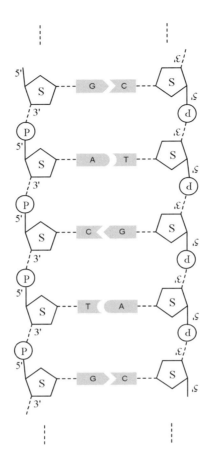

Fig. 8.4. Schematic diagram of the structure of the DNA polymer $5'-GACTG-3'$ according to the Watson–Crick model

in phases comprising 50–100 nucleotides. The strand (more precisely, strand fragment) of continuous replication is called the leading strand and the other one, the lagging strand. In the transcription process, the non-symmetry is related to the fact that only one of two strands is transcribed. In the context of transcription, the two strands of DNA are called the sense and antisense strand. Only one, the antisense strand, is copied to mRNA. The name "antisense" comes from the fact that amino acids are coded by the order of the nucleotides complementary to those in the (antisense) DNA strand.

Any sequence of DNA which one can find in a genomic database is always only one of the two strands; often it is the antisense strand of coding DNA. Note, that there is no way to tell which of the strands in Fig. 8.4 is the sense or antisense strand. The assignment is told by the context, which is

8.1 The DNA Molecule and the Central Dogma of Molecular Biology   219

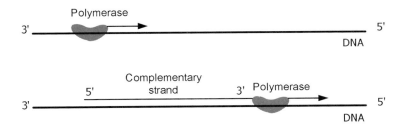

Fig. 8.5. Direction of the replication reaction. An enzyme, polymerase, slides along the DNA from 3' to 5', and the complementary strand is created along the upstream direction 5' − 3'

not depicted in Fig. 8.4, namely by positions and orientations of transcription control sequences in DNA.

For the functioning of living organisms, information from the DNA sequence must be applied effectively to the construction of proteins. There is a hypothesis concerning the flow of genetic information from DNA to proteins called the central dogma of molecular biology [58]. The diagram illustrating the central dogma, in Fig. 8.6, shows the main processes and their relations. The process of DNA replication already discussed allows the producing of copies of cells and reproduction of organisms. Two other basic processes are transcription of a fragment of DNA to RNA (ribonucleic acid), and translation leading to the construction of a protein based on the sequence of bases in the RNA. RNA is a polymer analogous and similar to DNA, composed of a sequence of repeating units. It also has the structure of a sugar–phosphate backbone with a sequence of organic nitrogen bases bonded along it. However, some components of RNA are different from those in DNA, namely: (1) sugar deoxyribose which appears in DNA is replaced in RNA by another sugar, ribose, and (2) the organic base thymine (T) that appears in DNA is replaced in RNA by another organic base, uracil (U). The chemical formulas of the two new compounds, ribose and uracil, are shown in Fig. 8.7. The main differences between RNA and DNA, following from the different chemical compositions, are the following. (1) RNA is usually a single-strand molecule. RNA is less stable and usually shorter than DNA. (2) The RNA molecule has a spatial structure that is more complicated and less regular than the DNA helix. The process of transcription involves creating a complementary RNA sequence corresponding to a fragment of the DNA. In transcription, only one (the antisense) DNA strand, is copied to a single-stranded complementary RNA molecule. The RNA carries a piece of information copied from DNA, which analogously to DNA, can be encoded as a sequence of bases, for example, the complementary RNA strand copied from the left strand of the DNA piece in Fig. 8.4 will have the following sequence of bases: $5' - CAGUC - 3'$.

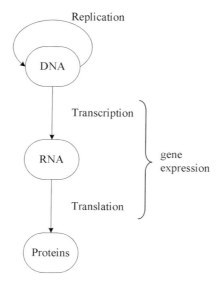

Fig. 8.6. Diagram illustrating the central dogma of molecular biology

By the mechanism of transcription, many copies of the RNA corresponding to particular DNA fragment can be created. After transcription, RNA molecules move from the nucleus to the cytoplasm where, in organelles of the call called ribosomes, they serve for the synthesis of proteins, in a process called translation. Proteins are fundamental structural and functional elements of living organisms. They are also long polymers of smaller molecules–amino acids, which have the ability to form chain structures. However, a crucial for their functioning is their spacial structure, which follows from very complicated molecular mechanisms. The templates for building proteins are encoded in DNA as linear sequences of codons (see the next section), each codon corresponding to one amino acid. The processes of transcription and translation are also called gene expression.

8.2 Genome Structure

None of the processes depicted in Fig. 8.6 follows spontaneously, but rather they are all controlled at many points and by many factors. DNA replication must be synchronized with many other processes of the cell cycle to allow for correct cell division. The process of gene expression cannot continue ceaselessly since it would lead to unnecessary excess of particular types of protein. A mechanism must then exist that limits gene expression under condition of excesses of the corresponding protein and initiates or boosts the gene expres-

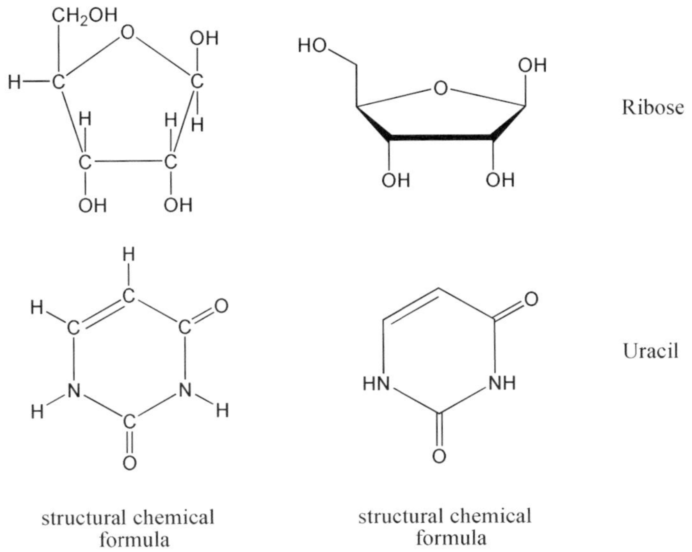

Fig. 8.7. Chemical formulae for two components that differentiate RNA from DNA. Sugar Ribose $C_5H_{10}O_5$ (*top*), and nitrogen base uracil $C_4H_4N_2O_2$ (*bottom*)

sion process under deficiency of the protein. Transcription and translation can be triggered or stopped by external (extracellular) factors, such as the presence of viruses, bacteria, or chemicals, or effect of the temperature, radiation, or cell signaling. The process of transcription in eukaryotes is more complicated than just copying the contents of DNA into RNA. It also involves (1) adding some molecules in front and after the copied region, and (2) splicing, which means transcribing exons (coding parts of genes) to RNA and leaving introns (noncoding parts of genes) untranscribed. Also, there are two types of RNA molecule involved in this process, mRNA (messenger RNA), which carries the information from the DNA to the cytoplasm, and tRNA (transfer RNA), a small RNA chain (74–93 ribonucleotides) that transfers a specific amino acid to a growing polypeptide chain at a ribosomal site.

In the following, we describe the composition of genome in association with its functional aspects.

Chromosomes. The genomes of eukaryotes consist of chromosomes, which are long linear molecules of DNA; wrapped around proteins, they form chromatin inside the nucleus of the cell. Chromosomes contain many genes and other functional elements of DNA. The numbers of genes in genomes differ between different organisms, for humans it is estimated that the total number of genes ranges between 20 000 and 30 000.

Genes. Genes are units of DNA sequences that contain information about the order of amino acids in proteins. The structure of genes differs quite

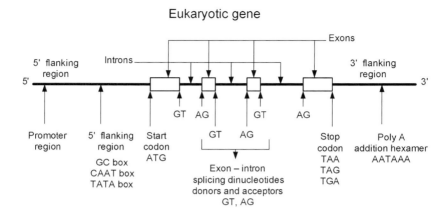

Fig. 8.8. Structure of a eukaryotic gene. Transcription from DNA to RNA proceeds along the direction from 5′ UTR (untranslated region) towards 3′ UTR. In the 5′ UTR flanking region of a gene, there are promoter regions and start codons. In the 3′ UTR flanking region of the gene, one finds a stop codon and polyadenylated (polyA) addition site

substantially between prokaryotic and eukaryotic organisms. The structure of a eukaryotic gene is shown in fig. 8.8. The sequence fragments of DNA outside the gene are called the untranslated regions (UTRs) or flanking regions of the gene. Transcription from DNA to RNA runs in the direction from 5′ towards 3′. The direction from 5′ to 3′ is also called the downstream direction, and the opposite direction, from 3′ towards 5′, is called the upstream direction of the gene. In the 5′ flanking region of the gene, control sequences that allow the transcription to be started, such as promoter regions or a $TATA$ box are placed. In the UTR neighboring the gene at the 3′ side polyadenylated (poly-A) sites are found, which control cessation of transcription. There are parts of the gene called exons and introns. Exons contain information that will eventually be used in protein synthesis. Introns are spliced out during transcription. Boundaries between exons and introns are signaled by DNA dinucleotides called donors and acceptors. Start and stop codons control the initiation and termination of the process of translation. All of these functional elements are related to the specific DNA sequences listed in Fig. 8.8 below the descriptions of the parts of the gene. Knowing these sequences is very helpful for example, in looking for genes in unannotated genomes, but it should be stressed that these codes are not rigid, in the sense that they come in different variants and can differ between organisms and between genes. This makes the analysis of DNA sequences of genes more complicated.

The genetic code. Cell organelles, namely ribosomes, which are complexes of RNA and proteins, located outside the cell nucleus, translate a linear sequence of mRNA bases to a linear sequence of amino acids. The genetic code

Table 8.1. Table of genetic codes

First base		Second base			
		U	**C**	**A**	**G**
	U	UUU (Phe/F)Phenylalanine UUC (Phe/F)Phenylalanine UUA (Leu/L)Leucine UUG (Leu/L)Leucine, *Start*	UCU (Ser/S)Serine UCC (Ser/S)Serine UCA (Ser/S)Serine UCG (Ser/S)Serine	UAU (Tyr/Y)Tyrosine UAC (Tyr/Y)Tyrosine UAA *Stop* UAG *Stop*	UGU (Cys/C)Cysteine UGC (Cys/C)Cysteine UGA *Stop* UGG (Trp/W)Tryptophan
	C	CUU (Leu/L)Leucine CUC (Leu/L)Leucine CUA (Leu/L)Leucine CUG (Leu/L)Leucine, *Start*	CCU (Pro/P)Proline CCC (Pro/P)Proline CCA (Pro/P)Proline CCG (Pro/P)Proline	CAU (His/H)Histidine CAC (His/H)Histidine CAA (Gln/Q)Glutamine CAG (Gln/Q)Glutamine	CGU (Arg/R)Arginine CGC (Arg/R)Arginine CGA (Arg/R)Arginine CGG (Arg/R)Arginine
	A	AUU (Ile/I)Isoleucine, AUC (Ile/I)Isoleucine AUA (Ile/I)Isoleucine AUG (Met/M)Methionine, *Start*	ACU (Thr/T)Threonine ACC (Thr/T)Threonine ACA (Thr/T)Threonine ACG (Thr/T)Threonine	AAU (Asn/N)Asparagine AAC (Asn/N)Asparagine AAA (Lys/K)Lysine AAG (Lys/K)Lysine	AGU (Ser/S)Serine AGC (Ser/S)Serine AGA (Arg/R)Arginine AGG (Arg/R)Arginine
	G	GUU (Val/V)Valine GUC (Val/V)Valine GUA (Val/V)Valine GUG (Val/V)Valine,	GCU (Ala/A)Alanine GCC (Ala/A)Alanine GCA (Ala/A)Alanine GCG (Ala/A)Alanine	GAU (Asp/D)Aspartic acid GAC (Asp/D)Aspartic acid GAA (Glu/E)Glutamic acid GAG (Glu/E)Glutamic acid	GGU (Gly/G)Glycine GGC (Gly/G)Glycine GGA (Gly/G)Glycine GGG (Gly/G)Glycine

is a dictionary for this translation process. Its basic principle was discovered by Francis Crick based on the basis only of the fact that there are 20 different amino acids (see Table 8.1). He assumed that the coding units were words made up of the letters of DNA "*A*", "*C*", "*T*", and "*G*", of some constant length. If the word length was 2, then the code capacity would be 4 (the number of different bases or letters) to the power of 2 (the word length) = 16, which is not enough to encode all 20 amino acids. For words of length 3 the code capacity is 64, which, by exceeding the lower limit of 20, allows the encoding of both all amino acids and the transcription/translation signaling sequences in DNA. A DNA word of length 3, the basic coding unit, is called a codon. The meanings of all codons, presented in Table 8.1, was discovered by Nirenberg and Matthaei in a series of experiments involving creating artificial RNA strands and observing the resulting amino acid sequences [209] (this work led to the Nobel Prize in Physiology or Medicine 1968). Although this code is called "universal" the genetic code varies between organisms in the meanings of some codons owing to the evolutionary processes [174].

8.3 Genome Sequencing

The sequencing of entire genomes of various organisms has become one of the basic tools of biology. Some of the main steps that led to the present massive capacity of DNA sequencing technology were (1) the application of restriction enzymes to cutting DNA into fragments, (2) an electrophoretic technique allowing one to separate DNA fragments of different lengths, (3) the Southern blot technique, (4) the polymerase chain reaction (PCR) which creates multiple copies of a given DNA piece, (5)the cloning of DNA, and (6) automatic DNA sequencers.

8.3.1 Restriction Enzymes

These enzymes, first discovered in the bacterium *Escherichia coli* [262] allow one to cut DNA molecule at particular motif sequences (4–12 base pairs long). Restriction enzymes are also called sequence-specific endonucleases or molecular scissors. In bacterial organisms, they serve the purpose of protecting against foreign DNA, by cutting it into pieces. (The host DNA is protected by methylation.) As an example, using the database of restriction enzymes ReBase [242, 333], we can find a restriction enzyme, in *E. coli*, named Eco RI, which cuts DNA as shown below:

$$\begin{matrix} & \downarrow & & & & & \\ 5' - G & A & A & T & T & C - 3' \\ 3' - C & T & T & A & A & G - 5'. \\ & & & & & \uparrow & \end{matrix}$$

In other words, any time the sequence $5' - GAATTC - 3'$ is encountered, the DNA is cut into two pieces with "scissor blades" marked by arrows. Note that the sequence $5' - GAATTC - 3'$ is DNA palindromic, its reversed complement $5' - 3'$ reads exactly the same as original.

By cleaving a given DNA double strand with restriction enzymes one obtains a collection of fragments of different lengths, called a fingerprint of the DNA, since they can be used in identifying DNA, by using electrophoresis to separate them by their sizes.

8.3.2 Electrophoresis

Electrophoresis technique uses agarose gels and an electric field to separate DNA strands by size (length). An agarose gel is a porous medium which acts like a sieve on DNA molecules; the longer the molecule, the slower it moves through the gel. DNA molecules are negatively charged, and when deposited on an agarose gel and subject to electric field, they will migrate towards the positive anode. Short molecules migrate faster than long ones, which results in the creation of a pattern of stripes on the gel, each stripe corresponding to DNA fragments of a specific length (see Fig. 8.9). The distance which a DNA fragment migrates in the agarose gel is inversely proportional to its length. The sensitivity of the electrophoretic technology allows one to separate DNA molecules with a resolution of one base in their length.

8.3.3 Southern Blot

Southern blot is a technique invented by Edward M. Southern [265] for the identification of specific DNA patterns in DNA samples. In this method, DNA

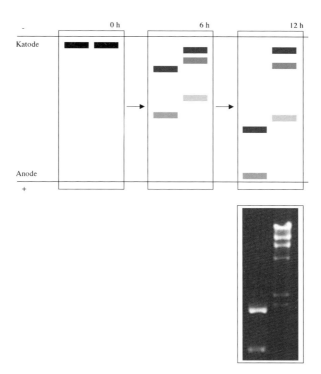

Fig. 8.9. *Top*: a schematic illustration of the separation of DNA molecules by their length by the use of agarose gel electrophoresis. Two samples of DNA molecules are exposed to an electric field at a time of $0h$. As time passes longer fragments move, more slowly and shorter fragments move faster towards anode. This creates a pattern of bands, which can be read for the lengths of the DNA molecules. *Bottom*: photograph of electrophoretic gel from a real experiment

is first treated with restriction enzymes and the resulting DNA fragments are separated by electrophoresis as described above. Next, the electrophoretically separated DNA molecules are denatured (which means separation of the complementary DNA strands by high temperature), blotted onto a nitrocellulose membrane, retaining their electrophoretic position, and hybridized with radiolabeled single-stranded DNA fragments with sequences complementary to those being sought. If present, the radiolabel is then detected by radiography.

8.3.4 The Polymerase Chain Reaction

The polymerase chain reaction invented by K. Mullis (which led to the Nobel Prize in Chemistry in 1993) allows amplification of a small initial amount of a specific sequence of DNA by factors of order 10^6–10^8. It proceeds by using

DNA primers and a polymerase enzyme in association with a carefully chosen schedule of temperature changes. DNA polymerase is an enzyme (protein) that controls the replication of DNA. DNA primers are artificial (synthesized in the laboratory) DNA sequences with a length 20–30 base pairs that bind (hybridize) to DNA at specific positions, owing to their complementarity to the DNA strands to be amplified, and allow initiation of a replication reaction. The scenario is presented in Fig. 8.10. Planning a PCR requires designing two primer sequences, one for each of the flanking regions of the DNA sequence to be amplified; in Fig. 8.10, these primers are named A and B. The PCR cycle starts from raising temperature to 96 °C which results in denaturation of the DNA (separation of complementary strands). Then the temperature is lowered to about 40–55 °C, which allows the primers to bind to the complementary DNA and, at a temperature of about 70 °C, the polymerase reads along the DNA strand and creates new copies. Each new cycle of the PCR approximately doubles the amount of DNA obtained from the previous cycle. A PCR process can consist of 20–30 cycles. The lengths of DNA fragments that can be amplified by using the PCR method are relatively short, of the order of 10 kb (kilobase pairs).

8.3.5 DNA Cloning

DNA cloning is a technique for amplifying DNA strands by using cellular mechanisms of DNA replication. In DNA-cloning technology, the first step is the isolation of the donor and vector DNA. The donor DNA is the DNA sequence to be amplified and the vector DNA comes from a host cell which will serve as a replication machine. The vector DNA can be a plasmid (a circular DNA structure present in many prokaryote cells). Both the donor and the vector DNA are digested by the same restriction enzyme, and ends of the DNA molecule are joined by the enzyme ligase. The recombinant plasmid DNA is then inserted into a host cell and cells replication machinery is used to obtain many copies of the donor DNA.

8.3.6 Chain Termination DNA Sequencing

Chain termination DNA sequencing [249], which has led to automatic DNA sequencers, uses the dideoxy nucleotides ddATP, ddCTP, ddGTP, and ddTTP to stop DNA chain reaction at positions corresponding to the complementary base. Dideoxy nucleotides, (see the bottom of Fig. 8.11), are analogous of normal deoxy nucleotides dATP, dCTP, dGTP, and dTTP (top of Fig. 8.11), which are used to build DNA molecules in the replication process. In contrast to normal (deoxy) nucleotides, dideoxy nucleotides, when inserted at the end of a DNA strand, cannot accept anther DNA element, which results in termination of the replication process. If a limited amount of ddATP (dideoxy adenosine triphosphoran, A) is added to a solution where a DNA replication reaction takes place, it will result in events of termination of DNA chain

8.3 Genome Sequencing   227

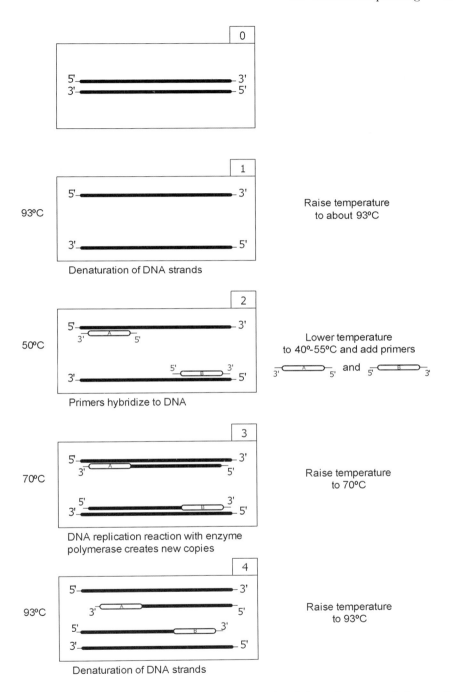

Fig. 8.10. Schematic representation of one cycle of the polymerase chain reaction. At 96 °C double strand DNA denaturates (two strands separate). At about 40-55 °C (this temperature depends on the length of the DNA primers) the primer strands bind to DNA. At about 70 °C the polymerase starts replicating the DNA. Cycles 1, 2, 3, and 4 are repeated 20–30 times, which leads to amplification of the initial amount of the DNA strand between the primers A and B, by a factor of order $10^6 - 10^8$

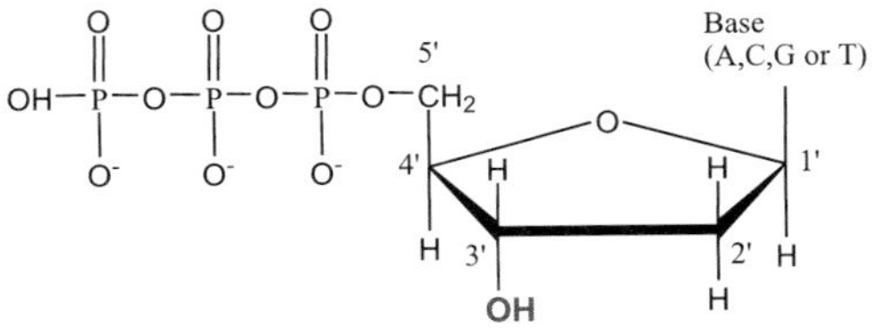

228   8 Genomics

Normal triphosphorate deoxynucleotide (2′ deoxynucleotide)

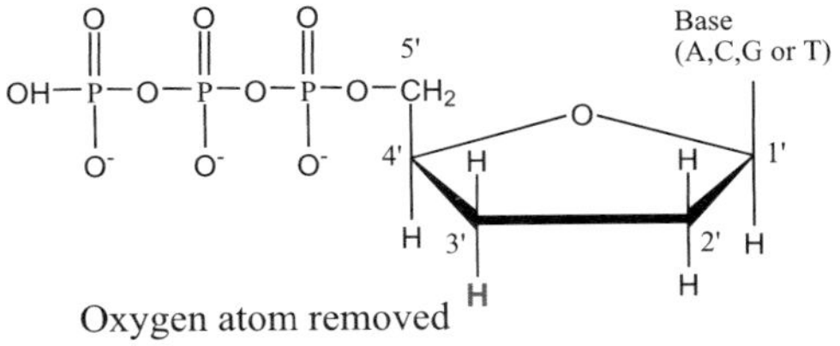

Oxygen atom removed
from OH group

Triphosphorate dideoxynucleotide (2′ 3′ dideoxynucleotide)

Fig. 8.11. Dideoxy nucleotides (*bottom*) versus deoxy nucleotides (*top*). If a dideoxy nucleotide binds to DNA at the end of the strand, then the replication reaction terminates

building at positions corresponding to its complement thymine (T). The electrophoretic gel technique for separation by size will then allow the lengths of T-terminated DNA fragments to be measured, in other words, reading of the positions of all letters T in the DNA sequence. Since this can be done for each of the bases, then the full contents of the DNA can be determined by reading the lengths of all terminated DNA strands. A subsequent improvement in automatic sequencing technology was the use of capillary electrophoresis and a detection method using a laser beam, to replace slab gel separation and reading.

8.3.7 Genome Shotgun Sequencing

Automatic DNA sequencers allow high-throughput reading of fragments of DNA sequences with lengths of several hundred base pairs. The lengths of the genomes of organisms are much longer. Therefore, the basic strategy used in genome sequencing, called shotgun sequencing, involves assembling long DNA sequences from short overlapping pieces. In shotgun sequencing, large numbers of overlapping DNA fragments several hundred base pairs long are read randomly from the basic DNA strand and their sequences are recorded. Then on the basis of the overlaps between the reads, the whole DNA sequence is reconstructed. This strategy is illustrated in Fig. 8.12. We have assumed

8.3 Genome Sequencing   229

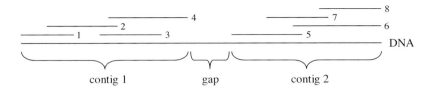

Fig. 8.12. Illustration of the shotgun method sequencing of a DNA strand

here that eight short DNA sequences, numbered 1–8, were read from the basic DNA strand; these are depicted above the basic DNA strand. The reads overlap, which makes reconstruction of the underlying sequence possible. A union of overlapping fragments is called a contig. In Fig. 8.12, there are two contigs. A DNA strand not covered by any of the reads is called a gap.

The complete DNA sequences of many organisms have been reconstructed. The most famous projects were two projects associated with sequencing of the human genome, by the Human Genome Consortium [126] and Celera Genomics [276]. Both were completed at almost the same time, in early 2001. The two projects took different approaches. The Human Genome Consortium divided the whole human genome (about 3 billion base pairs) into fragments of several hundred thousand base pairs each. The fragments were identified by fingerprinting with restriction enzymes and their positions in the genome were established by the use of sequence-tagged-sites (STSs) of known location. These fragments were cloned and distributed among participating laboratories, where they were sequenced by the shotgun method. Celera Genomics took an approach called the whole-genome shotgun (WGS) method. As explained by the name, they omitted the tedious phase of partitioning of the genome into smaller fragments, and assembled the whole genome from a very large collection of shotgun reads. Today as more genomes are being sequenced, it seems that the WGS approach is acquiring an advantage, because it reduces laborious experimental work and shifts the workload towards informatic side.

Appropriate bioinformatic tools and hardware must be used to store and organize the information obtained in sequencing projects for different organisms and appropriate mathematical methods are associated with the following tasks typically related to biomolecular and biological studies: (1) assembling DNA from reads, (2) looking for structure in and inferring information from a genomic database, and (3) comparing two or more DNA sequences or comparing a DNA sequence against a database. Projects may involve these problems or combinations of them. In our subsequent presentation, we cover some of these issues. We shall call task (1) the genome assembly, task (2) the genome annotation problem, and task (3) DNA alignment. Genome assembly and genome annotation are covered in the remaining part of this chapter; while DNA alignment was discussed in the Chap. 6, where problems involving aligning DNA, RNA, and amino acid sequences were treated together, owing to parallel techniques involved.

8.4 Genome Assembly Algorithms

Erroneous reads, false overlap detection, incomplete coverage, the non-homogeneous quality of reads, repetitive DNA structure, DNA polymorphisms leading to different reads from homologous chromosomes, the unknown orientation of DNA sequences, and, importantly, the enormous size of genomes make DNA assembly difficult and complicated. The quality of genome assembly depends to a large extent on the structure of the genomic sequence, notably signals such as repeats, polymorphisms, and nucleotide asymmetry, as well as structural motifs such as gene promoters, enhancers and suppressors, transcription-factor-binding sites, exon/intron splice junctions, and regions of homology between sequences. The existing approaches to DNA sequence assembly, for example using Atlas Genome Assembly [117], Arachne [20], Celera Assembler [203], Jazz [62], Phusion [201], PCAP [125], and Euler, [224] are multilevel and multistage and involve a substantial amount of heuristics. The assembly phases, such as the overlap phase, layout phase, and consensus phase [284], may need to be repeated or corrected several times. Genome assemblies can also use additional information that comes from molecular biology laboratories, such as PCR gap closure experiments, double-barrel data, and transposon-mapped sequencing, to improve the quality of reconstruction. The stage of development in this field is rather dynamic in the sense that, depending on the software-engineering aspects, different DNA assembly software packages can be used with different organisms and different types and sizes of data.

In the following we describe some mathematical problems that arise in DNA fragment assembly, as well as some ideas and algorithms that prove useful in their solution.

8.4.1 Growing Contigs from Fragments

How can the DNA strand in Fig. 8.12 be reconstructed from reads 1–8? The idea is to grow overlaps into contigs, as illustrated in Fig. 8.13. By starting from any read, for example, 1 and then adding reads $2, 3$, and 4, we obtain a contig including all reads. When an overlapping substring for two sequences has been computed (or estimated), these sequences can easily be merged. This stresses the need for developing fast and reliable methods for detecting overlaps between reads.

8.4.2 Detection of Overlaps Between Reads

Overlap detection can be done by several of the techniques described in previous chapters of this book. A fast and robust approach would be the following

(1) Apply the method of writing occurrences of l-mers in reads (the number l will often be $20 - 25$) in a hash accumulator array, recording the labels

8.4 Genome Assembly Algorithms   231

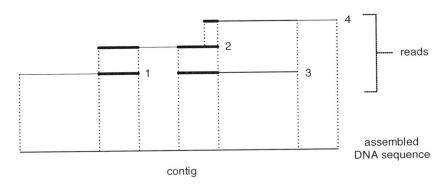

Fig. 8.13. Illustration of the idea of growing a contig from fragments of DNA. DNA reads are merged, on the basis of the detected structure of their overlaps

of the reads from which the l-mers are taken. Then, detect overlaps, as presented in Sect. 3.6 by searching through the hash accumulator array for entries marked with two (or more) read labels.

(2) Establish the exact overlapping substring by using methods of sequence alignment (see Chaps. 5 and 6). If two reads show enough similarity, then the overlap relation between them is established. By setting proper threshold values on the length of overlapping fragments, we can reduce the probability of false overlap detection. Overlap detection by sequence alignment allows one to establish the direction of the arrow showing the direction of contig growth by the overlap of the reads.

Overlaps between DNA fragments can also be established by using fingerprinting with restriction enzymes. [50, 264] If two overlapping DNA fragments undergo digestion by the same restriction enzymes, the patterns of the band lengths obtained (restriction enzyme fingerprints) should show similarity and can be used to establish the overlap relation between these fragments. The following quantitative result that allows us to estimate the probability of false overlap detection was derived in [270]. Let us assume that two *nonoverlapping* DNA fragments X and Y have been digested by the same restriction enzyme(s), which resulted in X being cut into n_L bands and Y being cut into n_H bands. What is the probability that the lengths of the bands in X and Y will exhibit a similar pattern as a result of random events? Assume $n_L < n_H$. The lengths of the bands in X and Y are compared, with some tolerance tol, i.e., they are considered equal if the difference between their lengths is $\leq tol$. The band lengths and the tolerance tol are measured as numbers of base pairs. We pick out one of bands in X. The probability that one of the bands in Y will accidentally be equal in length to that one can be approximated by the value $b = 2tol/\text{gelLength}$. The value of gelLength must again be scaled in terms of numbers of base pairs and serves as an estimate for the length of the shortest band. The probability that none of the n_H bands in Y will match the band picked out is then approximately equal to $p = (1-b)^{n_H}$. Finally, by the binomial formula, the probability of observing M or more matches when comparing X and Y is

$$\sum_{m=M}^{n_H} \binom{n_L}{m} (1-p)^m p^{n_L - m}. \tag{8.1}$$

If the above expression has a very small value for the observed number of matches M, the hypothesis that X and Y match by random coincidence can be safely rejected. Since b is an upper bound rather than the true probability of random match of two band lengths, (8.1) gives an upper bound for the true probability of observing M or more matches.

Overlap detection by restriction enzymes is direction-blind. One cannot decide which of the overlapping fragments X or Y comes first in the DNA strand. However, using some additional data on the structure of the genome to organize the estimated overlap graph, such as STSs, short regions in the DNA with known sequences and known positions, one can obtain a reasonable path for the overlapping DNA fragments and then sequence these fragments to get an estimate of the whole DNA sequence [52, 145].

Overlap detection by restriction enzymes is a basis for some methods named "map first, sequence later" (e.g., [142, 202]), where the (approximate) overlap structure is estimated first using restriction enzyme fingerprinting, and then (selected) reads are sequenced on the basis of the result of the first step. These methods use algorithms to obtain minimal spanning trees (forests) of the overlap graph. With some knowledge about the structure of the genome to be sequenced, such as which reads belong to the beginning and which to the end of the DNA strand, these methods allow one to find a reasonable path through the overlap graph and reconstruct the DNA by sequencing only reads that belong to that path.

8.4.3 Repetitive Structure of DNA

The reconstruction problem as presented above may seem fairly straightforward to solve by partially heuristic contig-growing algorithms, with possibly some difficulties due to erroneous reads and false overlap detection. Multiple coverage of the DNA can help in rejecting misassembled fragments due to errors and false overlap structure. However, some of the simple approaches to contig growth described above will instead fail for many examples of real DNA data, owing to the repetitive structure of DNA. Repetitive elements are among the most frequent features of a genome. Many of them display complicated evolutionary dynamics (for some models, see, e.g., [146]). There may be many repeats in the sequence [232], such that many parts of the sequence appear in it twice or more. The repetitive structure of the genome may cause overlaps to be detected between two DNA reads despite the fact that the

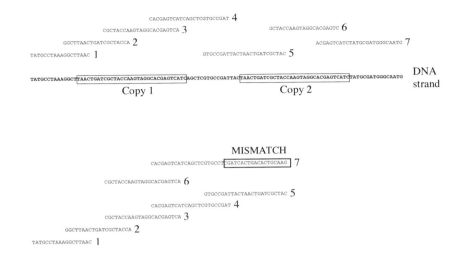

Fig. 8.14. Assembly of a DNA strand with repetitive structure. The true DNA strand contains two identical copies of sequences, marked by rectangular boxes, named copy 1 and copy 2. In the *upper par* of the figure this DNA strand is covered by seven reads, 1, 2, ..., 7. The *lower part* shows the process of DNA assembly by a heuristic method of growing a contig from reads. Reads are added in the order 1, 2, ..., 7. Note that reads 6 and 7 are positioned incorrectly, owing to the repetitive structure of the DNA strand, and that when read 7 is added a mismatch occurs

physical positions of these two reads are far away from one another. Let us assume the DNA strand structure shown in Fig. 8.14. There are two identical parts in this DNA, depicted by rectangular blocks and named copy 1 and copy 2. The coverage of this DNA strand by reads, numbered 1–7 is shown above the DNA strand. We can see that the sequence of overlap unions, resulting from merging successively reads 1, ..., 7, will lead to an erroneous structure of the contig with a "collapse" of two repetitive copies into one.

8.4.4 The Shortest Superstring Problem

One mathematical model of the problem of growing contigs from DNA reads is that of looking for the shortest superstring for a collection of strings (reads) [111, 142, 143, 260] (SSP problem) mentioned already in Sect. 5.3. SSP problem is NP-complete , but there are approximate approaches which allow one to find suboptimal solutions in polynomial time [271, 274]. Given the set of reads in Fig. 8.14,

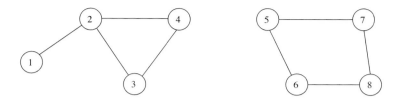

Fig. 8.15. The connectivity graph of the overlaps of the DNA reads corresponding to the DNA strand coverage in Fig. 8.12

$$s_1 : TATGCCTAAAGGCTTAAC$$
$$s_2 : GGCTTAACTGATCGCTACCA$$
$$s_3 : CGCTACCAAGTAGGCACGAGTCA$$
$$s_4 : CACGAGTCATCAGCTCGTGCCGAT$$
$$s_5 : GTGCCGATTACTAACTGATCGCTAC$$
$$s_6 : GCTACCAAGTAGGCACGAGTC$$
$$s_7 : ACGAGTCATCTATGCGATGGGCAATG \qquad (8.2)$$

whose overlaps can also be seen in Fig. 8.14 the solution to the SSP problem is

$$s = TATGCCTAAAGGCTTAACTGATCGCTACCAAGTAGGCA$$
$$CGAGTCATCAGCTCGTGCCGATTACTAACTGATCGCTA$$
$$CCAAGTAGGCACGAGTCATCTATGCGATGGGCAATG,$$

which indeed reconstructs the true underlying DNA sequence.

8.4.5 Overlap Graphs and the Hamiltonian Path Problem

The coverage of a DNA sequence by reads, for example reads 1–8 shown in Fig. 8.12, can be represented by an overlap graph with reads represented by vertices (nodes). Two vertices are connected by (undirected) edges if their corresponding reads overlap. An undirected graph representing the overlap structure in Fig. 8.12 has the topology shown in Fig. 8.15. It has a structure of two disjoint parts owing to the existence of a gap between two contigs. Several heuristic algorithms that enable reconstruction of a DNA sequence are based on the undirected overlap graph of the reads.

Let us now describe a more informative method, of defining a directed overlap graph. Again, the reads (sequences) s_i are the vertices of an overlap graph. If two reads s_1 and s_2 overlap, then they are connected by a directed edge (an arrow). The direction of the arrow is determined by the standard of

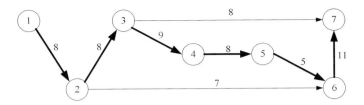

Fig. 8.16. Directed, weighted overlap graph corresponding to the DNA coverage in Fig. 8.14. Formulation of DNA assembly as a Hamiltonian path problem

reading DNA from left to right. Moreover, we equip arrows with weights. The weight of the arrow connecting two reads is the length of their overlapping part. The weighted, directed overlap graph for the coverage in the upper part of Fig. 8.14 is presented in Fig. 8.16. With the use of weighted, directed overlap graphs corresponding to DNA coverages, we can formulate the problem of assembling a DNA sequence as a version of the Hamilton path problem, [224, 244, 260, 290], presented in Sect. 5.3: find a maximal-total-weight path which visits all vertices of the graph, each exactly once. The Hamilton path problem is NP-complete, but there are many suboptimal methods which can give us an approximate solution in polynomial time [8, 272]. For the graph in Fig. 8.16, there is exactly one directed path which visits all nodes; $s_1 \rightarrow s_2 \rightarrow s_3 \rightarrow s_4 \rightarrow s_5 \rightarrow s_6 \rightarrow s_7$, marked in bold. Here, weighting the arrows by the overlap length does not lead to any change in the solution. However, weighting edges can be useful in the situation where overlaps are detected with errors, and then longer overlaps are more reliable than shorter ones.

8.4.6 Sequencing by Hybridization

Hybridization is a popular technique in biological/genetic laboratories. It involves labeling or dyeing biologically active agents (molecules) with radioactive atom isotopes or colored dyes, and then using these agents in some reaction or experiment. Labeled molecules can be monitored in the sense that their spatial positions or the number of labeled molecules bounded to the sample analyzed can be estimated, which can help in drawing conclusions concerning the structure or properties of the species under study, for example, the sequence of bases in a DNA strands. The idea of sequencing by hybridization [13, 128] is to construct a labeled set of probes and present them to the (single-strand) DNA molecule to be analyzed. This idea is presented in Fig. 8.17 where the probes are all 64 trinucleotides AAA, AAC, ..., TTT. The situation where a probe binds to the (complementary) DNA molecule can be detected and the conclusion is that a particular trinucleotide t_k is a substring in the DNA string.

TTAGCTTAAA

AAA	AAC	AAG	AAT	ACA	ACC	ACG	ACT
AGA	AGC	AGG	AGT	ATA	ATC	ATG	ATT
CAA	CAC	CAG	CAT	CCA	CCC	CCG	CCT
CGA	CGC	CGG	CGT	CTA	CTC	CTG	CTT
GAA	GAC	GAG	GAT	GCA	GCC	GCG	GCT
GGA	GGC	GGG	GGT	GTA	GTC	GTG	GTT
TAA	TAC	TAG	TAT	TCA	TCC	TCG	TCT
TGA	TGC	TGG	TGT	TTA	TTC	TTG	TTT

AAA	AAC	AAG	AAT	ACA	ACC	ACG	ACT
AGA	AGC	AGG	AGT	ATA	ATC	ATG	ATT
CAA	CAC	CAG	CAT	CCA	CCC	CCG	CCT
CGA	CGC	CGG	CGT	CTA	CTC	CTG	CTT
GAA	GAC	GAG	GAT	GCA	GCC	GCG	GCT
GGA	GGC	GGG	GGT	GTA	GTC	GTG	GTT
TAA	TAC	TAG	TAT	TCA	TCC	TCG	TCT
TGA	TGC	TGG	TGT	TTA	TTC	TTG	TTT

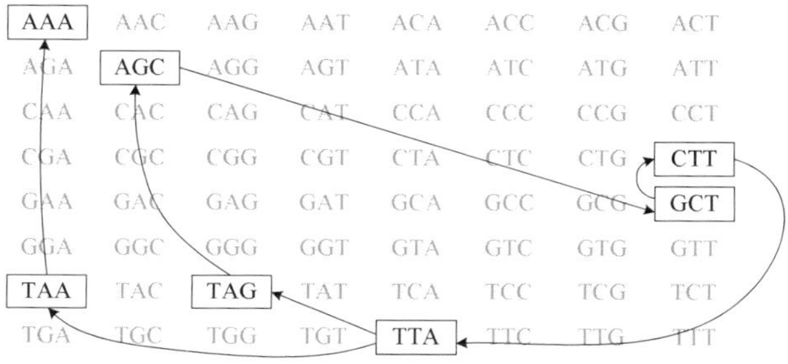

Fig. 8.17. Illustration of the idea of sequencing by hybridization. Assume that we are analyzing a DNA sequence complementary to the one written at the top, $TTAGCTTAAA$. The set of DNA probes is the array of all trinucleotides $AAA, \ldots, TTT$. In the *upper part*, rectangular frames show the trinucleotides which appear in the sequence $TTAGCTTAAA$. The sequence $TTAGCCTGAA$ can be reconstructed by using the suffix = prefix relation between DNA triples (see text for explanation). A graph depicting the suffix = prefix relations between the trinucleotides is shown in the *lower part*

Let us introduce the relation suffix = prefix. The prefix of a trinucleotide is the string of two letters at the beginning, for example, prefix(atg) = at. Analogously, the suffix of a trinucleotide is the string of two nucleotides (the dinucleotide) at the end, for example, suffix(atg) = tg. Two trinucleotides t_1 and t_2 are suffix = prefix related if suffix(t_2) = prefix(t_1). This relation is denoted by an arrow, i.e., $t_1 \rightarrow t_2$. An example of two trinucleotides that are suffix = prefix related is $ATG \rightarrow TGG$. A graph where the nodes are trinucleotides detected in the DNA strand and the arrows mark prefix-suffix relations between them, is given at the bottom of Fig. 8.17.

The true DNA sequence $TTAGCTTAAA$ can be reconstructed from the graph in the bottom of Fig. 8.17 by solving the *Euler path problem*: find a

8.4 Genome Assembly Algorithms

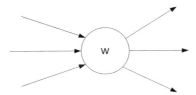

Fig. 8.18. Euler local balance condition. The number of incoming arrows (indegree(w)) must be equal to the number of outgoing arrows (outdegree(w)). In the above example, indegree(w) = outdegree(w) = 3

path that traverses all arrows of the graph, each exactly once. The Euler path problem can be solved very efficiently in linear time with respect to the number of nodes and arrows [244]. The existence of an Euler path can be verified by the local balance criterion: for each node w, except two terminal nodes, the number of incoming arrows (called indegree(w)) must be equal to the number of outgoing arrows (called outdegree(w)), i.e.,

$$\text{indegree}(w) = \text{outdegree}(w) \tag{8.3}$$

as shown in Fig. 8.18. The terminal nodes must satisfy

$$\text{indegree}(w) = \text{outdegree}(w) - 1 \tag{8.4}$$

at the beginning node and

$$\text{indegree}(w) = \text{outdegree}(w) + 1 \tag{8.5}$$

at the end node. A graph which satisfies (8.3)–(8.5) is called an Eulerian graph. The condition (8.3) is of course necessary for the existence of an Eulerian path, since every time a path arrives a node it must find an edge by which to leave it. Euler proved that (8.3) is also sufficient for the existence of an Eulerian path. The graph in Fig. 8.17 clearly satisfies (8.3)–(8.5). Using (8.4) and (8.5), we can identify TTA as the beginning node and AAA as the end node, we can easily find the Eulerian path

$$TTA \rightarrow TAG \rightarrow AGC \rightarrow GCT \rightarrow CTT \rightarrow TTA \rightarrow TAA \rightarrow AAA,$$

and the true sequence $TTAGCTTAAA$ can be reconstructed by traversing the graph and successively merging trinucleotides related by suffix = prefix.

It was possible to reconstruct the sequence of bases from the graph in Fig. 8.17 because the DNA strand was short enough and triplets did not appear too many times. For long DNA strands, it is most probable that all triplets will appear (each many times), leaving no possibility to assemble the true sequence. More precisely, the number of possible solutions would be very large and the true solution would be buried among them. One can imagine using longer strings as probes, but the exponential increase in their number would make the idea of presenting them to the DNA molecule, in an experimental framework impractical.

8.4.7 De Bruijn Graphs

A De Bruijn graph is a graph whose nodes are labeled by words (strings) over some alphabet and whose edges indicate some relations between the strings in nodes (e.g., those strings which overlap). The graph in Fig. 8.17 is a De Bruijn graph, since its nodes are labeled by trinucleotides and its edges connect suffix = prefix-related nodes. We have not labeled the edges of the graph in the lower part of Fig. 8.17, but it would be natural to use the dinucleotides shared by neighboring nodes for this purpose, as shown below:

$$AGC \xrightarrow{GC} GCT.$$

There are also other conventions for representing relations between strings as nodes and edges in De Bruijn graphs; for example, the edges between nodes corresponding to trinucleotides can be labeled by tetranucleotides, as shown below:

$$AGC \xrightarrow{AGCT} GCT.$$

8.4.8 All l-mers in the Reads.

As noted, the basic shortcoming of sequencing by hybridization is that if the probes to be hybridized to the DNA strand are short, then it is most likely that all of them will appear in the DNA strand many times, which will make the problem of reconstruction of the DNA sequence unsolvable. If we increase the length of the probes, aiming at higher specificity, then the size of the library of the probes will soon exceed the capacity of any computer system. Statistical evaluations show that DNA sequencing by hybridization with the set of probes given by all 8-tuples of A, C, T, G would allow reconstructing DNA strings of a length of order of 200 base pairs [219], which is far below the necessary efficiency. However, the idea of sequencing by hybridization can be modified in the following way. Let us try to collect all strings of length l, called all l-mers (or l-tuples), in the DNA strand. This is of course impossible, because we do not know in advance the sequence to be assembled. But instead of collecting l-mers from the underlying DNA, which is not available, we can obtain l-mers from the base sequences in the reads obtained from the DNA strand, which are given. Under the assumption of complete coverage and error-free reads, the set of all distinct l-mers in the reads is equal to the set of l-mers in the underlying DNA sequence. This set is also coverage-independent. If the length of the sequence to be assembled is G, then the number of distinct l-mers in this sequence is $N \leq G - l$. It can be seen that since there is no need to record all possible A, C, T, G l-mers, but rather only those which appear in the DNA strand, the problem becomes tractable even for large l.

8.4.9 The Euler Superpath Problem

The practical application of the above idea of using all l-mers in the reads for DNA assembly is contained in the Euler path algorithm developed in [224, 223]. We shall illustrate the approach by using the coverage of the DNA strand shown in Fig. 8.14 with strings $s_1, \ldots, s_7$ given in (8.2). Analyzing this example will help to explain how the effect of repeated structure in the DNA is taken into account in the assembly algorithms. The first step of the algorithm is to collect all l-mers in the DNA strand. In our evaluation of this example, let us assume that $l = 10$. For example, all 10-mers in the read s_1, $TATGCCTAAAGGCTTAAC$, are

$$TATGCCTAAA,$$
$$ATGCCTAAAG,$$
$$TGCCTAAAGG,$$
$$GCCTAAAGGC,$$
$$CCTAAAGGCT,$$
$$CTAAAGGCTT,$$
$$TAAAGGCTTA,$$
$$AAAGGCTTAA,$$
$$AAGGCTTAAC.$$

Since the length of the DNA strand in Fig. 8.14 is $G = 100$, there are no more than $N = G - l = 90$ different 10-mers. The actual number is $N = 74$, owing to the presence of two identical copies (copy1 and copy2) in the DNA. We recall that the set of all distinct l-mers is coverage-independent (assuming complete coverage and error-free reads).

In analogy to the previous treatment let us define the prefix of an l-mer w as the string of $l - 1$ bases at the beginning of w and the suffix an l-mer w as the string of $l - 1$ bases at the end. Two l-mers w_1 and w_2 are prefix–suffix related, or $w_1 \to w_2$, if prefix(w_2) = suffix(w_1). A graph with nodes given by l-mers in Fig. 8.14 and arrows showing their prefix-suffix relations is presented in Fig. 8.19. The number of nodes in this graph is lower than the actual number of different l-mers (74) in the DNA sequence in Fig. 8.14, for the sake of clarity. Also, owing to insufficient space, l-mers are not marked by numbers. Observe that the two copies copy 1 and copy 2 in the DNA strand now collapsed into one object. Nevertheless, it is still possible to reconstruct correctly the DNA string by following arrows in the graph. Some arrows will be traversed two times. Mathematically, the problem of reconstructing the DNA from the graph in Fig. 8.19 can be posed as "Find the minimal path in the graph in Fig. 8.19 which contains all arrows". This problem is sometimes called the Chinese postman problem [244] and it is a version of the Euler path problem. For the graph shown in Fig. 8.19, condition (8.3) is not satisfied since there are edges which must be visited twice so that the path traverses

240   8 Genomics

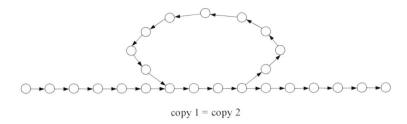

copy 1 = copy 2

Fig. 8.19. A graph of the l-mers in the upper plot of Fig. 8.14 with arrows showing prefix–suffix relations. Owing to the repetitive structure of the DNA, this graph contains a loop. The number of nodes in this graph is lower than the actual number of different l-mers (74) in DNA sequence in Fig. 8.14, for the sake of clarity

the whole graph. However, it is possible to introduce a multiplicity of edges (arrows), such that the local balances hold and an Eulerian path for this graph exists.

When analyzing practical problems, which are much more complicated than the one shown in Fig. 8.19, the authors of this method aimed at splitting the problem of the analysis of a large graph into a set of problems, each involving a smaller part of the graph. This led to the formulation of the *Eulerian superpath problem* [224, 222]: "Given an Eulerian graph and a collection of paths in this graph, find an Eulerian path in this graph that contains all these paths as subpaths".

By solving Eulerian path or Eulerian superpath problem, a repetitive structure of a DNA sequence can be deciphered and used to build a correct sequence of fragments in DNA. In [224] the linear time solution of the Eulerian Superpath Problem is presented.

8.4.10 Further Aspects of DNA Assembly Algorithms

What we have described above is some ideas about how mathematical algorithms can be applied to DNA assembly and which methods can be used. Heuristic methods of contig growth, run in low computational time, with respect to the DNA length. The Shortest superstring and Hamilton path problems are both NP-complete. The Euler path and Euler superpath problems, for the case of perfect reads, can be solved in linear time with respect to the size of the problem. However for data with erroneous reads the problem becomes NP-hard. In practical computations, NP-complete or NP-hard problems are always solved by using suboptimal algorithms with linear (polynomial) time, but with unpredicted output. The quality of the assembly process is, to some extent, a compromise between the computational complexity of algorithms and the available time.

Several assumptions must be satisfied to make the ideas of DNA assembly applicable. The structure of the DNA strand must have enough variety to allow mapping reads to their locations and detecting overlaps between reads. Also, the reads must be of satisfactory quality. The application of a technique for sequence assembly to data from some experiment will depend on the experimental aspects of how the experiment was planned and carried out. Also, important parts of algorithms, such as error correction and DNA comparison, can be realized in several different ways, which can lead to different outcomes for the whole structure. Some elements of the algorithms presented must be developed in more detail to make the algorithms work properly. The most important operation to be added is error correction, since errors are always present. Error (data inconsistency) detection and correction procedures should precede DNA reconstruction. In the following, we describe some more detailed aspects of the construction of DNA assembly algorithms in connection with the above remarks.

Error Correction

In what we have presented up to now we have assumed error free reads from the DNA strand, which in real data is never true. There are always errors in reads, which most often lead to predicting wrong bases at particular positions. Let us assume, in the case of a DNA coverage such as that shown in Fig. 8.12, that an isolated error appeared in the reads. This situation is shown in the upper part of Fig. 8.20, where the error (a single base change in the read) is marked by a black dot. The prefix–suffix overlap graph for the l-mers corresponding to the erroneous data is presented in the lower part of Fig. 8.20. We have a "bubble" resulting from the erroneous read, and now an Euler path that traverses all edges of the graph does not exist. Generally, the presence of errors in reads results in the appearance of bubbles and forks in the connectivity graph of l-mers.

From the above, we see that an error correction step should be applied to the data. We shall describe two approaches to error correction, or the problem of enforcing data consistency. In general, error correction methods rely on comparing the reads and obtaining information from the data, such as how many reads does an l-mer belong to, or which l-mers belong to a specified read, which is performed by the use of algorithms described in Chap. 3.

Error correction by solving the spectral alignment problem. This approach was developed in [225, 219]. Let us denote by Λ the set of all distinct l-mers obtained from the reads. Let us call any collection of l-mers from Λ, in other words, any subset of Λ, a *spectrum*. An l-mer is called *solid* if it belongs to at least M reads, where the parameter M is a predefined threshold for the algorithm. If an l-mer is not solid, it is called *weak*. We define the solid spectrum T as the set of all solid l-mers in Λ. A string s is called a T string if all its l-mers belong to T. The spectral alignment problem is "For a given string (read), find the minimum number of mutations that transform it into a T string". This problem can be solved by application of a dynamic programming method. The heuristic idea behind this method is that if a read contains an

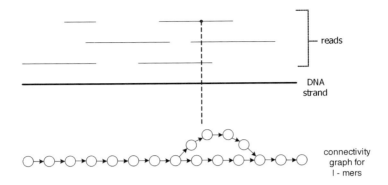

Fig. 8.20. *Top*: coverage of a DNA strand by reads, with one erroneous read containing a single base change marked by a black dot. *Bottom*: erroneous read results in a "bubble" appearing in the connectivity graph of l-mers

isolated mutation, then a family of weak l-mers appear in Λ. If all l-mers in the read are solid, then this read most probably does not contain an isolated mutation. Transforming reads into a T strings is performed successively until all l-mers and all reads are solid.

Error correction by searching for "orphans". Another approach [225], which has been proven to be efficient, uses the notion of "orphans" in the set of l-mers and pursues at the same time both error elimination and enforcement of data consistency. We denote again the set of all l-mers by Λ and the set of all reads by S. Suppose that one read has one isolated error (a mutation). How many l-mers in Λ will have this mutation? If the position of this mutation is at a distance $\geq l$ from the beginning and from the end of the read, then the number of l-mers containing this mutation is l. This is explained in Fig. 8.21 (where l is assumed equal to 10). For the string s_i, there are l ($l = 10$) l-mers covering the mutation. If the position of the mutation is at a distance $d < l$ from the beginning or end of the read s_i, then the number of l-mers containing the mutation is d. In order to make use of this observation, let us introduce some definitions. The *multiplicity* of an l-mer w, denoted by $m(w)$, is defined as the number of reads in S that contain this l-mer. Two l-mers w and v are called *neighbors* if they differ only at one base position (by one mutation). Finally, an l-mer w is called an *orphan* if (i) it has a small multiplicity $m(w) \leq M$, lower than a given threshold M, and (ii) it has only one neighbor v and $m(w) < m(v)$. The algorithm proceeds by screening the data read after read. For each read, all orphans are found on the basis of (i) and (ii) and an association between orphans and mutations in the read is established. If a mutation in the read is related to l orphans, it is removed by replacing w by v. After correcting all mutations related to l orphans, in the next stage, mutations close to the boundaries of the reads are removed on the basis of orphans for which $d < l$. There is a predefined

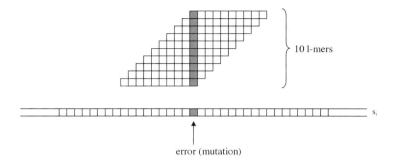

Fig. 8.21. Diagram plot explaining the number of l-mers containing an isolated error. Here we have assumed $l = 10$

limit of at most Δ corrections to be made in each of the reads. The condition which makes orphan removal procedure an efficient tool for error correction is that an orphan can be neighbor to only one other l-mer. Fulfillment of this condition makes error correction fully unambiguous.

Reads of Unknown Orientation

In shotgun sequencing experiments, DNA reads are copied from double-stranded DNA and it is not known whether a given read s_i comes from the 3'–5' or the 5'–3' DNA strand. There are two possible ways of dealing with this fact. The first possibility is to modify the definition of the overlap relation for reads, and to create appropriate algorithms for detection of the modified relation. Let us denote by $\bar{s}$ the reversed complement of the DNA string s (e.g., $\overline{actgtcc} = ggacagt$). We say that strings s_1 and s_2 have a forward or reverse overlap if either s_1 and s_2 or s_1 and $\bar{s}_2$ overlap. We then incorporate these forward and reverse overlap as an element of algorithms for DNA assembly. Another, probably simpler method to accommodate unknown orientation data is to augment the data strings with their reversed complements, i.e., to define
$$\check{S} = \{s_1, s_2, ..., s_n, \bar{s}_1, \bar{s}_2, ..., \bar{s}_n\}.$$
Now, in our data, we have two copies of every read, one in the forward and one in the backward direction, which corresponds to the real DNA structure. We can expect the DNA assembly algorithm to reconstruct two (disjoint) complementary DNA strands.

8.5 Statistics of the Genome Coverage

In this section, we focus on the statistical aspects of genome coverage and of the assembly algorithms presented in the previous section. A study on the statistical properties of the processes involved in genome assembly is crucial to making appropriate decisions concerning the lengths of reads to be used, the desired coverage folds etc. It is also of great importance in discovering the repetitive structure and estimating the heterozygosities of whole genomes [76, 163, 177, 220, 281].

The genome coverage for the shotgun strategy can be modeled as a binomial/Poisson stochastic process. From the properties of this process, it is possible to derive the statistics of the contig size, and in this way to determine the coverage needed to achieve an assembly of the desired quality. Analogously, in the case of probing reads with l-mers, it is possible to estimate the structure and size of the genomic sequence, even when sequence repeats are involved. This is accomplished by reconstructing the repeat structure, using a mixed Poisson distribution to model it and an expectation maximization algorithm to estimate parameters of the mixture. In the framework of this theory, it is also possible to estimate the total gap length and the stringency ratio.

Below, we provide an account of these topics. We also cover some other more specialized signals in sequences such as polymorphisms and nucleotide asymmetry.

8.5.1 Contigs, Gaps and Anchored Contigs

If the biological details are omitted, shotgun genome sequencing can be reduced to assembly of a sequence of total length G, from N reads (also called fragments) of equal length L. The fragments are randomly selected from the sequence G, and the assembly is feasible if there exists enough overlap between the fragments. To ensure this, the coverage $a = NL/G$ has to be greater than 1. Depending on the strategy used, G may represent the entire genome (in the WGS method) or else a subset of the genome. In Fig. 8.22, we present a coverage of a DNA sequence of length G by shotgun sequences of equal length L. Recall that a union of overlapping fragments is called a contig, and fragments of a DNA strand not covered by any reads are called gaps. In Fig. 8.22 there are three contigs, depicted below the line representing the DNA sequence. Of course the most desirable situation would be that one contig covers the whole DNA fragment to be analyzed and that there are no gaps.

One model for the random fragments is a binomial/Poisson stochastic point process, in which the coordinates of the left ends of the fragments are independent random variables uniformly distributed over G. Neglecting boundary effects, we obtain the result that the probability of the random event that the there are k fragments with left ends in the interval

$$(x, x - h)$$

has a distribution binomial$(N, h/G)$, or approximately Poisson(Nh/G). In order to obtain good quality of assembly, it is necessary that the contigs cover as much of G as possible. The number of contigs is equal to the number of

8.5 Statistics of the Genome Coverage

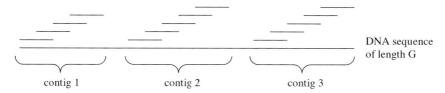

Fig. 8.22. Shotgun coverage of a DNA sequence of length G by clones of equal length L. Three contigs (unions of overlapping sequences) are depicted below the line representing the DNA strand

their rightmost fragments. Since it is easy to compute the probability that the fragment is rightmost (does not include the left end of any other fragment), one can readily obtain the following expression for the mean number of contigs

$$\begin{aligned} E[\#\text{contigs}] &= N \times \Pr[\text{fragment is rightmost in a contig}] \\ &= N \times \Pr[\text{fragment does not include the left} \\ &\qquad \text{end of any other fragment}] \\ &= N \times \exp(-NL/G) = (aG/L) \times \exp(-a). \end{aligned} \qquad (8.6)$$

Fig. 8.23 depicts the expected number of contigs, $E[\#\text{contigs}]$, as a function of the coverage a. $E[\#\text{contigs}]$ first increases but then decreases again, since smaller contigs coalesce with increasing coverage. A single (length G) contig is expected when the coverage reaches $a \approx 8$. For $a > 8$, the expected number of contigs becomes less than 1. This nonsensical outcome of (8.6) follows from the assumption that boundary effects can be neglecting neglected and calls for caution when using the expected full-coverage condition $a = 8$. If there is more than one contig then there are gaps between them. From Fig. 8.22, we have $\#\text{gaps} = \#\text{contigs} - 1$.

The formation of a contig can be considered as a point Poisson process, with intensity Nh/G, built from a sequence of left ends of shotgun fragments of equal size L, with the stopping condition: "stop when the interepoch distance is larger than L". By the interepoch distance we mean the distance between two successive left ends of fragments. Consequently, the number of fragments in a contig has geometric distribution with a parameter $\exp(-a)$, and the expected contig size, $E[S]$, can be expressed as

$$E[S] = E[\#\text{fragments} - 1]E[\text{interepoch distance}] + L$$
$$= \frac{1 - \exp(-a)}{\exp(-a)} \int_0^L x\lambda \frac{\exp(-\lambda x)}{1 - \exp(-\lambda L)} dx + L = L\frac{\exp(a) - 1}{a}.$$

$E[S]$ increases as smaller contigs coalesce. For the data in Fig. 8.23, $E[S] = G$ is expected when the coverage reaches $a \approx 8$.

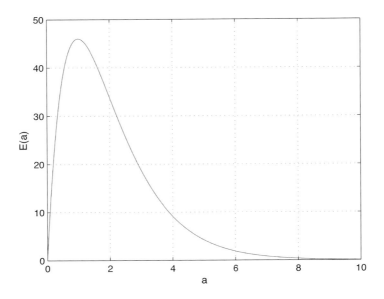

Fig. 8.23. Expected number of contigs as a function of coverage a. The parameter values are $L = 800$ and $G = 100\,000$.

8.5.2 Statistics with Minimum Overlaps Between Fragments, Anchored Contigs

The statistics of gaps (and contigs) will change if we introduce the assumption that there is some minimum overlap required between reads [163]. According to the Lander–Waterman theory [163], the expected number of gaps under this assumption is equal to

$$E(\#\text{gaps}) = N \exp[-\alpha(N-1)], \qquad (8.7)$$

where $\alpha = (L-T)/G$ is the effective fractional read (fragment) length, with T being the minimum overlap required. Consequently, the maximum number of gaps is equal to

$$E_{\max} = \exp(\alpha - 1)/\alpha. \qquad (8.8)$$

The Lander–Waterman theory also implies that the stringency σ, i.e., ratio of the number of gaps present to the expected maximum, can be expressed in terms of the effective coverage $\delta = N(L-T)/G$, as

$$\sigma = E/E_{\max} = \delta \exp(1 - \delta). \qquad (8.9)$$

In a recent paper [285] the prediction of the size of gaps in WGS projects was discusses. It was demonstrated that the above expressions underestimate the stringency (as it had been claimed earlier) and semiempirical estimates of stringency given the coverage, were provided, namely $\sigma_{emp} =$

$1.187\exp(-0.334\delta)$ for eukaryotes and $\sigma_{emp} = 0.701\exp(-0.211\delta)$ for prokaryotes.

We might require that the contigs be anchored, i.e., that each of them includes at least one anchor, which is a genomic site of known location (an element of a gene map). Let us define the coverage with M anchors as equal to $b = ML/G$ and suppose that the anchors are points which follow the binomial/Poisson process. We then have

$$E[\#\text{anchored contigs}] = Nb\frac{\exp(-a) - \exp(-b)}{b - a} \quad (8.10)$$

which reduces to the nonanchored case as $b \to \infty$, but usually is smaller.

8.5.3 Genome Length and Structure Estimation by Sampling l-mers

Let us first consider a simplified situation, illustrated in Fig. 8.24. We have a DNA sequence of unknown length G covered, by the shotgun method, by N DNA reads of equal length L. Instead of direct analysis of overlaps between fragments, we now take a different approach. We draw an l-mer randomly from the DNA strand. We denote the sequence of bases in this l-mer by w. The data which we base our analysis on, is the number of reads which the l-mer w belongs to, denoted by $x(w)$. In Fig. 8.24, we have $x(w) = 3$. We ask "What is the probability $p(w)$ that an l-mer w belongs to one, given, read?". By "l-mer w belongs to the read" we mean that both ends belong to the read. The number of bases which can hit the left end of the l-mer w when it belongs to the read is then equal to $L - l + 1$. The number of all possible choices for the left end of the l-mer w is $G - l + 1$, since again both ends must belong to a DNA sequence of length G. Therefore the expression for $p(w)$ is

$$p(w) = \frac{L - l + 1}{G - l + 1}.$$

Since there are N reads, $x(w)$, the number of reads containing w, is binomially distributed with parameters N and $(L-l+1)/(G-l+1)$, or approximately Poisson distributed with a parameter $\lambda = N(L-l+1)/(G-l+1)$. By drawing many l-mers randomly we can estimate the length G of the genome. We denote the drawn l-mers by $w_1, w_2, \ldots, w_{NW}$, where NW denotes the number of drawn l-mers. The log-likelihood function is (compare Chap. 2)

$$\text{Llik}(G) = \sum_{i=1}^{NW}\left\{-\frac{N(L-l+1)}{G-l+1} + x(w_i)\ln\left[\frac{N(L-l+1)}{G-l+1}\right] - \ln[x(w_i)!]\right\},$$

and the maximum likelihood estimate of G is

$$\hat{G} = l - 1 + \frac{N(L-l+1)}{NW}\sum_{i=1}^{NW} x(w_i). \quad (8.11)$$

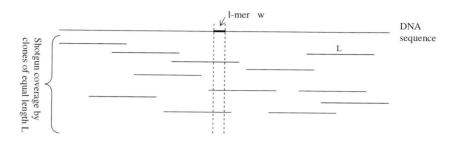

Fig. 8.24. Estimating the length and structure of a genome by sampling l-mers. The sequence of bases in an l-mer is denoted by w. The basic data assumed is how many fragments (reads) an l-mer belongs to, denoted by $x(w)$. In the figure, $x(w) = 3$

However, sampling an l-mer from the genome is impossible. Instead of sampling from the genome we sample an l-mer from the reads. In experiments where DNA is covered by N reads, we obtain an l-mer by selecting randomly one of N fragments of the shotgun coverage and drawing the l-mer from this fragment. This scenario makes the estimate (8.11) incorrect. There are two aspects which must be taken into account. The first is that, owing to sampling from reads, l-mers w with $x(w) = 0$ will never occur, which will lead to bias in the estimate. The second aspect is related to the repetitive elements in the genome G. If the genome G contains repetitive copies then the distribution of numbers $x(w)$ is going to be a mixed Poisson distribution rather than Poisson distribution. These aspects are discussed below.

As already mentioned in previous sections, there may be many repeats in the DNA sequence [232]. The rest of this section is devoted to extending the idea depicted in Fig. 8.24 and in (8.11) to the case of genomes with repetitive elements. A configuration with two identical copies of a fragment of DNA is shown in Fig. 8.25. From this figure we understand that if we choose one of the reads randomly and then we sample an l-mer w from it, then $x(w)$ will have either a Poisson distribution with parameter $N(L-l+1)/(G-l+1)$ or a Poisson distribution with parameter $2N(L-l+1)/(G-l+1)$, depending on whether w hits the nonrepetitive or the repetitive part of the sequence. On the basis of [177], we shall show here how the structure of repetitive elements in the genome, as well as the genome length, can be estimated on the basis of NW samples $w_1, w_2, \ldots, w_{NW}$ and the associated numbers of occurrences

$$x(w_1), x(w_2), \ldots, x(w_{NW}). \tag{8.12}$$

Because the fragments (reads) are randomly chosen from clone libraries, their positions in the original sequence are random, as is the number of occurrences of any l-mer in the sequence. Let us consider the distributions of these random variables first. Assume we know the coverage a of the genome by these N fragments. For a given l-mer w, that appears in the DNA sequence $n(w)$

8.5 Statistics of the Genome Coverage   249

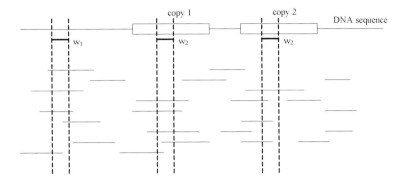

Fig. 8.25. A DNA sequence with repetitive structure. Assume that there are two identical copies in the DNA depicted by the boxes "copy 1" and "copy 2". From the explanation in the text we see that the number $x(w)$ of clones containing an l-mer w now has a mixed Poisson distribution

times (in Fig. 8.25, l-mer w_1 appears in the DNA sequence $n(w_1) = 1$ time, and l-mer w_2 appears in the DNA sequence $n(w_2) = 2$ times), how many times will it appear in N fragments? Analogously to the previous notation, we denote by $x_i(w)$ the number of fragments that cover the ith copy of w, where i ranges from 1 to $n(w)$. Note that the $x_i(w)$, $i = 1, \ldots, n(w)$, are independent identically distributed (i.i.d.) variables and have approximately Poisson distributions. The distribution of $x_i(w)$ is Poisson with a parameter $N(L-l+1)/(G-l+1)$. For any given l-mer w in the sequence, we do not have a vector $\{x_i(w), i = 1, ..., n(w)\}$, but $x(w)$, the sum of the elements. Owing to the additivity of the Poisson process, the distribution of $x(w) = \sum_{i=1}^{n(w)} x_i(w)$ is Poisson with parameter $n(w)N(L - l + 1)/(G - l + 1)$. For estimation, we can use observations from those l-mers as samples from a mixed Poisson distribution with intensities $a_1 N(L-l+1)/(G-l+1), a_2 N(L-l+1)/(G-l+1)$, ..., $a_K N(L - l + 1)/(G - l + 1)$, where $a_1, a_2, \ldots, a_K$ are associated with the repetitive structure of the genome.

Assume there are K families of l-mers (K components in the mixed Poisson distribution) in the original DNA sequence. The number of occurrences of any l-mer in the fragments is therefore a mixed Poisson random variable with intensities

$$\lambda_k = a_k a, \; k = 1, 2, \ldots, K \tag{8.13}$$

(where $a = N(L - l + 1)/(G - l + 1)$ is the coverage and a_k is an unknown integer) and with component probabilities (weights) $\alpha_k > 0$, $k = 1, 2, \ldots, K$, $\sum_{k=1}^{K} \alpha_k = 1$. This mixed Poisson parameter estimation problem can be solved with the use of the EM algorithm (see Sect. 2.6). We start from a random guess of the parameters λ_k^{start}, α_k^{start}, $k = 1, 2, \ldots, K$. Then recursions for successive approximations of λ_k, α_k, $k = 1, 2, \ldots, K$ are as follows:

250   8 Genomics

$$\lambda_k^{new} = \frac{\sum_i^{NW} n(w_i) \Pr[w_i \in \text{family k}|n(w_i)]}{\sum_i^{NW} \Pr[w_i \in \text{family k}|n(w_i)]} \quad (8.14)$$

and

$$\alpha_k^{new} = \frac{\sum_i^{NW} \Pr[w_i \in \text{family k}|n(w_i)]}{\sum_{j=1}^{K} \sum_i^{NW} \Pr[w_i \in \text{family j}|n(w_i)]}, \quad (8.15)$$

where

$$\Pr[w_i \in \text{family k}|n(w_i)] = \frac{\alpha_k^{old}}{\sum_{j=1}^{K} \alpha_j^{old} \left(\lambda_j^{old}/\lambda_k^{old}\right)^{n(w_i)} \exp(\lambda_k^{old} - \lambda_j^{old})}. \quad (8.16)$$

Similarly to Sect. 2.6, superscripts "old" and "new" have been added to denote two successive iterates. The coverage data is often obtained from hashing table (array), which leads to the following organization of data: #gr[0] is the number of l-mers found 0 times in reads, #gr[1] is the number of l-mers found once in reads, and #gr[n] is the number of l-mers found n times in reads, where $n = 1, 2, \ldots, MxGr$. The parameter $MxGr$ denotes the maximal count of reads containing an l-mer. Equations (8.14)–(8.16) can then easily be rearranged as follows:

$$\lambda_k^{new} = \frac{\sum_{n=0}^{MxGr} \#\text{gr}[n] n \Pr[w_i \in \text{family k}|n]}{\sum_{n=0}^{MxGr} \#\text{gr}[n] \Pr[w_i \in \text{family k}|n]} \quad (8.17)$$

and

$$\alpha_k^{new} = \frac{\sum_{n=0}^{MxGr} \#\text{gr}[n] \Pr[w_i \in \text{family k}|n]}{\sum_{j=1}^{K} \sum_{n=0}^{MxGr} \#\text{gr}[n] \Pr[w_i \in \text{family j}|n]}, \quad (8.18)$$

where

$$\Pr[w_i \in \text{family k}|n] = \frac{\alpha_k^{old}}{\sum_{j=1}^{K} \alpha_j^{old} \left(\lambda_j^{old}/\lambda_k^{old}\right)^n \exp(\lambda_k^{old} - \lambda_j^{old})}. \quad (8.19)$$

Equations (8.17)–(8.19) are equivalent to (8.14)–(8.16). However, in numerical computations (8.17)–(8.19) are more efficient, owing to avoiding multiple function calls.

Now let us recall that we have no data for w with $n(w) = 0$ because we cannot sample a w that does not belong to any of the reads. So the summations in (8.17)–(8.19) cannot be performed, because #gr[0] is not known. It is, however, possible to estimate the number of empty reads $n(w) = 0$ conditional on data. Denote by #gr[0] the unknown number of l-mers w for which $n(w) = 0$, and by #gr[> 0] the number of l-mers w for which $n(w) > 0$ (in other words, the number of all l-mers in our experiment). Knowing that $n(w)$ has a mixed Poisson distribution with parameters λ_k, α_k, $k = 1, 2, \ldots, K$, we obtain

$$\frac{\#\text{gr}[0]}{\#\text{gr}[0] + \#\text{gr}[> 0]} \simeq \sum_{k=1}^{K} \alpha_k \exp(-\lambda_k);$$

8.5 Statistics of the Genome Coverage

in other words, the estimate for $\#\mathrm{gr}[0]$ is

$$\#\mathrm{gr}[0] = \frac{\sum_{k=1}^{K} \alpha_k \exp(-\lambda_k)}{1 - \sum_{k=1}^{K} \alpha_k \exp(-\lambda_k)} \#\mathrm{gr}[>0].$$

A heuristic approach to adjusting the algorithm for the missing $\#\mathrm{gr}[0]$ might be as follows: augment each step of the recursions (8.14)–(8.16) or (8.17)–(8.19), with a step of estimation of $\#\mathrm{gr}[0]$ by use of the above equation and add $\#\mathrm{gr}[0]$ copies of l-mers w with $x(w) = 0$ to the data.

There are two conditions which our problem does not satisfy. They are the following.

(1) The numbers of occurrences of l-mers are not independent, although they have the same distribution. However, the dependence is not very deep, that is, the occurrences of one l-mer depend only on at most $L - l + 1$ others. Therefore we neglect the dependence in our computations. Assume that $w_1, w_2, \ldots, w_m$ are l-mers that all belong to the Poisson component n_k. Then

$$\frac{x(w_1) + \ldots + x(w_m)}{m}$$

approach $n_k \frac{N(L-l+1)}{G-l+1}$ when m is large.

(2) The intensities λ_k are not independent. They have a structure given by (8.13), and moreover they all depend on G, the parameter to be estimated. Assuming that the majority of the genome is a nonrepetitive strand we can base our estimates of $a_1, a_2, \ldots, a_K$ and G on the minimum of the values of the estimated $\hat{\lambda}_k$. We define $\hat{\lambda}_{\min} = \min_k \hat{\lambda}_k$. We then obtain the following estimates:

$$\hat{a}_k = \text{integer closest to } \frac{\hat{\lambda}_k}{\hat{\lambda}_{\min}}$$

and, analogously to (8.11),

$$\hat{G} = l - 1 + \frac{N(L-l+1)}{\hat{\lambda}_{\min}} \simeq \frac{N(L-l+1)}{\hat{\lambda}_{\min}}.$$

The proper choice of l is critical for the functioning of the method. We should let l be large enough that many l-mers in the original sequence are unique l-mers. That is, if the DNA is G base pairs long, l should satisfy $4^l > G$ if the sequence is generated by a uniform i.i.d. mechanism. On the other hand, we cannot let l be too large. For instance, if we let $l = L$, then there are N l-mers in all. And each l-mer appears once in the fragments, in general. Our estimate of a is then 1, which is incorrect. Moreover, the larger l is, the fewer the number of samples, and the less accurately we can estimate a and a_k for $k = 1, 2, \ldots, K$. Therefore, l must be large, but not too large. Some more comments on this are in [177].

Examples of Estimation of Repeat Structure and Genome Length

Li and Waterman [177] have provided examples of how their method performs. Their Example 3 assumes $G = 80\,000$, $L = 500$, and $a = 3$. There are two families of repeats. One is 6 kb long with two copies, whereas the other is 1 kb long with 12 copies. Repeats appear in tandem in the original sequence. In more than 95% of the simulations, the result obtained was that the estimated genome length was 73 104 bp; the estimated coverage was 3.283; and a unique sequence accounts for 82.7% of the genome. There is only one family of repeats, which had 12 copies and accounted for 17.3% of the sequence. This example illustrates difficulties that are inherent when short genomes and low coverages are considered.

8.5.4 Polymorphisms

The following is mostly based on [77]. Polymorphisms are differences in the variants of DNA sequences present in different individuals and in paternal and maternal chromosomes in a single individual. Most algorithms for large-scale DNA assembly make the simplifying assumption that the input data is derived from a single homogeneous source. The frequency of polymorphisms (single nucleotide polymorphisms (SNP), variable-length microsatellites and block insertions and deletions (indels)) depends on the evolutionary history of the species. For example, indels are twice as frequent in the human genome than in the *Drosophila* genome [203]. Discrepancies due to polymorphisms will cause false negatives in the overlap relations among the fragments and may result in confusion of the polymorphisms with evolutionary divergence in repeat regions of the genome. The Celera Assembler [127, 202] includes the A-statistic, which can help determine if a region containing a bubble seems repetitive. A sequence of overlapping fragments (assumed for simplicity to have equal lengths) of a genomic sequence can be depicted using a directed graph, which is linear for an unambiguous complete assembly. Polymorphisms will be represented as bubble-like structures (for examples, see [77]). Bubbles can be resolved in two ways. (1) A single path through the bubble can be designed, resulting in a consensus sequence. (2) Multiple alternative paths may be accepted and represented in the final assembly. An algorithm for bubble resolution (smoothing), which is a subroutine of the Celera Assembler, is described in [77]. The final validation of bubble smoothing is the resequencing of the putative polymorphisms.

8.6 Genome Annotation

Genome annotation means associating appropriate explanations with a sequence of DNA. As already said, a sequence of nucleotides in the genome of an organism has a lot of structure: it contains noncoding spaces between genes,

8.6 Genome Annotation 253

coding sequences, which are exons of genes; areas responsible for regulation of the transcription process, gene promoters, starting and stopping codons, dinucleotides for signaling splicing between exons and introns; and other features. If it is not already self-evident, the necessity for marking the functionality of DNA becomes evident when comparing the unannotated to annotated genes. From the NCBI GenBank depository [326], we have downloaded the nucleotide sequence of the gene, TEL1 on chromosome II of the organism, baker's yeast (*Saccharomyces cerevisiae*). The gene TEL1, in baker's yeast genome, is coding for protein kinase, Tel1p, primarily involved in telomere length regulation. It is a homolog of the human ATM gene, which has many functions in human genome. Below, we print a short sequence of the beginning and a short sequence at the end of this gene:

```
0001  tgagtttgta  cattactttt  cgtatttcta  taaacaaaaa  aaagaagtat  aaagcatctg
0061  catagcaatt  aataaaaagg  tgaccatccc  atatatataa  cactcaaatt  tgatggatcc
0121  gtggcttgct  gaatcaaatc  ttgtacgcta  gactctacac  ttagtccatt  acccataagc
      ⋮
8401  tttaaagttt  ctacaatccc  atgatcctcc  atcgtctatg  ttacactgat  ttccctttc
8461  tttgaaggct  ttttttcga   atttcctgct  tttttgcga   ggctttgaga  agtcaattag
8521  tcttgattat  tctattaact  tggaactaat  ttaccttgaa  aaatgtcaaa  atatgc....
```

The numbers on the left give a "local coordinate" for the linear structure of the gene. At the same address on the Internet databank one can also read the sequence of amino acids in the protein kinase Tel1p,

MEDHGIVETLNFLSSTKIKERNNALDELTTILKEDPERIPTKALSTTAEALVELLASE
HTKYCDLLRNLTVSTTNKLSLSENRLSTISYVLRLFVEKSCERF ....
⋮
LFEEEHEITNFDNVSKFISNNDRNENQESYRALKGVEEKLMGNGLSVESSVQDLIQQA
TDPSNLSVIYMGWSPFY

Knowing that the above sequence of nucleotides codes for a protein, it is a good exercise to try to decipher the order of amino acids in the kinase Tel1p by searching through the nucleotide string and using the genetic code shown in Table 8.1, and possibly methods for searching and comparing sequences presented in Chap. 3. Even if one knows the sequence of amino acids, as given above, identifying correspondences between codons and amino acid symbols can, initially, pose a problem, because this gene is transcribed from the complementary DNA strand. This means that the order of the amino acid sequence is reversed relative to the order of codons written in the above nucleotide sequence. The annotation information, which is again stored in the NCBI GenBank at the same address as the nucleotide sequence, tells this and allows solving the gene transcription problem. Using this information we can learn that transcription starts at nucleotide no. 8432 and ends at nucleotide

no. 69. With this information, assigning codons to amino acids becomes easy.
The assignment between codons and amino acids is presented below. At the
beginning of the gene

nucleotides forward 8401 tttaa*agt*tt c*tac*aatccc atg*atc*ctcc at...
RNA reverse   8401 aaauu*uca*aa g*aug*uuaggg uac*uag*gagg ua...
amino acids reverse   T   E   V I G   H D E   M...
⋮

and at the end of the gene,

nucleotides forward 0061 catagcaa*tt* *a*ataa*aa*agg *tga*ccatccc at*ata*...
RNA reverse   0061 guaucguu*aa* *u*ucuu*uu*acc *act*gguaggg uau*au*...
amino acids reverse   STP   Y F P   S W G   M Y...

In the above, we used alternating typefaces for easier identification of
codons. STP means a stop codon. Remember that when decoding, we read
RNA in the reversed direction. In the above assignments, by "forward" we
mean from left to right and by "reverse" from right to left.

8.6.1 Research Tools for Genome Annotation

Our knowledge about functional aspects of a genome comes not only from the
sequencing of it but also, to a substantial extent, from experiments performed
in molecular-biology laboratories, which aim at relating the contents of the
genome to its function. Our knowledge about the functional and structural
aspects of DNA has a hierarchy. Some genes in the human genome have been
studied for decades, the functions of the proteins they code for are known and
the tertiary structure of the protein has been found by crystallization and X-
ray diffraction measurements and stored in the database. However, some other
genes are not so well studied, and there are many regions in genomes which,
at the present stage of knowledge, are only declared to be putative genes, with
hypothetical sequences of amino acids in proteins and hypothetical functions.

A stream of useful information concerning genome annotation is coming
from comparing DNA sequences, searching for certain patterns in DNA strings
and computing appropriate statistics from the results of searches. These tasks
are done with the use of the algorithms, discussed in previous chapters involving string searching, sorting, indexing by suffix trees or the Burrows–Wheeler
transformation, and recording occurrences of patterns by hashing techniques.
Below we show some examples of the results of application of string-processing
algorithms to DNA annotation.

8.6.2 Gene Identification

The ideas which can be used for searching for genes in an unannotated genome
come from knowledge about the transcription process, and on the, related to

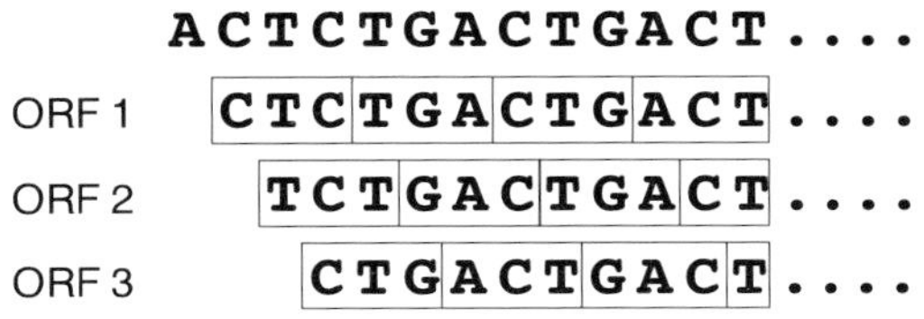

8.6 Genome Annotation   255

```
            ACTCTGACTGACT....
ORF 1       CTCTGACTGACT....
ORF 2        TCTGACTGACT....
ORF 3         CTGACTGACT....
```

Fig. 8.26. Three possible open reading frames (ORFs) for transcribing DNA from left to right

this process, structure of (protein-coding) genes. A diagram illustrating the structure of a protein-coding eukaryotic gene was presented in Fig. 8.8. A search for genes can be based on short DNA sequences, codons, dinucleotides or tetranucleotides occurring as specific elements of a gene. As already mentioned, these short DNA sequences come in differ variants and can differ between genes and between organisms, which makes gene searching more difficult.

Open Reading Frames

A concept very often used in the context of gene identification is that of open reading frames (ORFs). ORFs are DNA sequences (frames) between start and stop codons [85, 86, 173]. The start codon in eukaryotes is most often ATG, which also codes for the amino acid methionine (other possible start codons are TTG, CTG), and stop codons are TAA, TAG, and TGA (see Table 8.1. On the basis of this information, it is easy to search DNA for ORFs. Clearly, there will be many mistakes in identifying genes by ORF searching, but ORF searching can surely serve as a first step. In subsequent steps, regions identified as ORFs can be investigated further for their structure.

When reading codons from DNA there is an ambiguity of which exact position to start from. The use of ORFs is a possible solution to this ambiguity. There are six possible ORFs for a DNA sequence, three for each direction. Three possible ORFs for reading DNA from left to right are shown in Fig. 8.26.

Searching for ORFs can be performed with the help of many Internet software sources, for example, the NCBI Web site contains a service called ORF Finder, which estimates positions of ORFs in a given nucleotide sequence.

Hidden Markov Models for gene identification

More sophisticated methods for gene identification have been presented in papers by Burge and Karlin [43, 44]. In order to get familiar with the idea of their method, imagine traversing DNA strand as a Markov chain. The

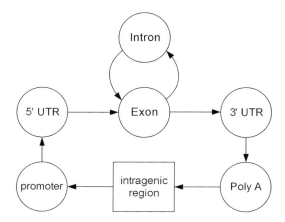

Fig. 8.27. Possible transitions between states, intregenic region, promoters, 5′ UTR, exon, intron, 3′ UTR, PolyA, visited when traversing a DNA strand.

possible states correspond to different functional units (regions) of the DNA (compare Fig. 8.8), namely intragenic, promoter, 5′ and 3′ UTR, exon, intron, and polyA regions. Looking at the structure of the eukaryotic gene in Fig. 8.8 we can imagine possible transitions between these states as presented in Fig. 8.27. The possible words emitted by the states of the Markov chain are those listed at the bottom of Fig. 8.8 and in the entries of Table 8.1. The problem is to reconstruct the states of the Markov chain on the basis of words emitted– the contents of the DNA strand. Provided the probabilities of transitions and the probabilities of emitting words are known, this problem can be stated as an application of the Viterbi algorithm presented in Chap. 2.

The diagram in fig. 8.27 is simplified compared with the analysis presented in [44] (compare Fig. 8.8 with Fig. 3 in [44]). We have made three main simplifications compared to [44]. (1) We have ignored the fact that genes can be transcribed along both directions of the DNA, in other words the sequence analyzed can correspond both to a gene and to its reversed complement. (2) In [44], there are more categories of introns and exons. There are introns and exons of phases 0, 1, and 2, which refers to their location relative to codons boundaries. These categories are related to the concept of reading frames. (3) In [44], the emitted words can be of different lengths. These differences are not only technical; they make the problem of state reconstruction more complicated in the sense that modification of the mathematical model is required. The main problems are phases and word lengths. For the sake of appreciation of the general idea and academic exercises, the model in Fig. 8.8 can still be an area for practice (see Exercise 14). A research with real DNA data requires a more detailed study of [43, 44]. Also there are open source programs employing models like those in [43, 44], see for example [313].

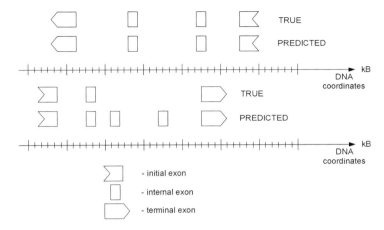

Fig. 8.28. Symbols used for presenting graphically the contents of a genome. The symbols can be reversed when transcription goes in the reverse direction (*upper diagram*)

When one is applying gene identification methods to DNA data, a most important problem is the quality of prediction of the positions and lengths of functional units in the DNA. Typically, verification of the correctness of identification is done by applying the method to a DNA fragment with a well-known function and which is well annotated. In such a situation, all functional units are known and we can compare the predictions of the gene identification algorithm with the true meaning of the DNA. Results of such comparisons are shown graphically using plots like the one presented in Fig. 8.28. The plot in Fig. 8.28 does not have a relation to any real gene in DNA; it only serves to illustrate the convention. Such plots can be used for comparing a true with an estimated structure, as well as for comparisons between different methods of gene identification.

8.6.3 DNA Motifs

There are many other methods similar to the describe above for searching for characteristic features of sequences of nucleotides. Examples are MEME (Multiple EM for Motif Elicitation) and MAST (Motif Alignment and Search Tool), [12, 324] which combine databases of highly conserved motifs in genomic sequences and tools for discovering them. Methodologies for discovering DNA motifs include algorithms described in this book, namely EM method and HMM methods.

8.6.4 Annotation by Words and Comparisons of Genome Assemblies

In [119], the Burrows–Wheeler transform was applied as a tool for annotation by words tool for analysis of assemblies of the human genome (compare Exercise 15 in Chap. 3). As already mentioned, the BW transform is a very efficient tool for both compression and search. The suffix array for the human genome constitutes approximately 12 gigabytes (3 billion 4-byte integers) of RAM. However, the BW string alone is sufficient to determine word counts and can be compressed to about 1 gigabyte of RAM. Furthermore, for the purpose of querying, all but a negligibly small portion of the compressed form can remain so throughout execution. Using the BW transformation, any region of the genome can be annotated with its constituent mer frequencies. Healy et al. [119] have depicted annotations of a 5-kb region of chromosome 19. For each coordinate and various word lengths, they determined the count of the succeeding word of the given length, in both the sense and antisense directions. The word lengths in their study were 15, 18, 21, and 24 bases. One of the most striking features of this region is the presence of narrow spikes in 15-mer counts. This is a virtually universal property of all regions of the human sequence examined, including coding exons and it is statistically significant (on the basis of a comparison with a random genome). Hypothetically, these spikes might result from an accidental coincidence of 15-mers in an ordinary sequence with 15-mers present in high-copy-number repeats.

Another application of the method was to monitor successive human genome assemblies. Healy et al. [119] annotated the December 2001 assembly of the human genome with probe l-mers ($l = 21$) and compared it with the June 2002 assembly. Unexpectedly, 1.2% of their probe words vanished from the June 2002 assembly (i.e., all of the constituent 21-mers went from copy number one in the original assembly, to copy number zero in the subsequent assembly). Systematic studies revealed both losses and gains of single-copy words between assemblies. Although there may be technical reasons to explain the dropout of some of these fragments, such as difficulty in assembly or a poor-quality sequence, it is also likely that, owing to insertion/deletion and order-of-sequence polymorphisms in humans, no fixed linear rendition of the genome is feasible. A detailed discussion is provided in [119].

8.6.5 Human Chromosome 14

The following examples illustrate the complex nature of the global-level organization of human chromosomes and demonstrate that the theory developed above is important for applications. Recently, edited sequences of human chromosome 14 and of the male-specific region of the Y chromosome (MSY) have been published [121, 259]. In chromosome 14, an ancient duplication involving 70% of chromosome 14 and a portion of chromosome 2 has been reported. This event is, however, only visible at the protein level and pre-dates the mouse–human separation. In [121] it was found that 1.6% of chromosome 14 consists of interchromosomal segmental duplications contained in fragments of 1 kb or more that show at least 90% sequence identity. A comparable value, based on a different comparison procedure, was reported earlier, and confirms that chromosome 14 has the lowest content of interchromosomal segmental duplications in the human genome. In a similar analysis that excluded repetitive DNA, it was found that internal duplications account for 1.1% of chromosome 14 and are clustered into four segments. The largest includes an 800 kb region adjacent to the centromere, which is also part of the segmental duplication shared with chromosome 22.

The male-specific region of the Y chromosome, MSY, differentiates the sexes and comprises 95% of the chromosome's length. Skaletsky et al. [259] determined, among other things, that the most pronounced structural features of the ampliconic regions of the Y chromosome are eight massive palindromes. In all eight palindromes, the arms are highly symmetrical, with arm-to-arm nucleotide identities of 99.94–99.997%. The palindromes are long, their arms ranging from 9 kb to 1.45 Mb in length. They are imperfect in that each contains a unique, nonduplicated spacer, 2–170 kb in length, at its center. Palindrome P1 is particularly spectacular, having a span of 2.9 Mb, an arm-to-arm identity of 99.97%, and bearing two secondary palindromes (P1.1 and P1.2, each with a span of 24 kb) within its arms. The eight palindromes collectively comprise 5.7Mb, or one quarter of the MSY euchromatin.

In addition to palindromes and inverted repeats, the ampliconic regions of Yq and Yp contain a variety of long tandem arrays. Prominent among these are the newly identified NORF (no long open reading frame) clusters, which in aggregate account for about 622 kb on Yp and Yq, and the previously reported TSPY clusters, which comprise about 700 kb of Yp. The NORF arrays are based on a repeat unit of 2.48 kb. Numerous further structural features of MSY and their evolutionary explanation are discussed in [259].

8.7 Exercises

1. Write a computer program to simulate shotgun coverage of a DNA artificial genome (DNA pseudocode) for small scales, e.g, 100–200 base pairs for the whole DNA sequence and 20 – 50 reads of lengths 10–50.
2. Write a program for detecting the overlap structure in the reads obtained in Exercose 1, and for drawing the overlap graph. What methods are you going to use in detecting the overlap structure of the reads? Compare to Chap. 3.
3. In the program in for Exercise 1, include a method to assign and control the repetitive structure of the artificial genome. What influence does repetitive structure have on the overlap graphs?

4. Try heuristic algorithms for growing contigs from reads, for data from Exercises 1 and 3.
5. Download a software package for solving Hamiltonian path problems from [340] and try to apply this package to the graphs obtained in Exercise 2 and Exercise 3.
6. Search for a program, on the Internet, for solving a superstring problem (see [340]). Apply this program to the data obtained in Exercises 2 and 3.
7. Include a mechanism of erroneous reads, by assuming base alterations, insertions and deletions. Observe influence of errors on the quality of the overlap graph and assembly process.
8. Try to advance the scale of the artificial genome in Exercise 1, to a size of approximately 10^3–10^4 base pairs for the whole DNA strand and 10^2–10^3 for the reads.
9. Use the Euler assembler [316] for the data obtained in Exercises 2, 3, 7 and 8.
10. The BW transform, used as an associative, contextual memory can be efficiently used for DNA assembly from reads. Try to develop an appropriate algorithm.
11. Derive (8.7)–(8.9) for expected number of gaps $E(\#\text{gaps})$, minimum number of gaps $E_{\min}$ and the stringency σ under a model of shotgun genome coverage with a minimum overlap T required for two consecutive reads [163].
12. Using the programs and data from Exercises 1, 3, and 7, generate l-mers which can be used for estimation of genome length and structure by fitting the mixed Poisson distribution to the data (8.12).
13. Write a computer program for searching for ORFs in DNA sequences, and try to apply it to real DNA data, for example, download a complete sequence of chromosome 1 in baker's yeast (*Saccharomyces cerevisiae*) [326]. Try to verify whether regions which your program classified as genes, coincide with data in the NCBI depository?
14. Write a computer program, based on the transition diagram in Fig. 8.8 for generation of artificial DNA sequences. Assume values for necessary parameters. Based on the generated DNA pseudocode, try to reconstruct states of the Markov process using hidden Markov model and the Viterbi algorithm.
15. Present your artificial DNA sequence, generated in Exercise 14 to the public domain gene identification program [313]. Download an example sequence of a gene (or a sequence containing more than one gene and intragenetic regions) from a gene bank [326] and present it to the gene identification program [313]. Compare the outcomes with analyses done with the program constructed in Exercise 14.

9
Proteomics

Proteomics is the branch of the biological sciences which deals with proteins. Proteins are basic constructional blocks and functional elements of living organisms. They are involved in all processes occurring in cells and tissues and therefore are linked to each other by numerous interactions. Proteomics can be viewed as the study of properties, interactions, and functions of proteins.

Proteins are assembled in cell organelles called ribosomes, the process of their production is called translation. Ribosomes are complexes of proteins and RNA. Two types of ribonucleic acid molecule are involved in this process, mRNA (messenger RNA), which consist of long strands of RNA, that carry a plan for the construction of the protein construction written in their codons, and tRNA (transfer RNA), which consists of short RNA molecules dedicated to the transport of a single unit of a protein, namely an amino acid. A guess could be made of the number of different proteins in one organism that one gene codes for one protein. However, this would lead to an underestimate owing to the number of ways in which different protein domains are used together within proteins, i.e., a larger number of protein architectures is made possible by the mechanism of alternative splicing, which involves the possibility of construction of different proteins or different variants of a protein by arranging the exons of one gene in different orders.

Proteins are molecules with a complicated three-dimensional structures, but they always have an underlying linear chain of amino acids as their primary structure. Information about the linear sequence of amino acids, the taxonomy and the functional aspects of proteins, and the organisms or the part of them in which they occur, as well as annotation data on known secondary and 3D structures of proteins is stored in proteominc bioinformatic databases [342, 329]. On the basis of criteria, such as functional aspects the activity of a protein or the part of an organism or compartment of a cell where a protein appears as a building element, proteins are classified into protein families. One database of protein families and domains is PROSITE [317]. This consists of biologically significant sites, patterns, and profiles that help to reliably identify to which known protein a given target protein belongs to.

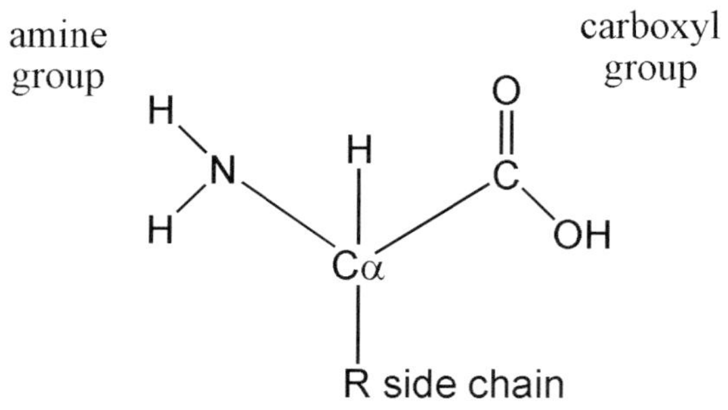

Fig. 9.1. The structure of an amino acid containing a C_α carbon atom, a C' carbon atom, an amine group NH_2, a carboxyl group COOH and side chain R. Amino acid differ in structures of their R side chains.

An example of a protein, the enzyme trypsin, obtained by using information from the Protein Data Bank (PDB) [329] and internet accessible molecular graphics program Ras Mol [332], is shown in Fig. 12.2. Trypsin is a serine protease that specifically cleaves at the carboxylic side of lysine and arginine. It can serve the purpose of digestion of proteins to polypeptides. Its atom coordinates are available in the Protein Data Bank under the symbol 2ptn.

9.1 Protein Structure

As said above, proteins are build from amino acids. All amino acids share a similar molecular structure, which allows them to be put in a chain. This molecular structure consists of a carbon atom called the central α carbon, an amine group NH_2, a carboxyl group COOH and a hydrogen atom H. The carbon atom in the carboxyl group is often labelled C'. The flanking amine and carboxyl groups can form peptide bonds with neighboring amino acids in the chain, and the α carbon is bonded to a side chain compound denoted by R, also called R group or side chain. The side chain is specific to each different amino acid. This scheme of construction is presented in Fig. 9.1.

9.1.1 Amino Acids

There are 20 different amino acids that occur in proteins, which differ from each other by the chemical structures of their R groups. The 20 amino acids were already tabulated in Table 8.1 explaining genetic code. Here we tabulate them once more, in table 9.1, where along with the names and codes we give some characteristic properties. In Fig. 9.2 we show the structural chemical formulas of side the chains of the amino acids.

9.1 Protein Structure

Table 9.1. Table of amino acids, abbreviations, symbols and basic properties

Amino acid	Abbreviation	Symbol	Properties
Alanine	ALA	A	Nonpolar, hydrophobic
Arginine	ARG	R	Polar, hydrophilic
Asparagine	ASN	N	Polar, hydrophilic
Aspartic acid	ASP	D	Polar, hydrophilic
Cysteine	CYS	C	Polar, hydrophilic
Glutamine	GLN	Q	Polar, hydrophilic
Glutamic acid	GLU	E	Polar, hydrophilic
Glycine	GLY	G	Polar, hydrophilic
Histidine	HIS	H	Polar, hydrophilic
Isoleucine	ILE	I	Nonpolar, hydrophobic
Leucine	LEU	L	Nonpolar, hydrophobic
Lysine	LYS	K	Polar, hydrophilic
Methionine	MET	M	Nonpolar, hydrophobic
Phenylalanine	PHE	F	Nonpolar, hydrophobic
Proline	PRO	P	Nonpolar, hydrophobic
Serine	SER	S	Polar, hydrophilic
Threonine	THR	T	Polar, hydrophilic
Tryptophan	TRP	W	Nonpolar, hydrophobic
Tyrosine	TYR	Y	Polar, hydrophilic
Valine	VAL	V	Nonpolar, hydrophobic

The properties of the amino acids are dictated by the chemical characteristics of their R groups. As can be seen in table 8.1 and figure 9.2 R groups of 20 amino acids differ in their size, structure, electrical properties and chemical reactivity. There are many characteristics by which one can describe properties of amino acids in proteins and each of them has some special and individual properties [18].

Important properties of the side chains, characterizing their interaction with other chemical compounds and the surrounding water molecules, are their polarities and charges. On the basis of their polarity at a pH between 6.0 and 7.0, the range corresponding to intracellular conditions, and their charge, the amino acids can be divided into several main classes.

Nonpolar Amino Acids

Nonpolarity is related to a distribution of electrical charges over R such that no asymmetry between positive and negative charges appears. Owing to the lack of polarity, such molecules have no (or little) affinity with polar water molecules and therefore they are also called "hydrophobic", "water hating", or "internal". As seen from Table 9.1, the group of non polar contains 8 amino acids, namely alanine, leucine, isoleucine, valine, proline, phenylalanine, tryptophan and methionine. Hydrophobic amino acids typically occupy the space inside the protein molecule, with no contact with surrounding H_2O molecules.

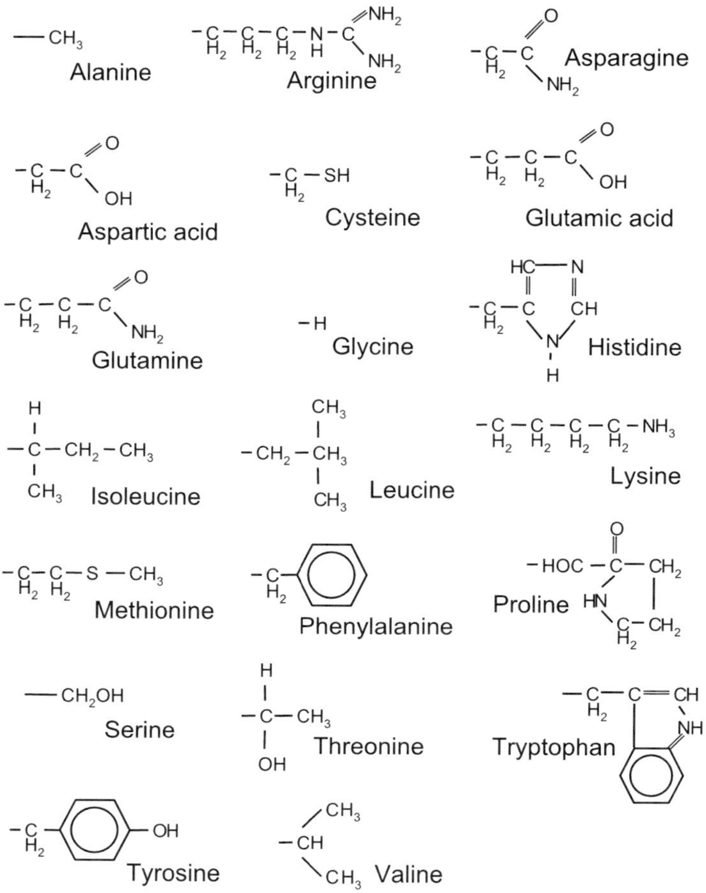

Fig. 9.2. Chemical formulas of the side chains of the 20 amino acids

Non polar amino acids are the most frequent components of proteins of many organisms.

Polar Uncharged Amino Acids

Polar uncharged amino acids are more soluble in H_2O. Owing to their polarity, these amino acids tend to be placed on the exterior of proteins, in contact with solvent H_2O and are also called hydrophilic amino acids. This group contains seven amino acids, namely serine, threonine, cysteine, proline, asparagine, glutamine and tyrosine.

Polar Charged Amino Acids

These amino acids have side chains which are both polar and charged and are very hydrophilic. They are located in the exterior parts of proteins and often belong to chemically active sites of proteins. There are five polar charged amino acids, namely lysine, arginine and histidine, which have positively charged R group, and aspartate and glutamate which have negatively charged R group.

Aromatic Amino Acids

Aromatic amino acids have side chains R containing aromatic carbon rings. The aromatic amino acids, namely phenylalanine, tryptophan, and tyrosine, are among the heaviest and biggest. Phenylalanine and tryptophan are nonpolar and hydrophobic. Tyrosine is the only one of the aromatic amino acids with an ionizable side chain. However, its polarity is not strong. Amino acids with polarities of this kind of strength are often called "indifferent". Aromatic amino acids have a tendency to be located in the interior of the protein molecule. The presence of aromatic nonpolar amino acids in the amino acid chain stabilizes the mechanical structures of a protein.

9.1.2 Peptide Bonds

Any two amino acids can form a bigger molecule, a dipeptide, by forming a peptide bond between them. The peptide bond is formed between a carbon atom in a carboxyl group COOH and a nitrogen atom in an amine group NH_2. This is shown in Fig. 9.3, which illustrates formation of a dipeptide, glycalanine, by the occurrence of a peptide bond between the amino acids glycine and alanine. Peptide bonds can lead to the formation of chains of amino acids, polypeptides, that are much longer than the dipeptide shown in Fig. 9.3. These chains define the linear, or primary, structure of proteins. By convention primary structures are represented as sequences of one-letter codes standing for the amino acids in the chains; the codes are shown in Tables 8.1 and 9.1. Chains of amino acids have a well-defined direction; the end containing a free amino group (left end of the dipeptide in Fig. 9.3) is called the N (amino) end and the end containing a free carboxyl group is called the C (carboxyl) end. The conventional representation of peptides by sequences of letters assumes that N end is at the left and C end at the right. So the two-letter representation of the dipeptide shown in Fig. 9.3 will be GA.

The peptide bonds, between nitrogen and C' carbon atoms are of the covalent type, quite strong. However, if required in some biological or molecular process peptide bonds can be broken by appropriate enzymes, such as trypsin, shown in Fig. 12.2.

The chain of carbon and nitrogen atoms connected by peptide bonds is called the backbone of the protein.

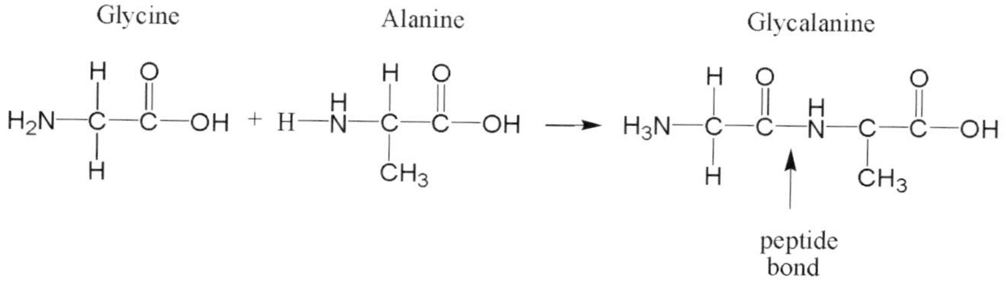

Glycine   Alanine   Glycalanine

$$H_2N-CH_2-COOH + H-NH-CH(CH_3)-COOH \longrightarrow H_3N-CH_2-CO-NH-CH(CH_3)-COOH$$

peptide bond

Fig. 9.3. Forming of the peptide bond between two amino acids glycine and alanine leads to the synthesis of the dipeptide glycalanine.

9.1.3 Primary Structure

Proteins are formed by one or more polypeptide chains. The primary structure of a protein is the order of amino acids in the backbone chains of the protein. The order of amino acids is a primary source of information about proteins. Bioinformatic databases contain a large number, of the order of 3×10^6 of amino acid sequences of proteins. On the basis of the linear order of amino acids, every newly discovered protein can be compared with a database, for similarities to already known sequences. Proteins which share similarities in their amino acid sequences will often also exhibit similar spatial structures, properties, and functions. Comparing amino acid sequences with database resources is therefore a basic initial step in research on proteins. It is done with the use of the algorithms and software discussed in Chap. 6. A good source of information on comparing protein amino acid sequences with database resources is a recent book [71].

9.1.4 Secondary Structure

By the secondary structure we understand certain structural forms which are shared by many proteins. Most often, the mechanisms of formation of secondary structures are related to the occurrence of hydrogen bonds between oxygen and nitrogen atoms. Below we discuss some of the secondary structures that often appear in proteins.

Alpha Helices

An alpha helix is a periodic structure where the protein backbone coils like a screw, and side chains of the amino acids stick out outside the helix. The number of amino acids per turn is approximately 3.6; each of the amino acids corresponds to a 100° rotation. The spatial stability of the alpha helix is

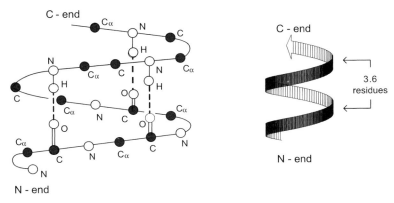

Fig. 9.4. *Left*: structure of a right handed alpha helix (α_R helix) in a polypeptide sequence. The side chains of the amino acids (sticking out of the helix) are not shown, for better clarity. Hydrogen bonds are represented by dashed lines. *Right*: symbolic representation of the α_R helix often used when illustrating structures of proteins.

maintained by hydrogen bonds between oxygen atoms in the CO group on amino acid number n and hydrogen atoms of the NH group in amino acid number $n+4$. Amino acid number $n+4$ is situated approximately above amino acid number n, when looking along the axis of the helix. This type of alpha helix is also called a 4-alpha helix. The structure of a 4-alpha helix is presented in Fig. 9.4. There are also other, similar helical structures 3_{10} and π alpha helices where number of amino acids per turn is, respectively 3 and 5. Two alpha helices are well seen in the left part of Fig. 12.2. Alpha helices are most often right-handed, like the one presented in Fig. 9.4. They are referred to by the abbreviation α_R-helices. Left-handed alpha helices, referred to as α_L-helices, are rare.

Beta Sheets

In contrast to the curly alpha helices formed by single amino acid chains, beta sheets are formed by pairs of expanded chains of amino acids. Again the structure is stabilized by hydrogen bonds between oxygen atoms in the CO group and hydrogen atoms of the NH groups on the other chain. The directions of amino acid chains can be either antiparallel, as shown in Fig. 9.5 or parallel as shown in Fig. 9.6. Beta sheets can also extend further in width, by incorporating more that two amino acid chains into the structure.

Other Structures

There are also other motifs shared by various proteins, which can be classified as second-order structural elements, for example hairpin loops, disulfide

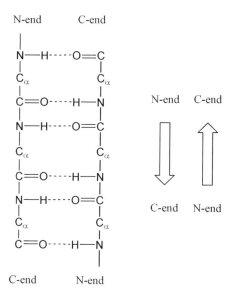

Fig. 9.5. *Left*: structural chemical formula of an antiparallel beta sheet. The side chains of the amino acids are again not shown. *Right*: Symbolic representation of an antiparallel beta sheet

bridges, and zinc fingers. They can have an important impact on the formation and stabilization of the shape of proteins. Investigating the possible occurrence and spatial position of these second-order motifs allows better understanding the structure of proteins [252].

9.1.5 Tertiary Structure

The tertiary structure is the spatial structure of a polypeptide chain. In principle, this structure is given by the spatial coordinates of the centers of all atoms in the protein. Exact data concerning tertiary structures is available for a fraction of the known amino acid sequences, which in total gives over 30 000 known conformations of proteins. Data, for proteins with solved spatial conformations, can be obtained from the Protein Data Banks [329].

The tertiary structure of a protein depends to large extent on the shape of its backbone, which in turn depends on the geometry of the peptide bonds and C_α atoms. The conventional definitions and notation concerning the geometry of amino acids and peptide bonds are presented in Fig. 9.7. This figure, at the top, shows a perspective representation of a fragment of a polypeptide chain (Kekulé representation, [29]). A characteristic property of the geometry of the chain is that atoms forming peptide bonds, i.e., the C and N and the surrounding four atoms C_α, H, O and, C_α, lie approximately in a plane.

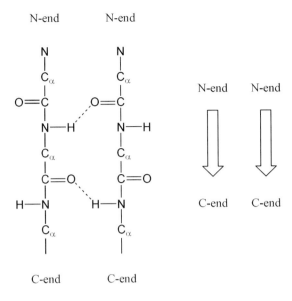

Fig. 9.6. *Left*: structural chemical formula of a parallel beta sheet. The side chains of the amio acids are left unmarked. *Right*: symbolic representation of a parallel beta sheet

This is depicted in Fig. 9.7 by drawing dashed-line rectangles enclosing the planar structures. The range of flexibility of conformational changes of the polypeptide structure is depicted in the lower part of fig. 9.7. The configuration of atoms C–C_α–N remains fixed and the conformational change is allowed by rotating planes depicted by dashed lines around axes C–C_α and C_α–N. The angles of rotation ψ and φ, measured in the range $-180°$ to $+180°$, are called dihedral angles.

Ramachandran Plot

A list of the dihedral angles, ψ_1, φ_1, ψ_2, φ_2, ..., ψ_N, φ_N, where N is the length of the amino acid sequence, describes the configuration of a polypeptide chains in the coordinate-free manner. A "dot" plot representing the values of the dihedral angles in the ψ–φ plane is called a Ramachandran plot of the polypeptide chain [237]. The geometry of a Ramachandran plot is presented in Fig. 9.8. Values of dihedral angles ψ and φ for polypeptide angles are not distributed uniformly. Some configurations of the angles ψ and φ angles are forbidden owing to spatial restrictions on the side chains residues of the amino acid sequence. Some areas in the Ramachandran plot can be associated with common motifs of the secondary structure, namely alpha helices and beta sheets. These areas are marked in the plot in Fig. 9.8

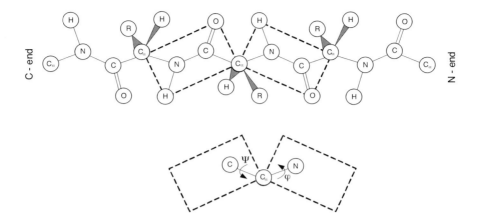

Fig. 9.7. *Top*: fragment of a polypeptide chain. The two atoms, C and N forming a peptide bond, and the surrounding four atoms C_α, H, O, and C_α lie, approximately in a plane, which is depicted by a dashed rectangle. *Bottom*: the configuration of atoms C–C_α–N remain constant in the polypeptide structure. Flexibility (conformational change) is allowed by rotating the planes shown by dashed lines around the axes C–C_α and C_α–N. Angles of rotation ϕ and ψ, measured in the range $-180°$ to $+180°$, are called Ramachandran angles or dihedral angles

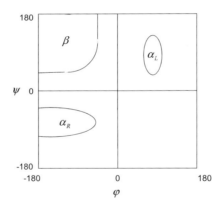

Fig. 9.8. Ramachandran plot. The areas marked by β, α_R and α_L are the regions corresponding to the basic secondary structures, namely antiparallel and parallel beta sheets (β), and right-handed (α_R) and left-handed (α_L) alpha helices

Spatial Positions of Side Chains

The next element which describes the tertiary structure of a polypeptide chain is the positions of the R groups relative to the backbone chain. There is a conformational flexibility involving rotations of the side chains of the amino acids relative to the backbone. Using the upper part of Fig. 9.7, we can illustrate a naming convention related to the positions of R groups relative to the backbone. It can seen in the upper part of Fig. 9.7 that the R groups stick out of the backbone chain in an alternating way: one points towards the front of the figure and the next one towards the back, and so forth. Such a spatial arrangement is called a trans configuration of amino acid side chains. The other possibility, when all R groups point to the same side of the figure, is called a cis configuration. Trans and cis configurations have different geometries and different properties.

9.1.6 Quaternary Structure

The Quaternary structure of a protein is described by the relative positions of the several polypeptide chains which make up the protein. This structure is determined by the shapes of the component polypeptides and by chemical interactions between the polypeptide chains.

9.2 Experimental Determination of Amino Acid Sequences and Protein Structures

Knowledge about the function and structure of proteins is developing at a rapidly increasing rate and here we describe some of the important experimental techniques behind this development. One experimental problem, with various applications, is the analysis of a solution containing a mixture of different proteins. Mixtures may contain several thousand different protein species, and mixture analysis involves identification of possibly many of their components. The associated techniques include electrophoresis, protein 2D gels, and Western blots. Another problem is, given a sample of polypeptides, to determine their sequence of amino acids. Chemical cleavage methods and methods of protein spectrometry can be used for this problem. Finally, the experimental techniques, which provide detailed molecular data about the tertiary structures of proteins are protein crystallography and X-ray diffraction, and protein NMR (nuclear magnetic resonance) spectroscopy.

The experimental techniques for the above problems rely very substantially on bioinformatic methods for representing, storing, and analysis of their outcome. We shall describe these bioinformatic techniques after presenting the experimental techniques for protein analyses.

9.2.1 Electrophoresis

Polypeptides are charged molecules of different sizes and molecular weights and so, similarly to DNA and RNA chains, they can be separated by polyacrylamide gel electrophoresis (PAGE). Polyacrylamide gel is a substance with properties analogous to the agarose gels mentioned in Chap. 8. Polyacrylamide has a more uniform sizes of pores (an advantage) but is more difficult technologically and is toxic (a disadvantage). The principle of the experimental technique, the same as that of DNA electrophoresis, is dragging charged molecules through a sieve of pores by an electrostatic field. Depending on their size and charge, the molecules will travel different distances, which is used to facilitate the separation task.

9.2.2 Protein 2D Gels

Analysis of mixtures of proteins often involves a large number of polypeptide species, which calls for more selective tools for their separation. One such tool is the use of 2D gels. Proteins have a complicated electrostatic structure; one molecule carries both positively and negatively charged atoms. The net electrical charge of a protein molecule depends on the value of the pH of the solution, determined by the ratio between acidic and alkaline ions. There is a value of the pH, at which the net electrical charge of a protein molecule is neutral. This value is called the isoelectric point (pI) of the protein. At a pH below their pI, proteins carry a net positive charge, and at a pH above their pI they carry a net negative charge. Therefore, in protein electrophoresis, depending on the pH value of the buffer of the gel, polypeptide molecules can migrate either to the positive electrode, if the pH of the buffer is below the pI of the protein, or otherwise to the negative electrode.

Proteins can be separated according to their isoelectric points with the use of a process called isoelectric focusing. In this process, used as the first step of obtaining a 2D gel image, proteins are arranged (focused) at their isoelectric points by moving along an axis with differential pH (i.e., along the direction of an immobilized pH gradient). In the second step of protein 2D gel separation, SDS gel electrophoresis is used along a second dimension. SDS is the abbreviation for sodium dodecyl sulfate, an anionic detergent which denatures proteins by sticking to them along the polypeptide backbone. It also confers a negative charge on the polypeptide, proportional to its length. The advantage of protein 2D gels method is its high specificity and selectivity. Large collections of protein mixtures resolved by 2D gel technique, for various organisms, are stored in protein databases [301, 342]. By precisely following the 2D gel protocols and using database data one can resolve thousands of proteins in one 2D gel experiment.

A protein 2D gel is a collection of spots in a plane, each spot corresponding to some polypeptide. Some computational problems related to reading 2D protein gels, are calibration of the coordinates of the gel and compensation for nonlinearities related to nonuniform distribution of materials. By correcting nonuniformities affecting the positions of spots on 2D gels, one can reduce the variation of the errors in the procedure.

9.2.3 Protein Western Blots

The Western blot (or protein immunoblot) is a technique for testing for the presence of specific antigens in a sample by presenting them to antibodies. The name "Western blot" is derived from an analogy to the main idea of the technique of the Southern blot assay by altering the word "Southern". Blotting of an electrophoretically separated sample onto nitrocellulose, retaining the electrophoretic positions, and reacting it with antibodies will result in the antibodies binding to specific proteins (antigens). Electrophoresis of known-molecular-weight standards allows for the determination of the molecular weight of each antigenic band to which antibodies may be produced.

9.2.4 Mass Spectrometry

A recent technique for the experimental analysis of properties of proteins is mass spectrometry. Mass spectrometry has been in use in physics and chemistry for a very long time. Mass spectrometry has become possible for proteins possible thanks to the technique of laser desorption and ionization [139]. Most often the sample preparation protocols in protein mass spectrometry involves digestion of proteins to shorter peptides by using an enzyme, for example trypsin. Protein species digested to shorter chains of polypeptides, are then transferred to the gas phase and at the same time ionized by laser radiation. In the transfer of the polypeptide chains to a gas phase a matrix of crystallized molecules is used, to prevent them from being destroyed. They then move along an evacuuated tube, pushed by an electrostatic field and their time of flight (TOF) is measured. This spectrometry technique is called MALDI (matrix assisted laser desorption ionization.) An illustration of this idea is shown in Fig. 9.9 at the top. At the bottom an example of a mass spectrum is presented. By measuring times of flight, it is possible to compute the ratios of m/z (mass to charge) of the polypeptide molecules.

An improvement in resolution of protein mass spectrometry can be achieved by combining the electrostatic field with a magnetic field [172]. As previously laser radiation desorpts and ionizes peptide chains, which then are trapped in strong magnetic field, where they undergo cyclotrone motion. The signal obtained is a mixture of many cyclic components, but can easily be resolved by application of the Fourier analysis. Frequencies of orbital motions carry information concerning (mass to charge) peptide chain m/z ratios.

Mass spectrometry may lead to the reading of amino acid sequences of polypeptide chains. In this method a protein species is obtained by sampling from a spot on a 2D gel. By comparing collections of measured values of mass to charge ratios with specialized databases, it is possible to find underlying

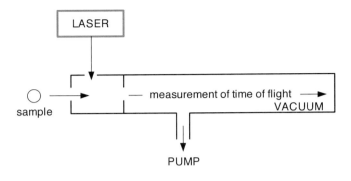

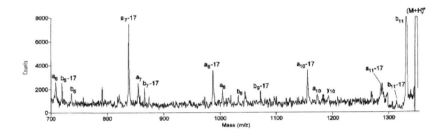

Fig. 9.9. *Top*: illustration of the principle of MALDI (matrix-assisted laser desorption/ionization). Laser radiation transfers polypeptide chains to the gas phase and causes their ionization. They then move in an electrostatic field and their time of flight (TOF) is measured. *Bottom*: Example of a MALDI spectrum

sequence of amino acids [30] in the target protein. Recovering the sequence of amino acids, in this method, involves using algorithmic tools similar to those applied in DNA shotgun assembly.

9.2.5 Chemical Identification of Amino Acids in Peptides

The composition of amino acids in peptides can be also resolved by use of amino acid analyzers [192], which first break the peptide bonds and then analyze the species of amino acids and their concentrations by chromatography. It is also possible to read the sequence of a peptide by a chemical reaction technique called Edman degradation. In this method, the amino-terminal residue is labeled and cleaved from the peptide without disrupting the other peptide bonds between the other amino acid residues. A major drawback is that the peptides that can be sequenced by Edman degradation method cannot have more than 50 to 60 residues.

9.2.6 Analysis of Protein 3D Structure by X Ray Diffraction and NMR

Our knowledge about the 3D structures of proteins comes from X-ray diffraction and nuclear magnetic resonance experiments. The X ray diffraction technique of estimating the structure of molecules involves crystallization and then shining X-rays onto the sample and using the scattering angles of the diffracted rays to obtain information about the spatial structure [218]. The diffraction technique was mentioned earlier in Chap. 8, in relation to determination of the structure of DNA. The native spatial conformations of proteins are therefore obtained by X-ray diffraction experimental techniques and the computed 3D coordinates of the atoms are stored in databses of protein 3D structures.

NMR exploits the magnetic properties of nuclei. NMR allows both identification of atoms and estimation of their spatial coordinates.

Despite rapid advances, the throughput of these experimental techniques still lags behind the potential of protein sequencing. Therefore, only a fraction of the known proteins have detailed descriptions of their tertiary and quaternary structures. Nonetheless, the PDB database [329] containing over 30,000, of known tertiary structures of proteins, is a large depository, which can support numerous researches projects concerning, for example, computational prediction techniques for proteins, methods of estimation of protein–protein, protein–ligand, and other interactions.

9.2.7 Other Assays for Protein Compositions and Interactions

There are also many other techniques for experimental analysis of presence of proteins in mixtures, analysis of protein interactions, analysis of amino acid sequences in proteins, and so forth. These techniques include protein expression mapping, high-throughput cloning of proteins, and protein–protein interaction mapping [214]. Similarly to the techniques described above, these methods also rely heavily on bioinformatic computational algorithms, many of which are described in this book.

9.3 Computational Methods for Modeling Molecular Structures

As said above, a lot of information can be inferred from the composition and frequencies of proteins in cells or cellular compartments. Also, the order of the amino acids carries a lot of information about the function of a protein. However, understanding the functioning of proteins at the molecular level relies basically on a knowledge of the tertiary and quaternary structures of proteins. The molecular mechanisms and processes in living organisms are made to happen by interactions, most of which depend on the spatial arrangement and chemical properties of atoms in the amino acid units in proteins. For a large numbers of molecular interactions, research has led to a quantitative description of energies and forces involved. For an even wider family of molecular interactions the mechanisms and processes are understood in qualitative terms.

There is an extensive effort to develop computational methods to predict structures and interactions involving proteins. This is motivated by (1) the importance of potential applications, involving understanding etiologies of diseases, responses to therapies, drug design etc., and (2) the increase in the data on the spatial structure of protein molecules, which can serve as a basis for estimating unknown structures of proteins, as well as a reference for grading the quality of new algorithms for predicting the structure and functions of proteins.

This area has traditionally belonged to computational chemistry and, recently, chemoinformatics. However, it is becoming very deeply interrelated with bioinformatics [71], mostly owing to the increasing need to search bioinformatic databases of sequences to support research in the prediction of the structure and function of proteins. Also, as will become evident, many of the mathematical and computational methods described earlier as well suited for bioinformatics are also very useful in the area of predicting spatial arrangements of proteins and interactions between protein molecules.

In the rest of this section, we overview some of these methods and approaches. We start from some rather general ideas and problems in computational chemistry, molecular force fields and computing the RMSD (root mean square distance) between molecules. Next we move on to approaches more specific to proteomics, namely protein structure estimation, prediction of active sites of proteins and estimation of interactions between proteins and ligands.

9.3.1 Molecular-Force-Field Model

At a basic level, molecular-modeling problems can be formulated by using the principles of quantum mechanics [166], which explain the time evolution of molecular-dynamical systems in terms of spatial quantum wavefunctions of the electrons and nuclei and the related probability distributions. However, owing to the prohibitive nature of computations related to quantum mechanical formulations, we go to the next level of approximation where spatial positions of electrons are not included as variables in the model, and the energies and force fields of molecular systems are expressed in terms of spatial positions of the centers (nuclei) of the atoms. By molecular mechanics, we understand the description at this level of approximation [166]. We imagine molecular mechanics in terms of a system of pointlike masses (atoms) moving in a force fields, which depends on the electrostatic interactions and chemical-bonding interactions between the atoms.

Molecular-force-field models combine the principles of quantum chemistry and empirical observations. There are many formulations, which differ in

9.3 Computational Methods for Modeling Molecular Structures

several details. Two well known standards for modeling molecular fields are CHARMM (Chemistry at Harvard using Molecular Mechanics) [42] and AMBER (Assisted Model Building with Energy Refinement) [217].

Below we present an expression for the energy of a molecule or a molecular complex [166]:

$$V(r^N) = \sum_{\text{bonds}} \frac{k_i}{2}(l_i - l_{i0})^2 + \sum_{\text{angles}} \frac{k_j}{2}(\theta_j - \theta_{j0})^2$$

$$+ \sum_{\text{torsions}} \frac{V_n}{2}[1 + \cos(n\omega - \gamma)] + \sum_{i=1}^{N}\sum_{j=i+1}^{N} \frac{q_i q_j}{4\pi\varepsilon_0 r_{ij}}$$

$$+ \sum_{i=1}^{N}\sum_{j=i+1}^{N} 4\varepsilon_{ij}\left[\left(\frac{\sigma_{ij}}{r_{ij}}\right)^{12} - \left(\frac{\sigma_{ij}}{r_{ij}}\right)^{6}\right]. \quad (9.1)$$

By r^N we mean a collection of the 3D coordinates of the N atoms in the molecule (or molecular complex). Using the atom coordinates obtained from r^N, one can compute the values of the variables $l_i, \theta_j, \omega, r_{ij}$ in (9.1) describing the coordinate-free spatial conformation of the molecule. The various terms in (9.1) represent different types of interactions. The first three of the five terms correspond to covalent bonds between pairs of atoms; the list of bonds must be specified for the structure analyzed. The last two terms in (9.1) represent nonbonding interactions; potentially these can occur between any pair of atoms in the molecular structure.

The geometrical meaning of the variables $l_i, \theta_j, \omega, r_{ij}$ is explained in Fig. 9.10. By convention, the angles θ_j and ω are measured in degrees (°) and the units used for distances l_i and r_{ij} are angstroms ($\mathring{A}$), where $1\mathring{A} = 10^{-10} m$. Equation (9.1) also includes the parameters $l_{i0}, \theta_{j0}, k_i, k_j, V_n, \varepsilon_{ij}, \sigma_{ij}$. These are indexed by a single index or a pair of indices, which is related to the fact that they depend on the types of atoms and bonds. They will be discussed in more detail later.

Partial Charges

The quantities q_i and q_j in the model (9.1) called partial atomic charges, are electrical charges placed at the geometrical centers of the atoms. The values of q_i and q_j are not positive or negative integer multiples of the electron charge. Rather, they are rational numbers, which approximate the distributions of electrons around the atomic nuclei resulting from geometries of electron orbitals and atomic configurations. There are many approaches to estimating the partial charges of atoms in molecules, combining experimental results with theoretical and computational techniques [42, 217]. A simple and easy-to-use is the iterative method of Gasteiger and Marsilli [95].

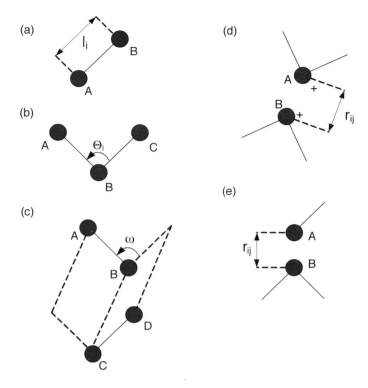

Fig. 9.10. Instances of atom configurations defining components of the molecular-force-field potential in (9.1)

Bond Lengths

The variables l_i describe the lengths of chemical bonds between atoms, as seen in Fig. 9.10a. A quadratic relation (Hooke's law) is assumed for the energy resulting from bond stretching or compression. The bond reference lengths l_{i0} and the force constants k_i depend on the atoms bonded and on the type of the bond.

Valence Angles

Valence angles θ_j are angles between straight lines AB and BC, where A, B and C are centers of atoms and the bonds between atoms are A–B and B–C, as shown in Fig. 9.10b, where the valence angle for three atoms is shown. Again a quadratic relation is assumed for the energy, and reference values θ_{j0} and the constants k_j depend on the atoms in the configuration A–B–C.

Torsion Angles

Torsion or dihedral angles ω are the angles between planes ABC and BCD determined by atom centers A, B, C, and D in a configuration with bonds A–B–C–D as presented in Fig. 9.10c. The constants V_n depend on the composition of atoms in the set A, B, C, and D. Parameter n is the multiplicity, in other words the number of maxima that the function representing the torsion component has as the torsion angle varies from $0°$ to $360°$. Finally, the constant γ describes the orientation of the bond at which the minimal value of the energy occurs. Parameters V_n, n, and γ depend on the atoms and on the configuration.

The energies related to changes of dihedral angles are lower than those associated with stretching of bonds and bending of valence angles. Compared with the rather stiff bond lengths and valence angles, there is a substantial conformational flexibility related to possible changes in the torsion angles. The variations of the shapes of molecules and complexes of them result, to a large extent, from interactions between the torsional and nonbonding components of their force fields.

Electrostatic Interactions

The energy of the electrostatic interactions, given by the fourth term in (9.1), is the usual energy of the Coulomb interaction between electric charges q_i and q_j (in units of Coulombs), placed at a distance r_{ij} apart. The constant ε_0 is the permittivity of free space or the vacuum permittivity equal to 8.854×10^{-12} F/m. The electrostatic interactions are long range since they decay proportionally to the inverse of the distance between the atoms.

Van der Waals Interactions

The last term in (9.1) represents the energies of the van der Waals interactions between atoms at a distance r_{ij}. This expression is also called Lennard–Jones 12–6 potential for the obvious reason for the "12–6" being the occurrence of powers with exponents 12 and 6. The power term with exponent 12 represents a repulsive force, and the other term, with exponent 6, an attractive force. As shown in Fig. 9.11, at some distance $r_{ij} = r^*$, specific to the interacting atoms, the energy of the van der Waals interaction becomes a minimum and the related force is zero. The parameters ε_{ij} and σ_{ij} are called the well depth and collision diameter, respectively. Geometric interpretations of these parameters are shown in Fig. 9.11.

The van der Waals interactions are short range since the attractive component decays as a power of the radius r.

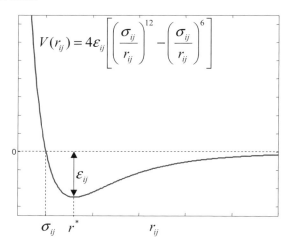

Fig. 9.11. Plot of the Lennard–Jones 12–6 potential representing the van der Waals interaction between atoms

Other Interactions

The estimation of parameters in molecular-force-field models is related to experimental research involving various measurements. The model validation concerns comparing the minimum-energy structures obtained from molecular field models, with the resolved conformations obtained, for example, from diffraction experiments. If the predictions of the modeling do not agree with the target structure, it may be necessary to make modifications to the structure of the force field and/or to include interactions other than those listed above, such as higher-order nonbonding interactions, "out of plane" terms for bonded interactions [166].

The Propane Molecule

In Fig. 9.12 a spatial model of a propane molecule, C_3H_8 is presented. For this molecule we have the following interactions:

Bonding interactions. In the first term in the sum in (9.1), there are ten bonds between atoms (two C–C bonds and eight C–H bonds). In the second term in the sum in (9.1), there are 18 valence angles (one C–C–C angle, ten C-C-H angles and seven H-C-H angles). In the third term in the sum in (9.1), there are 18 torsional interactions (12 H–C–C–H and six H–C–C–C torsion angles).

Nonbonding interactions. Potentially there are $11 \times 10/2 = 55$ possible nonbondig interactions between pairs of atoms in the propane molecule. However, we exclude interactions between atoms which are either directly bonded or two bonds distant (constrained by one valence angle). This is because the energies of nonbonding interactions are negligible compared with those related

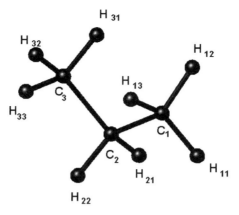

Fig. 9.12. Spatial model of a propane molecule, C_3H_8

to changes of bond lengths and valence angles and therefore with no influence on the geometry of the molecule. There are 27 nonbonding terms remaining, including 21 H–H interactions and six H–C interactions.

9.3.2 Molecular Dynamics

The molecular-force-field model (9.1) can serve various purposes, such as (i) predicting conformations of molecules, (ii) predicting geometries of polymers composed of several molecules, and (iii) formulating equations of motion in molecular dynamics. In (i) and (ii), the optimal conformation of a molecule or of a set of molecules is found by minimization of the potential energy $V(r^N)$. In (iii), one formulates the Newtonian equations of the dynamics of the multibody system by computing a force field from the potential-energy field $V(r^N)$ and then adding data on the masses of the atoms [42, 217].

9.3.3 Hydrogen Bonds

When the molecular force field model (9.1) is used for a system composed of multiple molecules, the relative positions of the molecules follow mainly from intermolecular interactions related to the electrostatic interactions between the partial charges of the atoms. In most situations the analysis of such intermolecular interactions is reduced to a finite number of interactions called hydrogen bonds. Hydrogen bonds, which have already been frequently mentioned in this book, can be represented with the use of the force field model (9.1) as dipole–dipole bonds resulting from electrostatic interactions. An example of a hydrogen bond between N–H and C=O dipoles in two amino acids is shown in Fig. 9.13. The partial atomic charges are [137] q_1^-, q_1^+ for

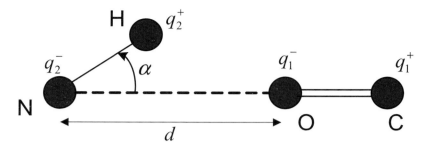

Fig. 9.13. A model of a hydrogen bond between N–H and C=O in two amino acids

(O=C) and q_2^-, q_2^+ for N–H, where $q_1 = 0.42e$, $q_2 = 0.2e$. One can intuitively understand the mechanism of attraction between N–H and C=O as following from a configuration where the attracting charges q_1^- and q_2^+ are closer to each other than the charges q_1^-, q_2^- and q_1^+, q_2^+ related to repelling forces. The bond configuration can be described by two parameters, a distance d and an angle α. Ideally for a hydrogen bond, $d = 2.9$ Å and $\alpha = 0$. However, electrostatic attraction between $N - H$ and $C = O$ can occur also for other values of d and α. From (9.1), the energy of electrostatic interaction between N–H and C=O in Fig. 9.13 is given by the following formula

$$E = \frac{q_1 q_2}{4\pi\varepsilon_0}\left(\frac{1}{d_{\mathrm{NO}}} + \frac{1}{d_{\mathrm{HC}}} - \frac{1}{d_{\mathrm{HO}}} - \frac{1}{d_{\mathrm{NC}}}\right), \qquad (9.2)$$

where d_{NO}, d_{HC}, d_{HO}, and d_{NC} are distances between atoms, which can be computed from d and α. Deciding on the occurrence of a hydrogen bond is done by setting a threshold for E in (9.2) [137].

9.3.4 Computation and Minimization of RMSD

A problem shared by many analyses related to the computational aspects of prediction of structures of molecules and complexes of them, is computing and/or minimizing the root mean square deviation (RMSD) of two particles or, more generally, of two sets of points. Here we define this problem and show two methods for its solution. One method uses the SVD technique described in Chap. 4 [136]. Another approach uses quaternions [124]. These methods are computationally equivalent [56], but in some circumstances one can be more convenient than the other.

Description of Displacements of a Rigid Body

Every displacement of a rigid body in three dimensions can be represented as a composition of rotation and translation. We denote by x a three-dimensional vector describing the coordinates of a point belonging to a rigid body before

9.3 Computational Methods for Modeling Molecular Structures

displacement, and by x' the vector of the three coordinates of the same point after displacement. Then the equation relating x' and x is then [156]

$$x' = Rx + t \tag{9.3}$$

where R denotes a 3×3 rotation matrix and t is a three-dimensional translation vector. The rotation matrix R is orthonormal, i.e., $R^T R = I$ (^{T} means transposition and I is the 3×3 identity matrix), and preserves the chirality of the space, i.e., $\det(R) = 1$. There are two basic ways of parametrization of a rotation matrix R, (1) describing R as composition of rotations around axes by three Euler angles, and (2) expressing R as resulting from rotation, by a given angle, around a specified axis of rotation [156]. The second of the two parametrizations is related to the representation of rotations by using algebraic operations on quaternions, which will be described later.

Definition of RMSD

Given two sets of points $Y = \{y_1, y_2, \ldots, y_N\}$ and $Y' = \{y'_1, y'_2, \ldots, y'_N\}$, with established correspondences y_1 to y'_1, y_2 to y'_2, ..., and y_N to y'_N, their RMSD is defined by

$$\text{RMSD}(Y, Y') = \sqrt{\frac{1}{N} \sum (y_i - y'_i)^2}. \tag{9.4}$$

The weighted RMSD (WRMSD), given by

$$\text{WRMSD}(Y, Y') = \sqrt{\frac{1}{\sum_{i=1}^{N} w_i} \sum w_i (y_i - y'_i)^2},$$

is also often used. However, by defining $z_i = \sqrt{w_i} y_1$ and $z'_i = \sqrt{w_i} y'_1$, WRMSD(Y, Y') can be transformed to RMSD(Z, Z') so we focus on the formulation in (9.4).

The Problem of Minimization of RMSD

This problem involves rotating and translating one structure (e.g., a rigid chemical molecule), called the model, such that its points become as close as possible to another structure, called the target. Mathematically, this can be expressed as

$$\min_{R, t} \text{RMSD}(Y, RX + t), \tag{9.5}$$

where, again, $Y = \{y_1, y_2, \ldots, y_N\}$ is a set of points, and by $RX + t$ we mean translated and rotated points of the corresponding set X, i.e., $RX + t = \{Rx_1 + t, Rx_2 + t, \ldots, Rx_N + t\}$.

9.3.5 Solutions to the Problem of Minimization of RMSD over Rotations

We start from the case where there is no translation, where we solve the problem of minimization of RMSD over all possible rotations R,

$$\min_R \text{RMSD}(Y, RX). \tag{9.6}$$

This problem can be approached analytically.

Solution by Singular Value Decomposition (SVD)

Here we present the application of the singular value decomposition discussed in Sect. 4.4.1 to solving (9.6). First we use the interpretation given in Sect. 4.4.2 to build up our understanding of the geometric structure of the problem and then we show the general solution.

For an X in (9.6) composed of three-dimensional column vectors x_1, x_2, ..., x_N, we use the matrix notation

$$X = [x_1 \ x_2 \ \ldots \ x_N] \tag{9.7}$$

and consider the economy-size SVD described in (4.44)

$$X = U_X \Sigma V^T, \tag{9.8}$$

where U_X and Σ are 3×3 matrices and V^T is a $3 \times N$ matrix. Observe that from the geometric interpretation in Sect. 4.4.2, the matrices Σ and V^T in (9.8) are invariant with respect to rotation of the vectors x_1, x_2, ..., x_N. In other words, by using (4.30) or (4.20), we can see that replacing X in 9.8) by RX, where R is any rotation matrix, will not change the matrices Σ and V^T

We form a matrix Y in a way analogous to (9.7),

$$Y = [y_1 \ y_2 \ \ldots \ y_N], \tag{9.9}$$

and assume that it is known for certain that the molecule with atom coordinates Y results from rotating the molecule X, i.e.,

$$Y = RX. \tag{9.10}$$

From the considerations above, we have

$$Y = U_Y \Sigma V^T, \tag{9.11}$$

where Σ and V^T are the same as in (9.8), and

$$U_Y = RU_X,$$

and so the matrix R can be computed as

$$R = U_Y U_X^T. \tag{9.12}$$

Equation (9.12) is not a good solution to the problem (9.6) since, in general, the assumption (9.10) is not satisfied owing, for example, to errors in the measurements of the values of the coordinates. Therefore the solution to the minimization problem (9.6) is obtained in a way different from that in (9.12) [56, 188, 136], which we show below.

Using (9.7) and (9.9), we can express $RMSD(Y, RX)$ in (9.6) as follows:

$$RMSD(Y, RX) = \text{Tr}[(Y - RX)(Y - RX)^T], \tag{9.13}$$

where Tr stands for the trace operator for a matrix (sum of the diagonal elements). Using the properties of a trace of a matrix [103], we have

$$\text{Tr}[(Y - RX)(Y - RX)^T] = \text{Tr}(YY^T) + \text{Tr}(XX^T) - 2\text{Tr}(RXY^T). \tag{9.14}$$

Observe that only the last term, $-2\text{Tr}(RXY^T)$, depends on the rotation matrix R. So minimizing (9.14) is equivalent to maximizing $\text{Tr}(RXY^T)$. Let us compute the SVD decomposition for XY^T,

$$XY^T = U_{XY} \Sigma_{XY} V_{XY}^T, \tag{9.15}$$

where Σ_{XY} is a 3×3 matrix of singular values and U_{XY} and V_{XY}^T are appropriate 3×3 orthonormal matrices. Substituting (9.15) in $\text{Tr}(RXY^T)$ yields

$$\text{Tr}(RXY^T) = \text{Tr}(RU_{XY}\Sigma_{XY}V_{XY}^T) = \text{Tr}(V_{XY}^T RU_{XY}\Sigma_{XY}), \tag{9.16}$$

where the last equality follows from the rules for applying the trace operator [103]. In (9.16), $V_{XY}^T RU_{XY}$ is an orthonormal matrix, and Σ_{XY} is diagonal with positive entries. It can be easily verified, either by using the algebraic properties of orthonormal matrices or by interpreting (9.16) geometrically as a rotation of the vectors defined by columns of Σ_{XY}, (see Exercise 9), that (9.16) attains its maximum iff

$$V_{XY}^T RU_{XY} = I, \tag{9.17}$$

where I stands for the 3×3 identity matrix. Solving (9.17) leads to

$$R = V_{XY} U_{XY}^T. \tag{9.18}$$

Summing up the above, the solution to the problem of minimization of the RMSD expressed in (9.6) is as follows:

1. Find SVD decomposition of XY^T as in (9.15).
2. Compute optimal R using expression (9.18).

When using the procedure above, we may encounter one rather minor difficulty. Namely, for some data structures, it may return as the optimal R an orthogonal matrix but not a rotation matrix, so that $\det(R) = -1$. Such solution does not preserve chirality; it gives a mirror reflection of the molecule as the solution. For reasonably conditioned data, this should not, however, happen. Also, we can modify the above algorithm to avoid mirror reflections; see [56] and Exercise 10.

Solution by Using Quaternions

Quaternions are a convenient algebraic structure for manipulating rotations in three-dimensional space. Not only they are applicable to the RMSD minimization problem, but they can also be used for many other aspects of molecular dynamics in three-dimensional space. Below we shall introduce quaternions and present their application to minimization of RMSD. In this part of the presentation, we shall make one notational change, compared with the rest of this chapter. Namely, vectors in three-dimensional space will be denoted by capital letters to distinguish them from scalars and quaternions, which will be represented by lower-case letters.

Notation. First we describe the notation and some basic properties concerning vectors in three-dimensional space. We denote vectors in three dimensional space by capital letters X, Y, Z, U, Q, P. Scalars and quaternions are denoted by lower-case letters. Scalar components of vectors are also denoted by lower-case letters, for example, $X = [x_1, x_2, x_3]^T$. (Transposition sign T means that we understand X as a column vector.)

The scalar (inner) product of two vectors is denoted by $\langle X, Y \rangle$ or $X^T Y$ and is given by

$$\langle X, Y \rangle = X^T Y = x_1 y_1 + x_2 y_2 + x_3 y_3. \tag{9.19}$$

For two vectors X and Y, we can also define the vector (also called outer, or cross) product, denoted by $X \times Y$,

$$X \times Y = \begin{bmatrix} x_2 y_3 - x_3 y_2 \\ x_3 y_1 - x_1 y_3 \\ x_1 y_2 - x_2 y_1 \end{bmatrix}. \tag{9.20}$$

The vector product $X \times Y$ is a vector perpendicular to the plane determined by vectors X and Y with its direction given by a right-hand screw rule. The vector product is not commutative, $X \times Y = -Y \times X$. We shall also need to use the cross product $X \times Y \times Z = (X \times Y) \times Z$. It can be decomposed along the directions given by the vectors X and Y, as follows:

$$X \times Y \times Z = \langle X, Z \rangle Y - \langle Y, Z \rangle X. \tag{9.21}$$

Rotation of a Vector X by an Angle δ Around an Axis Given by a Unit Vector U. This situation is presented in Fig. 9.14. The vector after rotation is denoted by X^R. We also denote by X^U the vector obtained by orthogonal projection of X onto U. From the fact that U is of unit length, we have

$$X^U = \langle X, U \rangle U. \tag{9.22}$$

We also define

$$\hat{X} = X - X^U. \tag{9.23}$$

9.3 Computational Methods for Modeling Molecular Structures

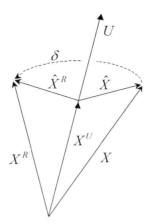

Fig. 9.14. Rotation of a vector X around an axis given by a unit vector U

The vector $\hat{X}$ undergoes a simpler, planar rotation by an angle δ. This rotation, in the plane normal to U, can be represented by a simple formula. To formulate this representation, we need to introduce an orthogonal coordinate system in this plane. One axis is parallel to $\hat{X}$. The other axis $\hat{Y}$, is obtained from

$$\hat{Y} = U \times \hat{X} = U \times (X - X^U) = U \times X. \qquad (9.24)$$

It is easily seen that $\hat{X}$ and $\hat{Y}$ are perpendicular to each other and that both vectors $\hat{X}$ and $\hat{Y}$ are of the same length. Therefore when we rotate the vector $\hat{X}$ in the plane $\hat{X},\hat{Y}$ by an angle δ, we obtain a vector $\hat{X}^R$ given by

$$\hat{X}^R = \hat{X}\cos\delta + \hat{Y}\sin\delta. \qquad (9.25)$$

Substituting (9.22), (9.23) and (9.24) in (9.25) yields

$$X^R = X^U + \hat{X}^R = (1 - \cos\delta)\langle X, U\rangle U + X\cos\delta + U \times X \sin\delta. \qquad (9.26)$$

In the above formula, we have expressed the result of rotation, X^R, as a vector sum of three vectors X, U and $U \times X$, with appropriate coefficients.

Quaternions. A quaternion q is a quadruple of real numbers a, b, c, d, represented as

$$q = a + bi + cj + dk, \qquad (9.27)$$

where the symbols i, j and k distinguish the coordinates of the quaternion. The algebra of summation and multiplication of quaternions fulfills all of the standard axioms, except commutativity of multiplication. The multiplication table for the symbols i, j, and k is

$$\begin{aligned} i*i &= -1, & i*j &= k, & i*k &= -j, \\ j*i &= -k, & j*j &= -1, & j*k &= i, \\ k*i &= j, & k*j &= -i, & k*k &= -1. \end{aligned} \qquad (9.28)$$

The norm of a quaternion $\|q\|$ is defined in a standard way,

$$\|q\| = \sqrt{a^2 + b^2 + c^2 + d^2}. \tag{9.29}$$

The conjugate of a quaternion q is

$$\bar{q} = a - bi - cj - dk \tag{9.30}$$

and by using the multiplication rules (9.28) we can prove (Exercise 11) that

$$q * \bar{q} = \|q\|^2. \tag{9.31}$$

The existence of the conjugate and the property (9.31) allow us to define the inverse and, consequently, division of quaternions:

$$\frac{1}{q} = \frac{\bar{q}}{\|q\|^2}, \quad \frac{q_1}{q_2} = q_1 * \frac{\bar{q}_2}{\|q_2\|^2}. \tag{9.32}$$

It can be proven (Exercise 11) that the norm and conjugate have the standard properties

$$\overline{q_1 * q_2} = \bar{q}_1 * \bar{q}_2 \quad \text{and} \quad \|q_1 * q_2\| = \|q_1\| \|q_2\|. \tag{9.33}$$

There are also representations of quaternions other than (9.27)-(9.28), for example, by pairs of complex numbers and by matrices. We shall use one more, a scalar–vector, representation, which is convenient for our further discussion. In this representation, a quaternion q has the form

$$q = q_0 + Q, \tag{9.34}$$

where q_0 is the scalar and Q is the three-dimensional vector of the quaternion q. The addition and multiplication rules for quaternions represented as in (9.34) are

$$(q_0 + Q) + (p_0 + P) = (q_0 + p_0) + (Q + P) \tag{9.35}$$

and

$$(q_0 + Q) * (p_0 + P) = (q_0 p_0 - \langle Q, P \rangle) + (q_0 P + p_0 Q + Q \times P). \tag{9.36}$$

In the above, $\langle Q, P \rangle$ and $Q \times P$ are scalar and vector products of three-dimensional vectors. It can be readily verified (Exercise 12) that the addition and multiplication rules (9.27) and (9.28) and (9.35) and (9.36) are equivalent. The conjugate $\bar{q}$ of the quaternion q in (9.34) is

$$\bar{q} = q_0 - Q. \tag{9.37}$$

Representation of rotations by using quaternion algebra. Let us consider the quaternion equation

$$q * x = y * q, \tag{9.38}$$

9.3 Computational Methods for Modeling Molecular Structures

which yields

$$y = \frac{1}{\|q\|^2} q * x * \bar{q}. \tag{9.39}$$

From (9.39), we see that scalars x_0, y_0 of the quaternions x and y are equal. What can be said about the vectors X and Y of the quaternions x and y? They are related by a rotation transformation around the axis given by the vector Q of the quaternion q (see Exercise 13).

The above fact allows us to state the following quaternion representation of rotations. Assume that q is a unit quaternion of the form

$$q = \cos\frac{\delta}{2} + U\sin\frac{\delta}{2}, \tag{9.40}$$

where U is a unit three-dimensional vector, i.e., $\|U\| = 1$. We then have

$$q * X * \bar{q} = X^R, \tag{9.41}$$

where by X we mean a quaternion with a scalar equal to zero, i.e., $X = 0 + X$. X^R is the vector resulting from rotating X around the axis given by the vector U of the unit quaternion q by the angle δ. This fact can be verified by using rules (9.35)-(9.37) in (9.41) and comparing the resulting expression for X^R with (9.26) (Exercise 13).

Solving the RMSD Minimization Problem by Using Quaternions. In the current notation the problem (9.6) of minimization of the RMSD over rotations can be stated as follows:

$$\text{RMSD} = \min_R \sum_{i=1}^{N} \|Y_i - RX_i\|^2 \tag{9.42}$$

where X_i and Y_i, $i = 1, 2, \ldots, N$ are the corresponding three-dimensional vectors and R is a rotation matrix. Using (9.41), we have

$$RX_i = X_i^R = q * X_i * \bar{q}, \tag{9.43}$$

and minimization over rotation matrices is transformed to minimization over unit quaternions. Substituting (9.43) in (9.42), we obtain

$$RMSD = \min_q \sum_{i=1}^{N} \|Y_i - q * X_i * \bar{q}\|^2 = \min_q \sum_{i=1}^{N} \|Y_i * q * \bar{q} - q * X_i * \bar{q}\|^2$$

$$= \min_q \sum_{i=1}^{N} \|(Y_i * q - q * X_i) * \bar{q}\|^2 = \min_q \sum_{i=1}^{N} \|(Y_i * q - q * X_i)\|^2. \tag{9.44}$$

The last problem is a minimization of a sum of squares of elements which depend linearly on the coordinates of the quaternion q. Therefore this problem can be formulated as a homogenous least squares problem and solved by using appropriate methods of static optimization from Chap.5.

9.3.6 Solutions to the Problem of Minimization of RMSD over Rotations and Translations

In the case where we have to minimize RMSD over both rotations and translations, the solution proposed by many authors involves centering both sets of vectors in space. Then translation is the vector defined by two centers and rotation is computed as described above. In the case where the problem is error-free or close to error-free, this approach will lead to the true minimum. However, when there are errors, this approach does not guarantee that the minimum will be reached. In that case iterations of centerings and minimizations over rotations are typically applied.

9.3.7 Solvent-Accessible Surface of a Protein

The interactions between a protein and other molecules (proteins or ligands) are determined by the shape of the outer surface of the protein and the electrostatic properties of adjacent atoms. The problem of modeling the outer shapes of proteins has been researched for a long time [167, 239, 254]. Computer software is available in the internet and appropriate algorithms are constantly being improved and developed [31, 176, 178]. In the context of describing the outer surface of a protein molecule, we can introduce two models, presented in Fig. 9.15, the solvent accessible surface, defined as the locus of the center of a solvent molecule as it rolls over the van der Waals surface of the protein (on the left-hand side of the figure), and the molecular surface, defined as the locus of an inward-facing probe sphere (on the right-hand side). The radius of the probe sphere can be assumed to have different values; its value is often taken as approximately equal to the radius of a water molecule. An algorithm for calculating the solvent-accessible surface, developed in [167], uses the idea of slicing the expanded atom volume into two-dimensional crossections along a single direction. Two-dimensional slices are easier to analyze. By aggregating the results of analyses of all of the slices, one obtains the solvent accessible surface of the whole molecule. In [254], another idea was introduced, of placing dots, with a fixed density, on the expanded atom surfaces. Dots shared by two or more expanded spheres are removed, and dots which are not inside any other expanded sphere are considered to be accessible to a solvent.

In many approaches to studying protein structure and function, the area of solvent accessible surface of a protein is computed and used as an index or parameter in modeling.

9.4 Computational Prediction of Protein Structure and Function

Computational prediction of the structure, properties, and functions of proteins is very important in computational chemistry, chemoinformatics, and

9.4 Computational Prediction of Protein Structure and Function

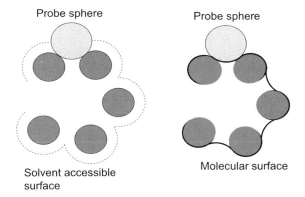

Fig. 9.15. Two-dimensional illustration of solvent-accessible surface (*left*) and molecular surface (*right*)

bioinformatics. The interest in this area is motivated by a wide spectrum of applications, in molecular biology, the medical sciences, pharmacology, and so forth. There are many different aspects and a variety of methods and approaches. Related problems involve predicting the secondary and tertiary structures of proteins on the basis of linear sequence of amino acids, estimating the positions of active sites in proteins, and predicting different types of interactions, both protein–protein and protein–ligand. Such problems may also involve inference about the relation between parameters such as the measured amino acid composition of samples and the outcomes of various types of diagnoses, the relation between these parameters and risks etc. In many of the proposed approaches there is a lot of three-dimensional geometry, since, as shown above, it is important for the modeling of many types of interactions. Also, appropriate statistical methods must be applied owing to biological and measurement-related variations. Finally, the results are based on large bioinformatic databases of measurements, structures, interactions, etc.

Below, some of the methods and approaches for the computational prediction of protein structure and function are presented.

9.4.1 Inferring Structures of Proteins

The linear sequences of amino acids are known for a large number of proteins, on the basis of reading sequences of codons from completed genomes of many organisms. The secondary and tertiary structures are known for only a fraction of these proteins. This fact naturally raises the problem of predicting the secondary and tertiary structures of proteins by using their primary structure. Basically, there are two main approaches to inferring the structures of proteins, by comparative modeling and de novo approach. In comparative modeling the query sequence (the sequence for which the structure is to be predicted) is aligned against a number of sequences from a database of known structures. These sequences are selected on the basis of a criterion of similarity to the query. In the de novo approach, the prediction is done solely on the basis of the sequence of amino acids, without using alignment as the initial step. The methods in the two groups share some common elements. After a query sequence has been aligned against similar sequences from a database, the differences are resolved by using molecular modeling and indices such as hydrophobicity and solvent accessible surface area. Also, de novo methods do not start from first principles but rather use short template sequences generated with the use of a protein database.

9.4.2 Protein Annotation

Resolving the spatial structure of a protein, by X-ray diffraction or NMR, allows one to study the relations between the order of amino acids in the protein and the parameters describing its properties, i.e., Ramachandran angles, secondary motifs, accessible surfaces, etc. This analysis may be called protein annotation. The relations obtained may be then helpful for predicting the structure of proteins.

A frequently used method for protein annotation and a related computer program [137] are based on identifying hydrogen bonds in the protein by computing and thresholding the energies of dipole–dipole electrostatic interactions (9.2). Finding all hydrogen bonds allows one to describe the secondary structures. The program [137, 311], takes as an input a standard file from the PDB database [329]. Besides the secondary structure, it also computes energies of hydrogen bonds, the Ramachandran and torsion angles, the area of the solvent-accessible surface, etc. We shall now illustrate it using the enzyme trypsin, presented previously in Fig. 12.2. Below, in the first row we have printed a fragment of the amino acid sequence of trypsin, and in the second row, we show the codes representing the secondary structure of the protein, obtained using the program from [137]:

```
PVVCSGKLQGIVSWGSGCAQKNKPGVYTKVCNYVSWIKQTIASN
EEEECCEEEEEEEECCCCCCCCCEEEEECCCHHHHHHHHHHC
```

In the second row E stands for a beta sheet, H for an alpha helix and C for coil. Often, more categories are used to annotate the regions referred to in the above by the one symbol. More symbols and explanations of them can be found in Table AII of [137].

9.4.3 De Novo Methods

In principle one can imagine computational prediction of protein structures by using procedures of minimization of the conformational energy (9.1) over the conformation parameters. However, not only do proteins contain very large numbers of atoms, but also the spatial folding of a protein is a consequence of both amino acid–amino acid interactions between amino acids in the protein, and interactions between amino acids in the protein and the surrounding (water) molecules. These latter interactions depend both on (nonbonding) interactions between water molecules and amino acid molecules, described in (9.1), and on the kinetic energy of the water molecules. Increasing the kinetic energy, i.e., raising the temperature of the solution, changes the protein conformation and eventually may lead to thermal denaturation of the protein. Most often, systems for experimental assays of protein structures and systems for predicting protein conformations are focused on constant temperature of $37°C$.

The above facts make the problem of de novo computation of protein conformations complicated. In order to sum, or average, the effect of the surrounding environment on the protein conformation, intuitive indices are used [14], following from observations that, in the optimal conformation, hydrophobic amino acids should be buried inside the molecule and the global shape of the molecule should be as compact as possible.

In the area of practically used de novo methods, yielding reliable low-resolution predictions of proteins spatial structures of proteins, we should mention algorithms given in [256, 48] and the related Rosetta Internet server [257, 336]. In the algorithm in [256], the amino acid sequence of a query protein is chopped into short (nine residue long) pieces. On the basis of its amino acid sequence, each of the nine-residue pieces is then associated with a probability distribution of possible spatial shapes. In principle, this step of relating short sequences of amino acids to 3D shapes and their probabilities, could be done by browsing through the PDB database. For increased efficiency, however, the authors of the algorithm in [256] are using another, more specialized database, HSSP [248, 319], of protein structure–sequence alignments. The next phase of the algorithm involves sampling the shapes of nine-residue pieces, on the basis of the probability distributions of the shape as a function of the sequence. This allows one to make guesses of the global shape of the target protein. Each of the guesses is scored by using the criteria that the hydrophobic amino acids should be buried inside the molecule, the hydrophilic amino acids should occupy outer surface of the molecule, and that the shape of the molecule should be as compact as possible. Developing mechanisms for sampling and scoring allows one to organize the procedure of conformation optimization in the form of a simulated annealing-version of the Metropolis–Hastings algorithm, described in Chap 2.

9.4.4 Comparative Modeling

Comparative modeling is a flexible and powerful tool, which can be used to predict structures, active sites, and functions of proteins. It is based on multiple alignments of the query sequence against sequences in a protein database, which reveal similarities in the sequences. It can be also helpful here to construct a tree for several sequences, based on similarity relations between aligned sequences.

Predicting Secondary and Tertiary Structures of Proteins by comparative Modeling

Inferring the secondary and tertiary structures of proteins by comparative modeling is based on the observation that proteins with similar sequences of amino acids also have similar secondary and tertiary structures [38, 300]. The observation [300] that when sequences are aligned, the regions of insertions and low sequence conservation often correspond to loop regions, whereas the regular secondary structures correspond to amino acid sequences conserved in alignments, has allowed improvements to be made in algorithms for prediction of the secondary structures of proteins. In the present algorithms for predicting of the structures of proteins, such as those in [59, 322, 245], the target amino acid sequence is aligned against a large database of sequences with known secondary structures. A specialized alignment tool, PSI BLAST [250], is often used, which was developed with the aim of increasing the sensitivity to distant sequence relationships. A simple idea for estimating the secondary structure of a target sequence is to find its nearest neighbors in a database and to choose the secondary structure of the majority of its neighbors as the prediction [247]. However, more sophisticated methods, such as neural networks and hidden Markov models, can improve the predictions [9, 231].

Ideas similar to the above can be developed for predicting proteins tertiary structures by using alignment of amino acid sequences [38]. A recent method for estimating the spatial structures of proteins is threading [101]. Protein threading is also known as fold recognition. The target sequence is drawn, or "threaded", along the backbone structures of a collection of template proteins with known spatial structures, collected from a protein library. A score index is calculated for each position of the query sequence. The construction of this index is crucial for performance of the algorithm, and many methods and approaches have been proposed in the literature. Typically, the index contains two types of components, some stemming from the similarity between the query and database sequences (quality of alignment) and some reflecting the energy of the protein molecule (quality of folding).

Prediction of Active Sites of Proteins by Comparative Modeling

Also, by analyzing multiple alignments of amino acid sequences, the locations of functionally important residues in proteins can be predicted [300]. Functionally important residues can be related to active sites of a protein, responsible for its interactions with other proteins or other molecules, as well as residues whose interactions are responsible for fixing the shape of the protein molecule. A method which takes advantage of this idea is the evolutionary trace method, developed and presented in [170, 171]. This method is based on using multiple alignments of a group of homologous proteins in a protein database, building a phylogeny tree (a sequence identity dendrogram), and using the sequence–tree hierarchy relation to predict functionally important residues. The name "evolutionary trace" emphasizes the possibility of retrieving useful information from databases, where sequences homologous to the protein of interest record past mutations.

9.4.5 Protein–Ligand Binding Analysis

An important area is the computational prediction of interactions between proteins (protein active sites) and ligands. Ligands are molecules which can bind to an active (receptor) site of a protein. The interactions between these receptors and ligands depend on their spatial molecular shapes and the distribution of their electrostatic charges. The problem of computational prediction of interactions between ligands and proteins is also called molecular docking. Knowledge about protein–ligand interactions and their prediction is important, for example, in pharmacology for the purpose of drug design [40]. There are many computational approaches to molecular docking [159]. Here we mention several of the directions.

One approach is related to constructing scoring or energy functions, originating from the energy formula (9.1). These functions are intended to describe the energy of protein–ligand interactions. The formula (9.1) cannot be directly applied owing to the need to take the contribution of the surrounding water molecules into account. Therefore empirical energy formulas [34, 279] contain heuristic components for representing the relevant interactions. Computing ligand–protein interactions is done under the assumption of given relative spatial positions of the interacting molecules and of fixed conformations. The computational tools for predicting interaction energies are then accompanied by algorithms for changing and updating the spatial positions and conformations. These algorithms often use variants of the MCMC or simulated-annealing methods presented in Chap. 2.

Another less precise approach, is based on detecting complementarity of shapes of molecules. Typically, electrostatic interactions are ignored at this stage of computations. Detecting shape complementarity can be done with an algorithm using the method of geometric hashing (described in Chap. 4) [211].

There are also many other approaches, using principal component analysis, neural networks, genetic algorithms, and so forth.

9.4.6 Classification Based on Proteomic Assays

There is a great potential for proteomic assays in the diagnosis of states of organisms and cells by proteomic measurements at the molecular level. The

296   9 Proteomics

composition of protein species present in organic liquids, as well as the proportions of them, obtained in various types of experiments, can be treated as features or patterns and used as input data in methods for classification and clusterization. Several studies in this direction have already been pursued. For example in [175, 226, 293] proteomic mass spectra were applied to the classification of the blood samples as diseased or healthy. For this classification and clusterization of protein assay measurements, one can use the algorithms described in Chap. 4.

9.5 Exercises

1. Atoms in the propane molecule shown in figure 9.12 are at the following spatial coordinates

Atom	x (Å)	y (Å)	z (Å)
C1	2.103	1.469	−0.406
C2	2.538	2.947	−0.406
C3	4.078	2.947	−0.406
H11	0.963	1.469	−0.406
H12	2.452	1.003	−1.386
H13	2.555	0.865	0.448
H21	2.189	3.413	−1.386
H22	2.085	3.551	0.448
H31	4.530	2.343	−1.261
H32	4.427	2.480	0.574
H33	4.399	4.041	−0.406

    Using the above data, compute the following parameters describing the chemical bonds in the propane molecule: the bond lengths $l_1, \ldots, l_{10}$, the valence angles $\theta_1, \ldots, \theta_{18}$, and the torsion angles $\omega_1, \ldots, \omega_{18}$.
2. Write a computer code to solve Exercise 1.
3. Write a computer program to compute propane atom coordinates, based on the values of the bond lengths, the valence angles and the torsion angles.
4. Compute the permittivity of the free space $\varepsilon_0 = 8.854 \times 10^{-12}$ F/m using the following units: $e^2/(kcal\ \text{Å})$.
5. Using (a) the data in Exercise 1, the (b) the program from Exercise 2, and (c) (9.1), draw plots which represent how the energy $V(r^N)$ of the propane molecule depends on the values of the torsion angles H–C–C–C. In computations, use the following (simplified) set of values for parameters:

parameter	unit	value
k (bond lengths)	$(kcal/(mol\ \text{Å}^2))$	300
k (angle-bending)	$(kcal/(mol\ deg))$	0.01
V (dihedral angles)	$(kcal/mol)$	3

Omit the van der Waals interactions and use the following partial charges of atoms:

Atom	$q\ (e)$
C1	-0.63907
C2	-0.43079
C3	-0.63907
H11	0.21124
H12	0.21124
H13	0.21124
H21	0.21426
H22	0.21426
H31	0.21124
H32	0.21124
H33	0.21124

6. Try solving Exercises 1–5 for other molecules.
7. Describe how an algorithm for minimization of the RMSD can be applied for computing the Ramachandran angles, given the spatial coordinates of the atoms of a peptide chain.
8. Download a PDB file describing a protein, for example, trypsin (2ptn). Try to use the method derived in Exercise 7 to compute the Ramachandran angles.
9. Prove the property leading to the assertion (9.17) that if Σ denotes a diagonal matrix with nonnegative elements,

$$\max_{U=\text{orthonormal matrix}} \text{Tr}(U\Sigma) = \text{Tr}(\Sigma)$$

is attained for $U = I$.

10. Modify the algorithm for minimizing the RMSD by using SVD, such that mirror reflections will be excluded (see [56]).
11. Prove the assertions (9.31) and (9.33).
12. Prove the equivalence between the addition and multiplication rules (9.27) and (9.28) and (9.35) and (9.36).
13. Prove the representation (9.41) using the rules (9.35)–(9.37) in (9.41) and comparing the resulting expression for X^R with (9.26).
14. We call M_q^L a matrix representation of the quaternion left multiplication $p * q$ if

$$\begin{bmatrix} y_0 \\ y_1 \\ y_2 \\ y_3 \end{bmatrix} = M_p^L \begin{bmatrix} q_0 \\ q_1 \\ q_2 \\ q_3 \end{bmatrix}, \qquad (9.45)$$

where $y = p * q$ and $y_0 \ldots y_3$ and $q_0 \ldots q_3$ are the components of the quaternions y and q. Derive values for the entries of the matrix M_p^L such that (9.45) represents quaternion left multiplication $p * q$. Derive, in an analogous way, a matrix M_p^R for representing the right quaternion multiplication $q * p$. Use the matrix representations of left and right quaternion multiplications to develop a detailed algorithm for solving the minimization problem (9.44).

15. Derive a matrix representation for the transformation $q * X_i * \bar{q}$ in equation (9.43). This representation corresponds to a parametrization of the rotation matrix R by the direction of the axis of rotation and the angle of rotation.

10
RNA

Ribonucleic acid (RNA), is a polymer of repeating units, namely ribonucleotides, with a structure analogous to single-stranded DNA. It has a backbone composed of sugars (riboses) and phosphate groups, with organic nitrogen bases bonded along it. The differences between RNA and DNA (see Fig. 8.7) concerning their components are (i) sugar deoxyribose which appears in DNA is replaced in RNA by another sugar, ribose, and (ii) the organic base thymine (T) that appears in DNA is replaced in RNA by another organic base, uracil (U). Compared with DNA, RNA molecules are less stable and exhibit more variability in their three dimensional-structure which underline their different function.

In living organisms, RNA arises in the process of transcription which involves creating single-stranded RNA, based on a DNA template according to the complementarity pairing rules A–U, C–G, $G - C$ and $U - A$. Similarly to DNA, the RNA chain has a direction, two ends of RNA are labeled $5'$ and $3'$. For example, an RNA strand copied from the left strand of DNA piece from Fig. 8.4 will have the sequence of bases $3'$–$UGACUG$–$5'$. By the mechanism of transcription, many copies of RNA corresponding to one piece of DNA can be created. The process of transcription is catalyzed by an enzyme RNA polymerase, which performs two functions: it unwinds the DNA (separates the two DNA strands), and slides along the DNA strand, forming RNA according to the complementarity rules. In eukaryotic organisms, there are various types of RNA polymerase, specializing in producing different types of RNA [118, 216, 287].

In the central dogma of molecular biology illustrated in Fig. 8.6, RNA serves mainly as a carrier of information from the DNA to the ribosomes, where it is utilized for protein construction. However, recent developments in molecular biology suggest that the importance of the various kinds of RNA molecule for the development and functioning of living organisms has not yet been sufficiently appreciated. In the forthcoming sections, we will mention these arguments and reference some related recent publications.

Experimental research on the structure and functions of RNA, proceeds along many of the directions that we have were already sketched for DNA and proteins. These include electrophoresis techniques for estimating the length of molecules, sequencing, blots, and X-ray diffraction and NMR methods for analysis of the three-dimensional structure. These methods must also be supported by appropriate developments in bioinformatics, to order, organize, and make accessible large amounts of data. Information on the sequences of ribonucleotides in RNA, functions of RNA molecules and their spatial structures are available in appropriate databases, coding RNA sequences are in gene banks [326], some recently established databases for non coding sequences of RNA can be found in [106, 181, 318, 334], and databases of spatial structures of RNA can be found in [321, 329]. The mathematical and computational aspects of research on RNA are also in most respects parallel to those for DNA and/or proteins, concerning inference on function by aligning sequences (as in the case of DNA and proteins) and studying and predicting the three-dimensional shapes of RNA molecules associated with efforts to relate the function of RNA to its spatial conformation.

10.1 The RNA World Hypothesis

The Processes involving replication, transcription, and translation are catalyzed by proteins. On the other hand proteins are constructed on the basis of the information written in DNA and carried by mRNA. This poses an evolutionary puzzle about how this functional organization evolved. Early theories gave prominence to amino acids and short peptides, as the earliest molecules in evolution. However, explaining the evolutionary scenario by the protein-first hypothesis suffers from serious problems related mainly to the lack of molecular mechanisms for the self replication of proteins. The present hypothesis, called the RNA world hypothesis, is that in the process of evolution, RNA molecules preceded DNA strands and proteins [98]. The scenario of self catalytic replication of RNA in the early stages of evolution is being reproduced in several laboratory experiments, see, for example, [277] and references therein.

10.2 The Functions of RNA

On the basis of on their functions, RNA molecules can be divided into two groups, coding and noncoding. The coding RNA is messenger RNA (mRNA) assembled by the machinery of (eukaryotic) cell during the processes of transcription, splicing, and polyadenylation. The order of the codons formed by the ribonucleotides in mRNA corresponds to the order of the amino acids in the linear content of proteins.

Noncoding RNA is transcribed from noncoding regions of DNA. Classically, there are two types of noncoding RNA, transfer RNA (tRNA) and ribosome RNA (rRNA). The transfer RNAs are short-chain RNA molecules (74–93 ribonucleotides) involved in transporting amino acid molecules to ribosomal sites, where the process of growing polypeptide chains occurs. Ribosome RNA (rRNA) is a component of ribosomes.

Recently there has been a great interest in RNA transcribed from noncoding regions of DNA, leading to the development of a much broader taxonomy of noncoding RNA. This direction of development is partly related to recent theories of the "dark matter of noncoding DNA and RNA". According to these theories, noncoding DNA is not "evolutionary junk" in the interpretation based on the classical molecular-evolutionary viewpoint, but rather has some important, not yet discovered, function in organisms [186, 187, 198]. One of the chief arguments supporting these theories is based on comparison of genomes of simple and complex organisms. The scale of complexity, seemingly, is not related to the number of genes nor to the number of chromosomes, but correlates better with the size of noncoding DNA. The conclusion is that non coding DNA and the non coding RNA transcribed from it are responsible for important functions of organisms, which are behind their diversification. Examples of the functions of noncoding RNAs are [186, 187, 198] regulatory functions in the process of gene expression, maintaining the telomeres, gene splicing, and chemical modification of ribosomal RNA. Organizing information about RNA families and their functions in a systematic way involves creating electronic databases that allow the relevant data to be deposited and downloaded. The Rfam database, [106, 334] contains curated lists of noncoding RNA families, classified by aligning their sequences.

10.3 Reverse Transcription, Sequencing RNA Chains

The standard direction of information flow is copying from the DNA template to RNA. However, the opposite direction, called reverse transcription, is also possible. This process is performed by the enzyme reverse transcriptase. In nature, reverse transcription is often met with in retroviruses, which appear to consist of two or more RNA molecules and attack the host cells by a self-replication strategy based on reverse transcription of their sequence to the host's genome. A very well known example is the HIV virus.

In the area of scientific research, reverse transcription is most often used for sequencing RNA by use of polymerase chain reaction (PCR) as in the case of DNA. The PCR can only be applied to DNA. By using reverse transcription, RNA can be first copied to DNA and then amplified (replicated) by means of the PCR.

10.4 The Northern Blot

The Northern blot is a technique analogous to the Southern blot described in Chap. 8. RNA from a specimen is separated by electrophoresis and fixed on a supporting plate. In the next step, single-stranded DNA fragments complementary to the specific mRNA being sought are labeled with radioactive atoms and then hybridized to the immobilized mRNA. If the specific mRNA is present, a radioactive band is detected. The name "Northern blot" was invented by altering the term "Southern blot" used for DNA assays.

10.5 RNA Primary Structure

The primary structure of RNA is the linear sequence of ribonucleotides. Comparative sequence analysis of ribonucleotides is a basic premise for the classification of RNA and prediction of its function and structure. RNA databases [334] contain a large number RNA sequences, enabling newly discovered sequences to be compared with existing RNA families.

10.6 RNA Secondary Structure

The shape of RNA is to a substantial extent formed by a series of hydrogen bonds that occur between the nitrogen bases in its backbone. These bonds result in the formation of characteristic motifs, shared by many RNA molecules, which can be represented graphically by using two-dimensional plots. Graphical representation of RNA including these motifs is called the secondary structure of RNA. The motifs encountered in RNA strands are, hairpin loops, internal loops, multibranched loops, bulges, and stems. An example of a secondary structure of an RNA strand is presented in Fig. 10.1. Hydrogen bonds occur between the complementary (Watson Crick) pairs of bases in RNA A–U and C–G. Additionally, a bond between G and U is also energetically favorable and is often encountered in RNA molecules.

10.7 RNA Tertiary Structure

The formation of the spatial (tertiary) structure of RNA molecules is related to the occurrence of hydrogen bonds between bases, other than those responsible for the formation of the secondary structure. These additional bonds contribute to the final 3D shape of RNA molecules. The tertiary structure of RNA can be established experimentally by X-ray diffraction of crystallized molecules or by NMR techniques. After the spatial shape has been computed, the hydrogen bonds that stabilize the spatial form of the molecule can be

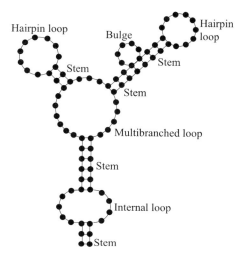

Fig. 10.1. Motifs in RNA secondary structure: stem, hairpin loop, bulge, multi-branched loop, and internal loop. Nitrogen bases are represented by small circles and hydrogen bonds between bases are depicted by thin, ladder-like line segments

found by analysis of the distances between the bases and their orientations [286, 19].

The motifs encountered in RNA tertiary structure and their occurrence in RNA species are stored in the SCOR database [153, 338]. Among the motifs in RNA spatial structure probably most important are coaxial bonds, which contribute to the formation of helical structures in RNA molecules. Other motifs include pseudoknots, kissing hairpin loops, and ribose zippers.

10.8 Computational Prediction of RNA Secondary Structure

The most reliable approach for prediction secondary structure, especially for long RNA sequences, is comparative sequence analysis discussed later in this chapter. This approach involves alignment of multiple RNA sequences and uses a covariance-type analysis aiming at identification of conserved base-pairing interactions in the RNA. Most of the secondary structures of long RNA sequences, accepted by experts and available via internet were obtained by using the method of comparative sequence analysis.

In cases where multiple RNA sequences either are not available or do not have enough diversity for comparative analysis, an alternative method of prediction of RNA folding is energy minimization. Although it is not as accurate as comparative analysis, it leads to useful predictions and can be applied, for example, when one wants to estimate quickly the second-order structure of a single RNA strand. Numerical minimization of the folding energy is performed by using the principle of dynamic programming. Below, we describe two basic algorithms in detail and also mention some other variants. The first approach, often called the Nussinov algorithm [210] simplifies the energy minimization problem by using the hypothesis that the more pairings there are between bases in an RNA strand, the lower the energy of the molecule. The second approach, [241, 281, 298, 299] assigns thermodynamic energies to motifs in the secondary structure. These energies depend on the size of the motifs, and the overall energy of an RNA strand is the sum of the energies of the motifs. The thermodynamic energies of the motifs have been tabulated [253] and they allow modeling the true ratios of the energy components. This enables, for example, the influence of the temperature on the shape of an RNA molecule to be modeled and studied.

10.8.1 Nested Structure

An important property of RNA secondary structures is that there are no crossings between bonds; in other words the structures are nested. The nested character of the secondary shape of RNA is presented in Fig. 10.2. In the upper part of this figure, a fragment of RNA of "hairpin" shape is presented and then in the lower part chain of bases is straightened out. Bonded bases are connected by half circles. The nested structure requires that the half circles cannot make crossings.

Observe that if there is at least one bond in an RNA chain then the secondary structure of this RNA string contains at least one hairpin loop. Contemplating the lower part of Fig. 10.2, one can notice an analogy between defining a nested structure of bonds between bases in RNA and arranging left and right parentheses in the correct order [104]. There is also a correspondence between nested secondary RNA structures and unlabeled trees, whose terminal leaves correspond to unbonded bases and whose topology reflects the structure of the bonds [281].

10.8.2 Maximizing the Number of Pairings Between Bases

As already said, the secondary structure of RNA is formed by series of bonds between bases. We assume that the number of bases in an RNA strand is N and we denote the sequence of the bases by $b_1, b_2, \ldots, b_N$. In this subsection we describe an algorithm for maximizing the number of pairings in the sequence $b_1, b_2, \ldots, b_N$ [210, 281]. It does not change the structure of the algorithm if we assume, more generally, that the score for a pair b_k–b_l is $s(b_k, b_l)$ and we maximize the sum of scores over all possible nested second-order foldings. This formulation becomes equivalent to maximizing the number of pairings when we take $s(A, U) = s(C, G) = s(G, U) = 1$ for energetically favorable pairs, and $s(b_k, b_l) = -\infty$ for all pairs b_k–b_l other than A–U, C–G, and G–U.

10.8 Computational Prediction of RNA Secondary Structure

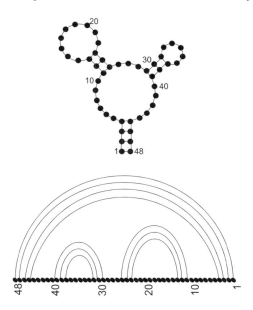

Fig. 10.2. Illustration of the nested property of the bonds between bases in the secondary structure of RNA. The *upper part* presents an example of the fragment of an RNA chain. In the *lower part*, the RNA chain is straightened out and bonded bases are connected by half circles

We introduce two triangular N-dimensional matrices $V(i,j)$ and $W(i,j)$, $i \leq j$, with the following meaning: $V(i,j)$ is the score of the best folding of the RNA subsequence $b_i, b_{i+1}, \ldots, b_j$, given that bases b_i and b_j form a bond, and $W(i,j)$ is the score of the best folding of the RNA subsequence $b_i, b_{i+1}, \ldots, b_j$ (no matter whether b_i and b_j are paired or not).

We shall state and explain recursions for $V(i,j)$, and $W(i,j)$. We start from the case where it is given that b_i and b_j form a bond. We then have

$$V(i,j) = s(b_i, b_j) + W(i+1, j-1). \tag{10.1}$$

In order to formulate a recursive relation for $W(i,j)$ one has to consider the following possibilities: (1) b_i and b_j form a bond, in which case $W(i,j) = V(i,j)$, and (2) b_i and b_j do not form a bond. In case (2), the nested property of bondings described in Sect. 10.8.1 guarantees that the strand $b_i, b_{i+1}, \ldots, b_j$ can be split into two strands $b_i, b_{i+1}, \ldots, b_k$ and $b_{k+1}, b_{k+2}, \ldots, b_j$ such that there are no bonds between them, and consequently their total score is $W(i,k) + W(k+1,j)$. Summing up (1) and (2) we obtain

$$W(i,j) = \max\{V(i,j), \max_{i \leq k < j}[W(i,k) + W(k+1,j)]\}. \tag{10.2}$$

The recursions (10.1) and (10.2) can easily be shrunk into one,

$$W(i,j) = \max\{s(b_i, b_j) + W(i+1, j-1), \max_{i \leq k < j}[W(i,k) + W(k+1,j)]\}. \quad (10.3)$$

Starting from $W(i,i) = 0$, $i = 1, 2, \ldots, N$, and $W(i, i+1) = 0$, $i = 1, 2, \ldots, N-1$ and then using (10.3) we can fill in all entries of $W(i,j)$ $i = 1, 2, \ldots N$, $i \leq j$.

The algorithm (10.3) has a complexity of order $O(N^3)$ since filling in each entry of $N \times N$ matrix requires running the index k over a range $O(N)$.

10.8.3 Minimizing the Energy of RNA Secondary Structure

Predicting the secondary structure of RNA only by maximizing the number of pairings between bases is an oversimplification. Results closer to the shapes found in experiments are obtained by assigning energies to the motifs of RNA secondary structure, i.e., stems, hairpin loops, bulges, internal loops, and multibranched loops, and searching for the structure with the lowest energy. This more accurate model assumes that RNA folding stems from an interplay between the stabilizing role of base pairings and the destabilizing effects of unpaired segments of hairpin loops, bulges, internal loops, and multibranched loops.

Hairpin RNA Structure

Let us start by describing an algorithm for minimizing energy of an RNA strand under some restrictions. Namely, we aim at minimizing the energy of RNA sequence of bases $b_1, b_2, \ldots, b_N$, assuming additionally that (i) b_1 and b_N are paired, and (ii) the possible secondary motifs are hairpin loops, stems, bulges, and internal loops. We call such a structure a hairpin RNA structure. Since multibranched loops are excluded, hairpin structures do not branch. An example of a hairpin RNA structure is shown in Fig. 10.3. As seen in this figure, a hairpin RNA structure can be thought of as a sequence of motifs, such as those shown, *stem*1, *iloop*1, *stem*2, *bulge*1, *stem*3, and *hloop*1. Each of the possible motifs is characterized by its size:

- $stem(k)$–a stem of k consecutive pairs of bases;
- $bulge(1,k)$, $bulge(k,1)$–right and left bulge of k bases,
- $iloop(k_1, k_2)$–an internal loop of k_1 bases on the left and k_2 bases on the right;
- $hloop(k)$–a hairpin loop of k bases.

The energies of motifs have been experimentally measured and their values are available in the literature and on the Internet [253, 344]. The energies of motifs depend not only on their size but also on the composition of bases, and it is necessary to specify the location of a motif relative to the sequence of RNA bases $b_1, b_2, \ldots, b_N$. It is convenient to introduce the concept of motifs of type 1 and type 2 (based on the presentation in [241]). A motif of type

10.8 Computational Prediction of RNA Secondary Structure

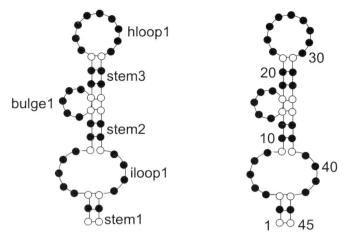

Fig. 10.3. A hairpin structure in RNA secondary folding. On the *left-hand* side, the sequence of motifs of the hairpin loop is marked *stem*1, *iloop*1, *stem*2, *bulge*1, *stem*3, *hloop*1. On the *right-hand* side, the numbers of ribonucleic bases are depicted, to be used in the text

1 is a hairpin loop. It is fully specified by a pair of indices i_s, j_s of bases, $i_s, j_s \in 1, 2, \ldots, N$, and is denoted by

$$M1(i_s, j_s). \tag{10.4}$$

The motifs of type 2 are stems, bulges, and internal loops. These motifs are defined by specifying indices of paired bases, $i_s, j_s, i_t, j_t \in 1, 2, \ldots, N$, where pairings are i_s–j_s and i_t–j_t, and is denoted by

$$M2(i_s, j_s, i_t, j_t). \tag{10.5}$$

In (10.4) and (10.5), the subscript s stands for "start" and the subscript t stands for "termination". In Fig. 10.3 the ribonucleotides corresponding to the indices i_s, j_s, i_t, j_t are marked in white. The hairpin loop *hloop*1 is a motif of type 1, and $hloop1 = M1(21, 31)$; the stem *stem*1 is a motif of type 2, and $stem1 = M2(1, 45, 3, 43)$, and the *bulge*1 is a motif of type 2, and $bulge1 = M2(12, 35, 18, 34)$. We denote energies of motifs (10.4) and (10.5) by

$$EM1(i_s, j_s) \tag{10.6}$$

and

$$EM2(i_s, j_s, i_t, j_t), \tag{10.7}$$

respectively Since the motifs in a hairpin RNA structure appear sequentially, arranging a dynamic programming recursion for minimization of its energy is particularly easy. We denote by $V(i, j)$ the lowest folding energy of the RNA hairpin structure, over the sequence $b_i, b_{i+1}, \ldots, b_j$, with b_i and b_j paired. We can then state the following recursion for $V(i,j)$:

$$V(i,j) = \min \begin{cases} EM1(i,j), \\ \min_{i_t, j_t}[EM2(i,j,i_t,j_t) + V(i_t,j_t)]. \end{cases} \quad (10.8)$$

In (10.8), the indices run over the ranges $1 \leq i +$ minimal size of $M1 < j \leq N$ and $i < i_t +$ minimal size of $M1 < j_t \leq j$, and the values of $V(i,j)$ on the diagonal and neighboring positions are initialized at $+\infty$, $V(i,i) = V(i,i+1) = \ldots = V(i, i +$ minimal size of $M1) = +\infty$.

The order of complexity of the algorithm (10.8) is $O(N^4)$, because filling in each of the entries of $V(i,j)$ requires $O(N^2)$ operations, following from minimization over two indices i_t, j_t.

RNA with Multibranched Loops

In our notation multibranched loops are motifs of types 3, 4, and so forth. Assume that in the RNA sequence $b_1, b_2, \ldots, b_N$ ending bases are paired and that possible motifs are type 1 (10.4), type 2 (10.5), and type 3, denoted analogously to (10.4) and (10.5) by

$$M3(i_s, j_s, i_t, j_t, i_q, j_q) \quad (10.9)$$

and having an energy

$$EM3(i_s, j_s, i_t, j_t, i_q, j_q). \quad (10.10)$$

We denote by $V(i,j)$ the minimal energy of the RNA strand $b_i, b_{i+1}, \ldots, b_j$ with b_i and b_j paired. The recursion analogous to (10.8) for $V(i,j)$ is

$$V(i,j) = \min \begin{cases} EM1(i,j), \\ \min_{i_t, j_t}[EM2(i,j,i_t,j_t) + V(i_t,j_t)], \\ \min_{i_t, j_t, i_q, j_q}[EM3(i,j,i_t,j_t,i_q,j_q) + V(i_t,j_t) + V(i_q,j_q)]. \end{cases}$$
(10.11)

In (10.11), updating $V(i,j)$ requires minimization over four indices, so the overall complexity of the recursion (10.11) is $O(N^6)$.

Comparing (10.11) and (10.8), one can see that adding loops with more branchings, given by motifs of type 4 and higher, will lead to recursions of successively higher complexity. In practical calculations related to the minimization of RNA folding energies, expressions for energies such as (10.10) are not, however, used because there are not enough experimental data describing the exact energies of multibranched loops. Instead, the energies of multibranched loops are approximated by sums of components related to generating a multibranched loop (M), closing base pairs (P), and unpaired bases inside a loop (Q). So e.g., the energy of multibranched loop $M3$ from (10.10) will be approximated by

$$EM3(i_s, j_s, i_t, j_t, i_q, j_q) = M + 3P + Q(i_t - i_s + i_q - j_t + j_s - j_q), \quad (10.12)$$

10.8 Computational Prediction of RNA Secondary Structure

where M, P, and Q are appropriate coefficients, and $Q(.)$. Analogous formulas hold for motifs of type 4 and higher.

Using the approximation (10.12), we can simplify the recursion for minimizing the energies of RNA foldings with multibranched loops. Assume that in the RNA sequence $b_1, b_2, \ldots, b_N$ ending bases are paired and we allow motifs of all types. Denote by $V(i,j)$ the minimal energy of the RNA strand $b_i, b_{i+1}, \ldots, b_j$ with b_i and b_j paired, and by $W(i,j)$ the minimal energy of the strand $b_i, b_{i+1}, \ldots, b_j$ inside a multibranched loop. Then, for $V(i,j)$, we have a recursion

$$V(i,j) = \min \begin{cases} EM1(i,j), \\ \min_{i_t, j_t}[EM2(i,j,i_t,j_t) + V(i_t, j_t)], \\ M + P + \min_k[W(i+1,k) + W(k+1, j-1)]. \end{cases} \quad (10.13)$$

In the above, the first and second row are the same as in (10.8). The third row stems from inserting a multibranched loop into the RNA secondary structure. The components M and P are related to the energy of creation of the multibranched loop and to the energy of the closing pairing i–j. The term in the third row is also called a bifurcation, because the secondary structures related to $W(i+1,k)$ and $W(k+1, j-1)$ will fold independently one of another. The recursions for $W(i,j)$ are as follows:

$$W(i,j) = \min \begin{cases} P + V(i,j), \\ Q + W(i+1,j), \\ Q + W(i,j-1), \\ \min_k[W(i,k) + W(k+1,j)]. \end{cases} \quad (10.14)$$

In the above the first row is related to closing the multibranched loop by a pairing i–j, the second and third rows represent leaving the ith and jth base, respectively, inside the loop unpaired, and the fourth row is related to adding a new bifurcation. The computational complexity of the algorithm (10.13) and (10.14) is $O(N^4)$. Before starting the recursions, one must initialize $V(i,j)$ and $W(i,j)$ so that the diagonal and neighboring positions are initialized to $+\infty$, i.e., $V(i,i) = V(i,i+1) = \ldots = V(i, i + \text{minimal size of } M1) = +\infty$, $W(i,i) = W(i,i+1) = \ldots = W(i, i + \text{minimal size of } M1) = +\infty$.

External Bases

Up to now, when discussing minimizing the folding energy of RNA, we assumed that the two terminating bases of the strand were paired, which is not the most general situation. The ending bases $b_1, \ldots, b_p$ and $b_{N-p}, \ldots, b_N$ of the RNA strand $b_1, b_2, \ldots, b_N$ may not form pairings. We shall call the unpaired bases $b_1, \ldots, b_p, b_{N-p}, \ldots, b_N$ of the RNA strand $b_1, b_2, \ldots, b_N$ external bases. Generalizing the algorithm (10.13) and (10.14) to the case of external bases is possible and involves adding one more score matrix $W^E(i,j)$, with the same recursion scheme as in (10.14)

310   10 RNA

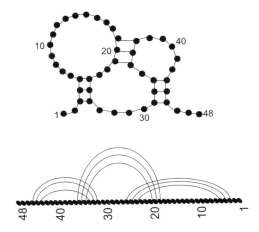

Fig. 10.4. Illustration of the secondary structure in RNA called a pseudoknot. The *upper part* shows the shape of a pseudoknot. The *lower part* illustrates the nonnested character of a pseudoknot

$$W^E(i,j) = \min \begin{cases} P^E + V(i,j), \\ Q^E + W^E(i+1,j), \\ Q^E + W^E(i,j-1), \\ \min_k [W^E(i,k) + W^E(k+1,j)]. \end{cases} \quad (10.15)$$

The difference between (10.14) and (10.15) is in the values of the constants P, Q and P^E, Q^E. For external bases, reasonable values for the parameters are $P^E = Q^E = 0$.

10.8.4 Pseudoknots

In Fig. 10.4 we present a motif of RNA structure, called a pseudoknot, which has not yet been discussed. Pseudoknots are not found in short RNA chains but they can form in longer RNA molecules. The RNA folding is stabilized here by additional pairings, which do not form a nested structure. Deriving dynamic programming algorithms for the analysis and prediction of RNA secondary structures with pseudoknots is possible [241], however, the time complexities of these algorithms increases substantially. Although it is possible to present them in the planar layout, pseudoknots are instead classified as tertiary motifs of RNA, since their occurrence often contributes to forming a nonplanar spatial structure of RNA.

10.9 Prediction of RNA Structure by Comparative Sequence Analysis

Comparative sequence analysis involves aligning a target RNA sequence with a block of RNA sequences of known structure. Then, using the correspondences obtained we can infer the secondary and/or tertiary structure of the target RNA sequence. The idea, analogous to that used in comparative modeling of proteins, is that similar sequences of ribonucleotides lead to similar secondary and tertiary structures of molecules. This idea is related to the paradigm that homologous RNA species result from an evolutionary relationship and that functionally homologous regions will adopt similar structures.

In comparative analysis of RNA species, sequences are searched for compensatory base pair changes. If, in the course of evolution, a base pair has changed, then a compensatory mutation should have occurred on the complementary string, allowing the molecule to maintain its spatial structure. The existing software for alignment-based structure prediction of RNA [55, 206], enables a group of sequences to be aligned with a new target sequence, and using regions of high sequence conservation of the group as predictors of secondary-structure motifs in the target sequence. The growing number of sequences in RNA databases will result in the possibility of quickly adding new RNA sequences to structured databases of homologous RNA molecules.

10.10 Exercises

1. Assume that RNA chain has length N and the bases are numbered $1, 2, \ldots, N$. There are K bonds between bases, depicted as follows:

$$\begin{aligned} i_1 &- j_1 \\ i_2 &- j_2 \\ &\vdots \\ i_K &- j_K. \end{aligned} \quad (10.16)$$

   How can one determine whether these bonds have a nested structure? Write a computer program for solving this problem.
2. Develop a computer program with graphics for drawing secondary structure of RNA of length N, given a list of nested bonds between bases, as in (10.16).
3. Develop a computer program for maximizing the number of pairings, on the basis of the algorithm described in Sect. 10.8.2.
4. Download a short tRNA sequence from the GtRNA database [181, 318]. Use the program from Exercise 3 to predict its secondary structure. Compare the structure obtained to that available in the GtRNA database.
5. Present the above RNA sequence to one of the RNA secondary-structure prediction servers [346].

6. Develop a computer program for minimizing the free energy of an RNA sequence using one of the algorithms described in Sect. 10.8.3. Use it to data from Exercise 4.

11
DNA Microarrays

Gene expression, which includes two processes, namely transcription of data from a DNA template to RNA, and translation, involving the construction of proteins on the basis of the information about the linear sequence of their amino acids encoded in the RNA, lies behind all functions of cells in organisms.

The idea of DNA microarray technology is to monitor gene expression processes by measuring levels of RNA species in biological samples, for example tissue cells or blood. RNA molecules in the samples are labeled by using appropriate techniques and presented to an array of spots, where complementary-DNA (cDNA) fragments corresponding to known coding DNA sequences are placed. It is also possible to copy RNA sequences, by a reverse transcription mechanism, back to a DNA strand, label the DNA with fluorescent dyes and hybridize it to a complementary-DNA probe fixed on the microarray. The measurement of RNA levels is based on the fundamental property of nucleotide sequences, already mentioned in this book, of binding (hybridizing) to their complements. If the level of the RNA product corresponding to the DNA placed at a spot X in the microarray is high in the sample being analyzed then we should observe a high fluorescence signal at spot X. The complementary-DNA sequences placed in the spots of microarrays are designed and synthesized on the basis of our the knowledge about the content of genomes of organisms. The number of spots in a DNA microarray is comparable to the number of known genes in the organism studied, and can reach tens of thousands in DNA microarrays dedicated to the human genome. Appropriate technology allows the precise positioning and stabilizing of complementary-DNA probes on glass or plastic plates and then, after hybridization of the labeled target molecules, estimation of the level of RNA in the sample by reading the intensity of the signals from the dots of the DNA array.

Experiments with DNA microarrays are performed to help study issues in biology and clinical practice, regarding cellular mechanisms, the functions of genes and proteins, the structure of gene networks and pathways, relating the risk of being affected by diseases to gene expression profiles, etc. Gene expression profiling has been successfully used in many medical research programs concerning monitoring cellular process, measuring the response of cells or tissues to therapeutic agents, classification or detection of disease symptoms and many other problems; such studies have been presented, for example, in [3, 23, 51, 102, 133, 238].

Gene expression experiments lead to the creation of huge data sets consisting of tens of thousands of RNA species, corresponding to known or putative genes. However, most of the genes whose expression profiles are generated in microarray experiments may be unrelated to the processes or phenomena being studied. Therefore an appropriate methodology must be developed for inference based on gene expression levels obtained from DNA microarrays to filter out irrelevant information. The amount of data makes it intractable manually, and appropriate bioinformatic tools must be applied to review and organize the information. The mathematical modeling approaches used must be consistent (i) with the aim of the study, i.e., they must help in verifying the hypotheses behind the experiment, and (ii) with the specific character of microarray data. A study based on gene expression usually involves analyzing issues such as the following:

(i) Are there differences between the gene expression profiles obtained in experiments A and B ?
(ii) Is gene X overexpressed or underexpressed in experiment A versus experiment B?
(iii) Is there a correlation between the gene profiles in experiments A and B?
(iv) Is there a group of genes that is always overexpressed (or underexpressed) under the experimental conditions of A?
(v) If there is an environmental, temporal, or spatial factor behind the experiment, are there genes that follow the pattern of this factor?

An important aspect of the analysis of microarray data is the possibility of repeating the experiment or of collecting multiple samples under different experimental conditions or situations. Some issues, such as (iv) or (v) above can be efficiently resolved when experiments are repeated many times, but become very difficult otherwise.

The process of inferring useful information from expression profiles involves several steps where models and mathematics, along with heuristics and intuition, are necessary for choosing between algorithms and between values of numerous parameters. In this chapter, we review some approaches to the analysis of large data sets consisting of gene expression profiles and illustrate them using publicly available data sets obtained from DNA microarrays. We present some basic techniques for data normalization, and the related statistics of expression levels and logarithms of expression levels. We present a maximum likelihood method for modeling probability distributions of logarithms of expressions, and we show how it can be applied to infer useful information. Modeling probability density functions by Gaussian mixtures allows one to study clustering properties that arise from similar values or patterns of change of gene expressions. Mixture model analysis can be enhanced by incorporating information about repetition of the measurement into the construction of the likelihood function. We overview some methods of dimensionality reduction. We also demonstrate or mention methods for class prediction and class discovery, namely hierarchical clustering, the K-means method, and linear and nonlinear classifiers. By analyzing distances and grouping data, hierarchical clustering explores the wealth of information encoded in the positive or negative correlations of expression values induced by simultaneous increases or decreases in gene expression.

11.1 Design of DNA Microarrays

The two main techniques for the measurement of gene expression measurement are (i) complementary-DNA arrays and (ii) oligonucleotide arrays. In both cDNA and oligonucleotide microarrays, labeled (dyed) target RNA or DNA molecules bind to immobilized complementary-DNA probes. The cDNA technology is cheaper and it is possible to implement and develop it on a laboratory scale. Therefore there are many cDNA standards, and cDNA chips dedicated to many research programs are manufactured by genomic laboratories and small enterprises, as well as university laboratories. The oligonucleotide technology is more involved and expensive and there are only a few industrial manufacturers of microarray chips and scanners that are used in scientific research; the most widely known is Affymetrix [1, 303].

In the cDNA technology, DNA strands (100–5000 base long) are presynthesized and placed on a glass or plastic plate by microrobots called DNA arrayers or spotters. The probes are deposited on the plate by a method similar to inkjet printing. The surface of the plate itself must be prepared appropriately to allow attraction and stabilization of the probes. The sequence of DNA probes is established using data from the DNA databases. RNA material from experimental samples is isolated and then reverse transcribed to DNA with the use of reverse transcriptase (see the Chap. 10). The cDNA strands obtained are labeled with fluorescent dyes and then presented to the DNA array, where a hybridization process occurs. Usually, two different fluorescent dyes (Cy3, orange or green, and Cy5, dark red) are used to label the samples. One color is used for the case samples, corresponding to a biological experiment, and another color is used for the reference or control RNA samples. The case and control samples, labeled with different dyes, are mixed and presented to the cDNA on the microarray spots, where the hybridization process takes place. The effective information obtained in a cDNA experiment is the fluorescence intensities, measured with the use of a scanning device. Owing to imperfect making of cDNA probes, reading the dye intensities is prone to errors from many sources, such as nonuniform intensities of the dyes in the spots and irregularities in the spot shapes. In order to reduce the influence of these errors, digital image-processing techniques are applied as a preprocessing stage in data acquisition. Images of cDNA arrays are often published to accompany the results of their analysis. Therefore they are widely available on the Internet. An example of an image corresponding to one cDNA array, from [3] and available at the Web site supporting the data in that paper, is presented in Fig. 12.3. The study in the paper referred to was devoted to identifying types of B-cell lymphomas (human lymphatic cancers) on the basis of the cDNA gene expression profiles. The cDNA microarrays were specially designed and included 17856 cDNA clones chosen from DNA libraries related to B cells and lymphomas [3]. In order to achieve the aim of the study, both malignant and normal tissue and blood samples were collected and analyzed with the use of the manufactured cDNA microarrays. Fig. 12.3 shows the full image of a cDNA array corresponding to a sample of blood B cells (labeled *lc7b023*), which consists of cDNA spots organized into rows and columns. Normally, cDNA microarray apparatuses are equipped with computer software that allows for the intensities of the fluorescence signals in the cDNA spots to be translated into numerical values; for example [337] was used for the images in [3]. Using that program, one can estimate the fluorescence intensities of both the red Cy5 and the green Cy3 spectral components.

Oligonucleotide microarrays are composed of DNA strands synthesized in situ on the solid support. The oligonucleotide probes have a length of 20–50 bases and each gene is represented by 20–25 probes corresponding to its exonic (coding) fragments. The technology of assembling oligonucleotide microarrays is a combination of solid-phase chemical synthesis [169] and photolithography technology similar to that used for manufacturing LSI and VLSI electronic circuits [16, 90]. In order to control the growth of the oligonucleotide strand, chemical reagents are applied to block reactive groups on the bases and on the deoxyribose ring (benzoyl and isobutyryl are used to block the reactive NH_2 groups on the bases and dimethoxytrityl is used to block the $5'$ position on the deoxyribose ring). The blocking reagent at $5'$ position of the deoxyribose ring is removed before the phase of adding the new base to the DNA sequence. In the process of manufacturing an oligonucleotide array, tens of thousands of different oligonucleotide strands are synthesized at the same time on a plate. This is achieved by steps where access to spots on the plate is opened and closed by the use of light-activated photolithographic masks, synchronized with the stages of synthesis of solid-phase single-strand oligonucleotide DNA. A diagram explaining these technological steps is presented in Fig. 11.1. By using light-sensitive masks, oligonucleotide spots are either masked or made available to the reagents, with a time schedule controlled by a computer algorithm.

The targets for the immobilized probes on the microarray plate are fluorescent, labeled RNA molecules. After hybridization, the signal intensity is measured with the use of a digitally controlled laser-optic system. The signal corresponding to one gene results from averaging and comparing signals from many probes; the value available to the user is proportional to the level of

11.1 Design of DNA Microarrays   317

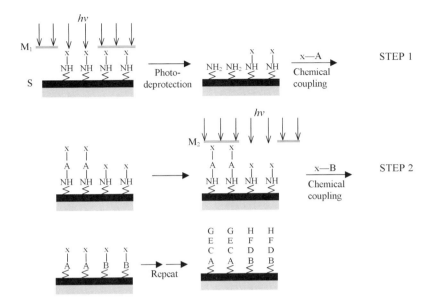

Fig. 11.1. The process of synthesis of DNA strands applied to make oligonucleotide DNA microarrays. By using light-sensitive masks, oligonucleotide spots are either masked or made available to the reagents. STEP1: the spot for base A is prepared by illuminating the plate through a M_1, and base A is fixed (chemically coupled) to the solid support. In STEP2, all spots are again protected. The spot for base B is activated by a procedure analogous to that in STEP1, using mask M_2, and base B is fixed to the support

the RNA species. Some controls, or flags, are also added that report on the quality of the signal.

The basic difference between the two microarray formats is in the lengths of the complementary-DNA probes. In oligonucleotide arrays the probes are of constant length in the range of 20–50 base pairs, while in cDNA arrays the lengths of the cDNA strands differ between spots. In oligonucleotide microarrays, in principle, the readouts of fluorescence intensities are comparable between different spots. In contrast, in cDNA microarrays, the different probe lengths change the reaction rates between spots. Therefore experiments performed using cDNA always require the samples to be compared to controls. Each spot containing DNA clones is exposed both to targets from the sample RNA and from the control RNA. As already mentioned, the case samples and control samples are labeled with fluorescent dyes of different colors (red and green), and their RNA levels are compared by measuring the intensities of the corresponding dyes.

In the Affymetrix oligonucleotide microarrays, one complementary-DNA probe is also associated with two spots called PM and MM, where PM stands for "perfect match" and MM for "mismatch". The PM probe is a DNA strand that corresponds uniquely to a gene. The MM probe has one nucleotide in the middle altered. It is intended that PM versus MM comparison will increase the precision of measurements by eliminating errors resulting from cross-bindings and background hybridization.

11.2 Kinetics of the Binding Process

The hybridization reaction can be represented by the following scheme [165]:

$$R + L \underset{k_r}{\overset{k_f}{\rightleftarrows}} C,$$

where R denotes the number of oligonucleotide strands available for reaction, L is the molar concentration of free target RNA samples, and C stands for the number of bound complementary complexes. The coefficients k_f and k_r are the forward (binding) and reverse (unbinding) reaction rates. The units for the rates are $\text{mol}^{-1}\text{time}^{-1}$ for k_f and time^{-1} for k_r. Different units of measurement are used for different molecules: L is measured as a molar concentration, while R and C are numbers of molecules. The different units underline the nature of the experimental setup. If we denote the molar volume of the solution interacting with the probe by V and the Avogadro's number by N_A, then we can compute the number of free target RNA molecules in the solution as LVN_A. The kinetics of the process of hybridization is governed by the law of mass action. The rate of forward binding of target molecules to immobilized probes is proportional to the product of the concentrations of the target strands and free, probes and the unbinding process is a first-order reaction with a rate proportional to C. This results in the following balance of flows:

$$\frac{dC}{dt} = k_f RL - k_r C. \quad (11.1)$$

We assume that at the beginning, i.e., at $t = 0$, there are no hybridization complexes, i.e., $C(0) = 0$, and we use the notation $L(0) = L_0$ and $R(0) = R_T$, where the subscript T stands for the total number of oligonucleotides available for hybridization. Since one RNA strand binds to one oligonucleotide, resulting in one binding complex, the following equalities for the flows hold

$$\frac{dR}{dt} = VN_A\frac{dL}{dt} = -\frac{dC}{dt},$$

which results in $R(t) + C(t) = R_T$ and $VN_A L(t) + C(t) = VN_A L_0$. So $L(t)$ and $R(t)$ can be expressed in terms of $C(t)$ and substituted in (11.1), leading to

11.2 Kinetics of the Binding Process

$$\frac{dC}{dt} = k_f[R_T - C(t)]\left[L_0 - \frac{C(t)}{N_A V}\right] - k_r C(t)$$

or

$$\frac{dC}{k_f(R_T - C)(L_0 - C/N_A V) - k_r C} = dt.$$

The left-hand side of the above equation can be expanded into two first-order fractions, which allows one to find the analytical solution. Often it is possible to approximate the above dynamics to only one exponent. Specifically, one of the following two asymptotic situations may hold. (i) There is a large excess of particles in the immobile probe over the potential number of binding targets, i.e., $R_T \gg C$, or $(R_T - C)/R_T \approx 1$, which results in

$$C(t) = \frac{R_T L_0}{K_D + R_T/N_A V}\left[1 - \exp\left(-\frac{t}{\tau_R}\right)\right], \qquad (11.2)$$

where $K_D = k_r/k_f$ and

$$\tau_R = \frac{1}{k_f(K_D + R_T/N_A V)}.$$

(ii) There is a large excess of free RNA strands with respect to the number of binding complexes, i.e., $L_0 \gg C/N_A V$, or $(L_0 - C/N_A V)/L_0 \approx 1$, which leads to

$$C(t) = \frac{R_T L_0}{K_D + L_0}\left[1 - \exp\left(-\frac{t}{\tau_L}\right)\right], \qquad (11.3)$$

where

$$\tau_R = \frac{1}{k_f(K_D + L_0)}.$$

Microarray experiments are very often planned in such a way that there is a large excess of particles in the immobile probe over the potential number of binding targets, i.e., case (i) holds. From (11.2) one can see that, after equilibrium has been reached, or at a predefined instant of time, the intensity of the fluorescence signal measured at a microarray spot is proportional to the level of the corresponding RNA species in the analyzed sample, i.e., $C \sim L_0$.

Along with the process of binding labeled target RNA or DNA molecules to their corresponding complementary-DNA sequences, a processes of cross-hybridization may occur. Cross-hybridization is the binding of target molecules to non-corresponding DNA regions [60]. It differs from complementary hybridization in its coefficients k_f and k_r. Dai et al. [60] have noted that the forward hybridization coefficients k_f have comparable values for complementary hybridization and cross-hybridization, whereas the coefficient k_r of the reverse process is at least of one order of magnitude, higher for cross-hybridization than for complementary hybridization. Therefore in a first-approximation model, the influence of cross-hybridization on the measured values of gene expression can be neglected.

11.3 Data Preprocessing and Normalization

Owing to imperfections in the assembly processes and the high density and large number of spots on DNA microarray chips, measurements of expression levels are contaminated, to a substantial extent, by measurement noise. Also, there is a large variation in the concentration levels of different RNA macromolecules resulting from their biological functions in cells. Signals are not present or are low at some of the microarray spots owing to the absence of the corresponding RNA species in the analyzed sample. On the other hand, some of the RNA molecules that appear in very small amounts are nevertheless crucial for the proper functioning of many molecular mechanisms. Therefore, it is important to employ preprocessing and normalizing steps on the raw expression data, with the aim of eliminating some errors and reducing the variation of the measurement noise. The preprocessing and normalization procedures are based on the hypothesis that there is a systematic error between experiments, which can be removed (or reduced) by averaging or scaling [1, 16, 35, 130, 234, 235, 280]. The aims of the normalization steps are (i) to label measurements of low reliability, which introduce mostly noise accompanied by a very low or no useful signal, and (ii) to estimate of the useful signal by averaging or applying other transformations of this type to the components of the measurements. Several possible approaches to reducing the variation of errors are mentioned in the literature. They can be grouped into two classes: (i) normalization based on a model of the transcription process, such as normalization to the total or ribosomal RNA, normalization to housekeeping genes, normalization to a reference RNA, or normalization by spiked in control RNA sequences; and (ii) normalization by applying empirical scaling functions that transform the distributions of the expression values to the desired shape. We can also distinguish between normalization applied to one scan of a microarray chip and normalization resulting from averaging over repetitions of an experiment under the same or similar conditions.

Despite the preprocessing steps built into the microarray software by their manufacturers, the researcher often needs to add additional rules for data analysis. When analyzing data from several DNA microarray chips, one may encounter the situation where a gene is marked present on one chip and absent on another one. Also, for some microarray standards, it may happen that the values of gene expression returned by the microarray software are negative. Some rules must then be introduced to infer useful information from data of these types. After this, data on the expression of genes in microarrays are often visualized by plotting histograms of RNA levels over different subsets of the genes spotted on the plate.

DNA microarray experiments involve both measuring and comparing RNA levels between samples taken from cell lines at different times, between different cell lines, between different individuals, and so forth. So, the statistical description of the interplay between random and systematic elements in the samples may be complex and may require a considerable research effort.

Even in very carefully planned biological experiments, many sources of error are rather poorly recognized. Therefore, despite their potential advantages, models of transcription processes are often ignored, and, instead, methods that belong to class (ii) above are used to reduce the systematic errors. The operations performed on microarray data include logarithmic transformations (or more general nonlinear transformations, [96]), centering transformations, and variance standardization transformations. These transformations allow one to eliminate some systematic errors without bothering with precise models of the mechanisms that cause them. The approaches that belong to class (ii) can be called black-box modeling, since no (or almost no) hypotheses are introduced regarding the hybridization process or its parameters.

An important issue in the analysis of DNA microarray data is standardization of the data processing procedures, which paves the way towards comparing microarray experiments between different studies. One element of ensuring repeatable results is to standardize the normalization procedures. If the studies to be compared use the same normalization procedures, their results should be comparable at the level of the final estimates of RNA levels. However, despite the existence of some recommendations regarding normalization methods, there is still no single standard for these methods. Therefore, a possibility, often applied when experimental data are published, is making available fluorescence intensities signals at the level of the probes (see Sect. 11.3.1). This makes it possible to redo statistical analyses from the raw data to the final conclusions and allows easier comparisons between different studies. The normalization procedures published in the literature are supported by a lot of publicly available software that performs normalization procedures for microarrays; some of the popular programs are [306, 335].

Generally, it is commonly believed that the preprocessing and normalization steps are very important and have an impact on the overall results of studies involving the use of DNA microarrays. The lack of an adequate normalization step can lead to misleading conclusions. Below we describe some approaches to normalization procedures for DNA microarrays.

11.3.1 Normalization Procedures for Single Microarrays

First we describe the normalization procedure, included as a part of the software developed by manufacturer of DNA microarrays scanner, Affymetrix. This procedure can be applied to a single microarray. The algorithm for preprocessing and normalizing the measurements of RNA intensities is being modified as products are upgraded; here we present the version called MAS 5.0 [1, 306, 303]. The files of expression level data produced by DNA microarray scanners are organized such that the measured expression levels are marked additionally by labels that describe their level of reliability. In the DNA microarray chips manufactured by Affymetrix, each gene is represented by K probes (also called spots) placed on the surface of the plate. For example, in Affymetrix human-genome chip HG U133 [1], $K = 11$, L the length of the cDNA sequence = 25, and the number of genes is 22000. Each probe contains a pair of two cDNA sequences denoted by PM (Perfect Match) and MM (Mismatch). The PM sequence is a sequence L bases long sampled from the exonic part of the gene. MM sequence is equal to the PM sequence at all bases except the 13th (the one in the center) which is altered to another base. The PM and MM signals from the K probes are used to decide whether the measurement should be labelled P for present, A for absent or M for marginal. To make the decision, discrimination scores R_k, $k = 1, 2, ...K$ are computed with the use of the following formula:

$$R_k = \frac{PM_k - MM_k}{PM_k + MM_k},$$

where PM_k and MM_k, $k = 1, 2, \ldots, K$ are the intensities of the PM and MM fluorescence signals, respectively. By definition, the discrimination scores R_k are in the interval $(-1, 1)$. The null hypothesis (see Chap. 2) is that the gene is absent, which can be denoted by $H_0 = H_{\text{Absent}}$. This means that there are no RNA strands in the sample that correspond to the probe cDNA sequence, and we should expect fully random binding of RNA sequences to both PM and MM. The alternative hypothesis is that the gene is present, i.e., $H_A = H_{\text{Present}}$. In this situation the PM spot should attract more RNA strands than the MM spot, owing to its higher affinity. To decide between H_{Absent} and H_{Present}, or, more precisely, to accept or reject $H_0 = H_{\text{Absent}}$, a rank statistic is used. The values $R_1, ..., R_{11}$ are assigned a plus or minus sign by the criterion that a plus is assigned if $R_k >$ Threshold, otherwise a minus is assigned. Here Threshold is a predefined number; the default is Threshold $= 0.015$. Then, the positive values of R (those to which a plus sign has been assigned) are sorted and ranked in ascending order, the negative values of R are sorted in descending order, and the one-side Wilcoxon signed-rank statistical test is applied [297], (see Chap. 2) to compute the detection p-values. Although the rank test has a slightly lower power than the paired t-test [297], its advantage is robustness against non normality. Commonly, when statistical tests are applied, the critical value for p is taken as $p_{\text{critical}} = 0.05$. A similar, but slightly modified approach is adopted here. The following three subintervals are used to categorize the expression of genes on the basis of the computed p-values:

$$\text{if } p \in \begin{cases} (0, 0.04) \text{ then the gene is labeled } P \text{ (Present)}, \\ (0.04, 0.06) \text{ then the gene is labeled } M \text{ (Marginal)}, \\ (0.06, 1) \text{ then the gene is labeled } A \text{ (Absent)}. \end{cases} \quad (11.4)$$

The final estimate of the expression level of the gene is computed from component measurements PM_k and MM_k, $k = 1, 2, \ldots, 11$, as follows:

$$X = \frac{1}{\#S} \sum_{k \in S} (PM_k - MM_k), \quad (11.5)$$

11.3 Data Preprocessing and Normalization

where S is the subset of probes for which $PM_k - MM_k$ is within three standard deviations of the average.

To illustrate the labeling P, M, and A, let us look at the analysis of the data in [102], available electronically at an accompanying Web site. The experimental study in that paper is based was based on Affymetrix microarrays of type Hu6800, with 7129 spots, including cDNA strands corresponding to 6817 human genes and a number of control spots corresponding to control RNA species that could be used for calibration purposes. Gene expression profiles were obtained from patients affected by two types of leukemia, acute lymphoblastic leukemia (ALL) and acute myeloid leukemia (AML). The initial training set included 38 patients (27 ALL and 11 AML). The measurements of RNA levels were labelled P, M, and A according to the decision rule (11.4) and assigned numerical values based on differences between PM and MM intensities as described in equation (11.5). In total, the training set contained $38 \times 7129 = 270902$ measurements (spots). Among them there were 187892 spots labelled absent A, 78632 spots labelled present P and 4378 spots labelled marginal M. So, it seems that majority of spots contain no useful data. However, we must take into account the fact that the data are not 270902 independent spots but, rather, 7129 spots each repeated 38 times. Typically, if there are multiple measurements (experiments) of the expression for each gene, then the genes labeled A in all experiments are excluded from further analysis.

11.3.2 Normalization Based on Spiked-in Control RNA

Normalization methods which belong to class (i) above, "normalization based on a model of the transcription process", need information or modification of the design of the experiment and/or knowledge about the properties of the cDNA sequences deposited on the array spots. An example of a mathematical model that leads to corrections to the numerical values of measurements of RNA levels was presented in [113]. This involves using spiked-in control RNA species. Spiked-in control RNA species are RNA strands of known, controlled levels in the probes, with base sequences that correspond to sets of spots used in the DNA microarray chips. Let us assume that M spiked-in control RNA species have been added to each of N oligonucleotide microarray chips in a biological experiment and that each sample contains the same level of spiked-in control RNA number i. The matrix of the $M \times N$ measurements of spiked-in control RNA expression levels is

$$\begin{bmatrix} x_{11} & x_{12} & \cdots & x_{1N} \\ x_{21} & x_{22} & \cdots & x_{2N} \\ \vdots & \vdots & \ddots & \vdots \\ x_{M1} & x_{M2} & \cdots & x_{MN} \end{bmatrix},$$

where x_{ij} stands for the expression level of the ith spiked-in control in the jth microarray chip. The basic hypothesis, following from the design of the experiment, is that changes in the expression levels of the ith spiked control between chips are a consequence of the systematic errors resulting from the chip-manufacturing process. We assume the following multiplicative model for these data:

$$x_{ij} = m_i \cdot r_j \cdot e_{ij}. \tag{11.6}$$

By m_i, we denote the true expression level of the ith spiked-in control. The jth microarray chip is characterized by a multiplicative modifying factor r_j and e_{ij} is a random multiplicative error. After applying the logarithmic transformation, $y_{ij} = \log(x_{ij})$, $\mu_i = \log(m_i)$, $\rho_j = \log(r_j)$, $\epsilon_{ij} = \log(e_{ij})$, the model (11.6) becomes additive:

$$y_{ij} = \mu_i + \rho_j + \epsilon_{ij}. \tag{11.7}$$

We now assume that ϵ_{ij} is normally distributed with zero mean and a variance σ_i^2. In other words, we make one more assumption, that the variance of the logarithm of the error ϵ_{ij} does not depend on the array number, j, but only on the spiked-in control species number, i. In the model (11.7), the y_{ij} are observations and μ_i, ρ_j, σ_i are parameters to be estimated. Under the hypothesis of independence between different spiked-in controls and between errors in different chips, the following log-likelihood function is associated with the data:

$$L = \log \prod_{i=1}^{M} \prod_{j=1}^{N} p(y_{ij}, \mu_i + \rho_j, \sigma_i^2) \tag{11.8}$$

where $p(.)$ is the normal probability density function

$$p(y_{ij}, \mu_i + \rho_j, \sigma_i^2) = \frac{1}{\sqrt{2\pi\sigma_i^2}} \exp\left[\frac{(y_{ij} - \mu_i - \rho_j)^2}{\sigma_i^2}\right]. \tag{11.9}$$

Substituting (11.9) in (11.8) and taking derivatives with respect to the parameters μ_i, ρ_j and σ_i leads to the system of equations

$$\hat{\mu}_i = \frac{1}{N} \sum_{j=1}^{N} (y_{ij} - \hat{\rho}_j), \tag{11.10}$$

$$\hat{\rho}_j = \frac{\sum_{i=1}^{M} (\hat{\sigma}_i^2)^{-1} (y_{ij} - \hat{\mu}_i)}{\sum_{i=1}^{M} (\hat{\sigma}_i^2)^{-1}}, \tag{11.11}$$

$$\hat{\sigma}_i^2 = \frac{1}{N} \sum_{j=1}^{N} (y_{ij} - \hat{\mu}_i - \hat{\rho}_j)^2. \tag{11.12}$$

After the above system of equations has been solved, $\hat{\mu}_i$, $\hat{\rho}_j$ and $\hat{\sigma}_i$ are the maximum-likelihood estimates of the means and variances. The form of (11.10)–(11.12) dictates an easy method for their solution by iteration. One problem, which arises in the case of combinations or mixtures of continuous distributions is that for some data, the maximization of (11.8) may diverge to

11.3 Data Preprocessing and Normalization

infinity and, in iterations we will have $\hat{\mu}_i - \hat{\rho}_j \to y_{ij}$ and $\hat{\sigma}_i^2 \to 0$. A method to avoid this is to introduce a prior distribution of $\hat{\sigma}_i^2$ such that the probability density is zero at $\hat{\sigma}_i^2 = 0$, i.e., $p(\hat{\sigma}_i^2 = 0) = 0$. Hartemink et al. [113] assumed that the prior distribution for joint distribution of variances was a Wishart distribution [158], which, with additional assumptions about symmetry and independence, leads to a joint probability density of $\hat{\sigma}_1^2, \hat{\sigma}_2^2, ..., \hat{\sigma}_M^2$ in the factorized form

$$p_W(\hat{\sigma}_1^2, \hat{\sigma}_2^2, ..., \hat{\sigma}_M^2) = \prod_{i=1}^{M} C(\alpha, t) \left(\frac{1}{\hat{\sigma}_i^2}\right)^{(\alpha-3)/2} \exp\left(-\frac{t}{2\hat{\sigma}_i^2}\right), \quad (11.13)$$

where α and t are predefined constants and $C(\alpha, t)$ is an appropriate scaling factor. A plot of the family of functions

$$p(\sigma^2) = \left(\frac{1}{\sigma^2}\right)^{(\alpha-3)/2} \exp\left(-\frac{t}{2\sigma^2}\right), \quad (11.14)$$

with $\alpha = 5$ and several choices for t, is presented in Fig. 11.2. From the plot and (11.14), we see that the distribution in (11.13) satisfies $p_W(\hat{\sigma}_i^2 = 0) = 0$, provided that $t > 0$. Maximization of the log-likelihood

$$L_1 = \log \prod_{i=1}^{M} C(\alpha, t) \left(\frac{1}{\hat{\sigma}_i^2}\right)^{(\alpha-3)/2} \exp\left(-\frac{t}{2\hat{\sigma}_i^2}\right) \prod_{j=1}^{N} p(y_{ij}, \mu_i + \rho_j, \sigma_i^2)$$

instead of the L given by (11.8), leads to equations (11.10) and (11.11); (11.12) is replaced by

$$\hat{\sigma}_i^2 = \frac{\sum_{j=1}^{N}(y_{ij} - \hat{\mu}_i - \hat{\rho}_j)^2 + t}{N + \alpha - 3}. \quad (11.15)$$

It can be verified that the likelihood L_1 does not diverge to infinity and that the iterations following from (11.10), (11.11), and (11.15) (such that (11.12) is replaced by (11.15)) cannot result in $\hat{\mu}_i - \hat{\rho}_j \to y_{ij}$, $\hat{\sigma}_i^2 \to 0$.

From the estimate $\hat{\rho}_j$ computed from (11.10)–(11.12) there follows the optimal scaling factor for the jth chip, denoted by s_j,

$$s_j = \frac{1}{\hat{r}_j} = \exp(-\hat{\rho}_j) = \prod_{i=1}^{M} \left(\frac{\hat{m}_i}{x_{ij}}\right)^{w_i}$$

where $\hat{m}_i = \exp(\hat{\mu}_i)$ and the weights w_i (weights) computed as follows:

$$w_i = \frac{(\hat{\sigma}_i^2)^{-1}}{\sum_{k=1}^{M}(\hat{\sigma}_k^2)^{-1}}.$$

Finally, we conclude that all expression levels in chip number j should be multiplied by the factor s_j.

Methods similar to that presented above can also be used for normalization based on other data from a microarray, for example, the expression of housekeeping genes, which should remain at constant levels across experiments.

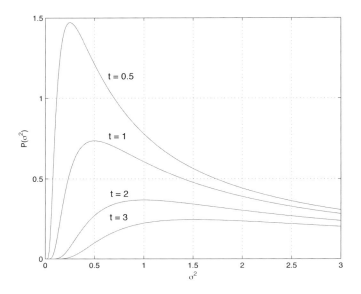

Fig. 11.2. Plots of the family of functions given by (11.14). We have assumed the parameters $\alpha = 5$ and $t = 0.5, 1, 2, 3$

11.3.3 RMA Normalization Procedure

An often applied method for the normalization of results of DNA microarray experiments is robust multiarray analysis (RMA) [35, 130, 335]. This approach proceeds in several steps and assumes a model for the expression profiles analogous to that presented in (11.6) and (11.7). It can be used both for single measurements and for normalization based on repetition of experiments on measurements of gene expression in DNA microarrays. In order to apply RMA normalization, one must have access to the data from separate probes in the microarray measurement.

We assume that a microarray scanning experiment has been repeated J times, and introduce the notation

$$PM_{ijk} \text{ and } MM_{ijk},$$

where, as in (11.5), PM and MM stand for reads of the perfect-match and mismatch intensities, and the indices are used as follows: the index $i = 1, 2, \ldots, I$ represents the genes in the samples, the index $j = 1, 2, \ldots, J$ represents different samples, and the index $k = 1, 2, \ldots, K$ represents the number of the probe. A characteristic feature of RMA is that it relies only on the perfect match probe intensities PM_{ijk}. The RMA algorithm proceeds in the following steps.

Background correction. The following exponential–normal model is assumed for PM_{ijk}

$$PM_{ijk} = bg_{ijk} + s_{ijk}, \qquad (11.16)$$

where bg_{ijk} stands for the background noise and s_{ijk} represents the hybridization signal. The background noise bg_{ijk} results from optical noise and non-specific binding and is modeled by a normal distribution with an array-specific mean level $E(bg_{ijk}) = \beta_i$. The hybridization signal s_{ijk} is assumed to be distributed exponentially. Using these assumptions a filter is constructed which is intended to reduce the background noise. This filter uses the model of a mixture of normal and exponential distribution (see Chap. 2).

Quantile normalization. Quantile normalization between different microarrays is applied to the background-corrected hybridization signals. This equalizes quantile-quantile plots [297] and histograms of expression profiles. Quantile normalization of two samples x and y of equal length involves:

(i) Sorting both x and y in decreasing order, which leads to vectors x_{sorted} and y_{sorted}. Sorting is an appropriate renumeration, so we can write

$$x_{sorted} = \mathrm{renum}_x(x),$$

and

$$y_{sorted} = \mathrm{renum}_y(y).$$

(ii) Computing the mean of x_{sorted} and y_{sorted}

$$z = \frac{1}{2}(x_{sorted} + y_{sorted});$$

and

(iii) Computing the quantile-normalized vectors by an operation of restoring the original order, applied to the mean z. By restoring the original order we mean applying the inverse operators renum_x^{-1} and renum_y^{-1} to the vector z, i.e.,

$$x_{quantile\text{-}normalized} = \mathrm{renum}_x^{-1}(z),$$

and

$$y_{quantile\text{-}normalized} = \mathrm{renum}_y^{-1}(z).$$

Quantile normalization of more than two samples is defined in the way analogous to the above. A variant of quantile normalization based on replacing the operator mean by the operator median is also possible.

Additive model for normalization. For the signals Y_{ijk} defined as the background-corrected, quantile-normalized and log-transformed values of PM_{ijk}, we now use the following additive model:

$$Y_{ijk} = y_{ij} + \alpha_{ik} + \varepsilon_{ijk}$$

where y_{ij} is the final normalized log transformed expression level, α_{ik} is the probe affinity effect, and ε_{ijk} is random Gaussian noise. The above model is fitted to the preprocessed data and the final, normalized values of the logarithms of the expression signals, y_{ij}, are obtained.

11.3.4 Correction of Ratio–Intensity Plots for cDNA

An example of normalization of cDNA microarray data based on a model of the transcription process is a procedure based on ratio–intensity plots. This procedure leads to corrections to the values of the logarithm of the ratio R/G [234]. The method uses the hypothesis that the statistics of the log-ratio $\log(R/G)$ should be independent of the logarithm of the intensity product, $\log(RG)$, where R is the fluorescence intensity of the red dye and G is that of the of green dye. Let us illustrate this procedure by use of the data shown in Fig. 12.3, concerning the expression profile of the sample $lc7b023$ described in [3]. For these data R is denoted by $Ch2$ and G by $Ch1$. A scatterplot of the logarithms of the products of intensities $\log_{10}(R \cdot G)$ versus the logarithms of the ratios $\log_2(R/G)$ is presented in Fig. 11.3 on the left. Data points are $[\log_{10}(R_k \cdot G_k), \log_2(R_k/G_k)]$, where k ranges from 1 to the number of clones. Each data point is represented by a plus sign. A systematic bias is clearly seen. We have estimated this bias by means of a third-order polynomial relation, i.e.,

$$Y_k = F(X_k) = a_0 X_k^3 + a_1 X_k^2 + a_2 X_k + a_3 \qquad (11.17)$$

where $Y_k = \log_2(R_k/G_k)$, $X_k = \log_{10}(R_k \cdot G_k)$, and the above equation is understood in the least squares sense (see Chap. 5). The parameters a_0, a_1, a_2 and a_3 can be estimated by using a simple least-squares algorithm. The resulting estimated bias relation is plotted as a black curve in the left-hand plot in Fig. 11.3. Removing the bias, i.e., subtracting the estimated bias from the logarithms of ratios, which means taking $Y_{k\ corrected} = Y_k - F(X_k)$, leads to normalized (corrected) values of $\log_2(R_k/G_k)$. A plot of $\log_{10}(R_k \cdot G_k)$ versus $\log_2(R_k/G_k) - F(\log_{10}(R_k \cdot G_k))$ (corrected), is shown in the right-hand part of Fig. 11.3. The simple version of the least-squares method, which we used to remove the bias in the data does not take into account the non-uniform density of data points along the $\log_{10}(R \cdot G)$ axis. A more adequate statistical analysis would be the use of a locally weighted linear regression [53] to remove the ratio–intensity bias.

11.4 Statistics of Gene Expression Profiles

The common paradigm in inference based on gene expression profiles is that information on the process behind the DNA microarray experiment is encoded in the ratios of RNA levels between different experiments. The researcher is interested in the factor by which the RNA concentration has increased or decreased from one measurement to another. In order to change the ratios to a linear scale, a logarithmic transformation is applied as a preprocessing stage. This approach has been confirmed by the statistics of gene expression. Histograms of RNA levels, or their ratios resemble an exponential distribution. After the logarithmic transformation transform distributions become similar to normal, or can be modeled by several normal components.

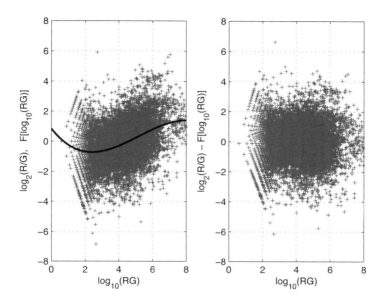

Fig. 11.3. *Left*: scatterplot of logarithms of intensities $\log_{10}(R \cdot G)$ versus logarithms of ratios $\log_2(R/G)$ for the dataset *lc7b023* from [3]. An estimated correction curve based on (11.17) is shown in black. *Right*: scatterplot of normalized log ratios

The typical statistics of RNA levels in oligonucleotide DNA microarrays can be illustrated well by using again the data in [102]. From the training set consisting of 38 patients (27 ALL and 11 AML), we have chosen one ALL patient (sample number 1) and one AML patient (sample number 38). In Fig. 11.4, we show histograms of RNA levels for these patients, (sample number 1 and sample number 38). We have excluded erroneous (negative) measurements of expression. The histogram bars are scaled as relative frequencies; in other words, their areas add up to one. In the upper plots in Fig. 11.4 the values on the horizontal axis are the RNA levels, and in the lower plots, base-2 logarithms of the RNA levels. One can see that the probability density functions in the upper plots resemble exponential functions and their logarithmic transforms are normal-like. More precisely, distributions of the logarithms of RNA levels are bimodal (or multimodal) and so they should be modeled by mixtures of two or more normal distributions rather than by a single normal distribution. We shall give more details concerning this approach in Sect. 11.4.1 later in this chapter.

We can also illustrate, analogously to the above, the statistics of the data obtained from cDNA microarray chips. Again we use, as an example, the expression profile of the sample *lc7b023* from [3]. Signals corresponding to the RNA levels, denoted by $Ch1$ and $Ch2$, are the background-corrected mean fluorescence intensities of pixels within the ellipses of the spots [73, 337],

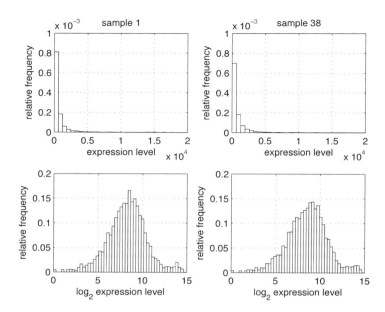

Fig. 11.4. *Upper plots*: histograms of expression levels for sample 1 and sample 38 from the data in [102]. *Lower plots*: corresponding histograms of logarithms to base 2 of the expression levels

computed according to the following equations:

$$Ch1 = CH1I - CH1B$$
$$Ch2 = CH2I - CH1B$$

where the index 1 means the green (532 nm) and the index 2 means the red (635 nm) component of the spectrum, $CH1I$ and $CH2I$ are the fluorescence intensities averaged over the pixels of the spot, and $CH1B$ and $CH2B$ are the average fluorescence intensities of the background pixels. Replacing the mean values $CH1B$ and $CH2B$ by medians can lead to better robustness of the measurements against noise. The data file created for the measurements on the cDNA chip contains quality control parameters for each spot specifying, for example, number of pixels in the spot with intensities greater than background. These parameters allow on to test statistical hypotheses related to the presence or absence of gene expression products in the sample. Spots which generated measurements that did not allow rejection of the hypothesis of the absence of the corresponding gene or which led to values of $Ch1$ or $Ch2$ less than zero, are removed from the analysis. The basic signals used in our inference based on these cDNA microarray experiments are the ratios $Ch2/Ch1$. In Fig. 11.5 histograms of the ratios $Ch2/Ch1$ (in the upper plot) and their logarithms $\log_2(Ch2/Ch1)$ (in the lower plot) are presented. Again the dis-

11.4 Statistics of Gene Expression Profiles   331

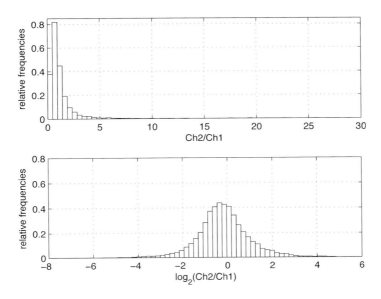

Fig. 11.5. Histograms of the ratios $Ch2/Ch1$ (in the *upper plot*) and their logarithms $\log_2(Ch2/Ch1)$ (in the *lower plot*) for the dataset *lc7b023* form [3]

tribution of the ratios $Ch1/Ch2$ is exponential-like, while, after logarithmic transformation the distribution of $\log_2(Ch2/Ch1)$ is similar to normal.

11.4.1 Modeling Probability Distributions of Gene Expressions

We now focus our attention on the probability distribution of the logarithms of gene expression levels. We denote the logarithm of the expression level of the nth gene by x_n. From the plots in Figs. 11.4 and 11.5 it can be seen that although they are similar to normal probability density functions, the logarithms of expression levels from [102] have more than one mode. Therefore it seems reasonable to model such distributions by mixture densities.

Several researchers have proposed using mixtures of distributions to solve various issues in the interpretation of DNA microarray data [41, 99, 100, 189, 193, 194, 213, 227]. The paper [189] overviews methods for obtaining parameters of mixtures of different distributions and proposes the use of mixtures of factor analyzers for unsupervised classification of colon and leukemia gene expression datasets. In [41] a decomposition of expression level probability density functions into Gaussian components is used to set thresholds to classify expression levels as "change", "no change", "overexpressed", "underexpressed", etc. Thresholds for logarithms of gene expression levels and for SLRs (logarithms of ratios of expression levels) are often set intuitively. For example, a gene is considered overexpressed if the base-2 logarithm of its expression level exceeds the average by 1, or, equivalently its expression is twice as much as the average. However, when gene expressions are classified into groups, for example, "no change", "overexpressed" and "underexpressed", based on boundaries computed by the use of estimated probability density functions corresponding to different clusters, the results are often biologically more sound. This can be verified by using information on ontologies of genes (Sect. 11.10). In [193] and [194], different variants of Bayesian-mixture based clustering procedures were studied. In [99] a hierarchical agglomerative clustering method was used for an initial guess of the mixture parameters and the mixture model was combined with a dimensionality reduction technique for the analysis of cutaneous melanoma data. In [100], a mixture model was used to determine the differential expression of genes in the presence of mixed cell populations. In [213], mixture modeling was applied to the problem of missing measurements in DNA microarrays.

The approach of using probability density function decomposition to grouping genes into coexpressing clusters can also be used successfully in the case where DNA microarray experiments are conducted to compare gene expression profiles under different experimental conditions. In that case it seems natural to impose the requirement that genes remain clustered into the same component over all experiments. However, the components can change their parameters between different experimental conditions. Also, it is possible that the repetitive structure represented by (11.22) may be combined with variability resulting from changing experimental conditions. Decomposition into components according to different patterns of expression can be used, for example, to study the functions of genes involved in the cell cycle, [51, 194] (see also Exercise 6). In [227] decomposition into Gaussian mixture was used to study time-course data on gene expression profiles of cancer cells after irradiation.

The probability distribution of a normal mixture model, $p(x)$, is given by the formula below (compare (2.77)):

$$p(x) = \sum_{k=1}^{K} \alpha_k p_k(x) \qquad (11.18)$$

In the above equation x denotes the natural logarithm of the expression level; α_k, $k = 1, 2, \ldots, K$, are weighting coefficients $\sum_{k=1}^{K} \alpha_k = 1$; and $p_k(x)$, $k = 1, 2, \ldots K$, are probability density functions of normal components, i.e.,

$$p_k(x) = p_k(x, \mu_k, \sigma_k) = \frac{1}{\sqrt{2\pi}\sigma_k} \exp\left[-\frac{(x-\mu_k)^2}{2\sigma_k^2}\right], \qquad (11.19)$$

where μ_k is the expectation and σ_k^2 is the variance of the kth normal component. The parameters to be adjusted are the number of components, the expectations and variances of each component, and the weighting coefficients.

They can be estimated by the maximum likelihood method. Denoting by x_n the logarithm of the expression level of the nth gene, we express the likelihood function as follows:

$$L(x_1, \ldots, x_N) = \prod_{n=1}^{N} \sum_{k=1}^{K} \alpha_k p_k(x_n). \tag{11.20}$$

To estimate the parameters, $\alpha_1, \alpha_2, \ldots, \alpha_K, \mu_1, \mu_2, \ldots, \mu_K, \sigma_1, \sigma_2, \ldots, \sigma_K$, we can use the iterations of the EM algorithm given in expressions (2.87)-(2.89).

A decomposition of the form (11.18) is natural for modeling the probability density function for one microarray measurement experiment. However, it does not incorporate information obtained by measurement repetition. When gene expression measurements are repeated, it becomes natural to make the assumption that all measurements of the expression of one gene belong to the same component of the mixture distribution. Let the number of repetitions be R, and denote by

$$\bar{x}_n = [x_n^1 \; x_n^2 \; \ldots \; x_n^R]$$

the vector of the logarithms of the expression obtained in repeated measurements, for gene number n. We have used an overbar in order to distinguish more explicitly non repeated and repeated measurements. Under the hypothesis that all repeated measurements of one gene belong to one Gaussian component, the probability density function for $\bar{x}_n$ becomes

$$p(\bar{x}_n) = \sum_{k=1}^{K} \alpha_k \prod_{r=1}^{R} p_k(x_n^r, \mu_k, \sigma_k), \tag{11.21}$$

where $p_k(x, \mu_k, \sigma_k)$ is the pdf of the normal distribution given in (11.19). The likelihood function analogous to (11.20) but accounting for repeating measurements has the form

$$L(\bar{x}_1, \ldots, \bar{x}_N) = \prod_{n=1}^{N} \sum_{k=1}^{K} \alpha_k \prod_{r=1}^{R} p_k(x_n^r, \mu_k, \sigma_k) \tag{11.22}$$

where R stands for the number of repeated measurements. The likelihood (11.22) can be maximized by techniques analogous to those discussed before. The EM iterations leading to an increase of the likelihood function (11.22) assume the following form (Exercise 3):

$$p(k|\bar{x}_n, p^{old}) = \frac{\alpha_k^{old} \prod_{r=1}^{R} p_k(x_n^r, \mu_k^{old}, \sigma_k^{old})}{\sum_{\kappa=1}^{K} \alpha_\kappa^{old} \prod_{r=1}^{R} p_\kappa(x_n^r, \mu_\kappa^{old}, \sigma_\kappa^{old})} \tag{11.23}$$

for updating the conditional probabilities, and

$$\mu_k^{new} = \frac{\sum_{n=1}^{N} x_n p(k|\bar{x}_n, p^{old})}{R \sum_{n=1}^{N} p(k|\bar{x}_n, p^{old})}, \quad k = 1, 2, \ldots, K, \tag{11.24}$$

and

$$(\sigma_k^{new})^2 = \frac{\sum_{n=1}^{N}(x_n - \mu_k^{new})^2 p(k|\bar{x}_n, p^{old})}{R\sum_{n=1}^{N} p(k|\bar{x}_n, p^{old})}, \ k = 1, 2, ..., K. \quad (11.25)$$

for updating the means and variances.

We treated the training set in the data concerning 27 ALL and 11 AML patients in [102] as 38 independent repeats of measurements of gene expression and used the iterative formulas (11.23)–(11.25) to perform a decomposition of the probability density functions into 11 Gaussian components. We excluded genes with expression levels less than zero (erroneous measurements) owing to necessity of taking logarithms. More precisely, we included in the analysis only genes with strictly positive values of their expression levels in all 38 scans. The number of Gaussian components, 11, was chosen arbitrarily. The resulting decomposition, described by estimated values $\hat{\alpha}_1, \hat{\alpha}_2, \ldots, \hat{\alpha}_{11}$, $\hat{\mu}_1, \hat{\mu}_2, \ldots, \hat{\mu}_{11}$, and $\hat{\sigma}_1, \hat{\sigma}_2, \ldots, \hat{\sigma}_{11}$, is shown graphically in Fig. 11.6. The upper left plot shows the probability density functions of all 11 components. The plots are scaled by the weights $\hat{\alpha}_1, \hat{\alpha}_2, \ldots, \hat{\alpha}_{11}$ (the areas under curves $1, 2, \ldots, 11$ are equal to $\hat{\alpha}_1, \hat{\alpha}_2, \ldots, \hat{\alpha}_{11}$, respectively). The other plots show, separately, the probability density functions of the components along with histograms of the logarithms of the expressions of the genes which belong to those components. By "gene n belongs to the kth component" we mean that $k = \arg\max p(\varkappa|\bar{x}_n, \hat{p})$, where $p(k|\bar{x}_n, \hat{p})$ is the probability density function given by (11.23) with $\alpha_1^{old}, \alpha_2^{old}, \ldots, \alpha_K^{old}, \mu_1^{old}, \mu_2^{old}, \ldots, \mu_K^{old}$, $\sigma_1^{old}, \sigma_2^{old}, \ldots, \sigma_K^{old}$ replaced by $\hat{\alpha}_1, \hat{\alpha}_2, \ldots, \hat{\alpha}_{11}, \hat{\mu}_1, \hat{\mu}_2, \ldots, \hat{\mu}_{11}, \hat{\sigma}_1, \hat{\sigma}_2, \ldots, \hat{\sigma}_{11}$. In other words, gene n belongs to the kth component if $p(\varkappa|\bar{x}_n, \hat{p})$ has its maximal value at $\varkappa = k$.

Introducing the requirement of that all values for genes should be positive resulted in confining the number of genes analyzed to 2568. Comparing the number of genes in the Hu6800 Affymetrix microarray 6817 with the number of genes included in the analysis, only 2568 we can see that the demand that a gene must give positive expression levels in all scans can be quite tough for some datasets and may lead to the elimination of many measurements. It would be both possible and reasonable to relax our data filter. However, we restricted the analysis to 2568 genes since it was intended mainly to serve for illustration of the methodology. We can understand the 11 normal components as 11 clusters of coexpressing genes; the numbers of genes in these components are 174, 31, 75, 344, 487, 209, 257, 381, 196, 358, and 56. From the plots in Fig. 11.6 we can see, by the quite good agreement between the histograms of the logarithms of the expression levels of the genes clustered into components and their pdf envelopes, that the model (11.21) seems to fit the analyzed microarray data quite well.

An issue to be solved in designing computational approaches is how many components K should appear in the sums (11.18) and (11.21). When we use a mixture model only for approximation of the probability density function,

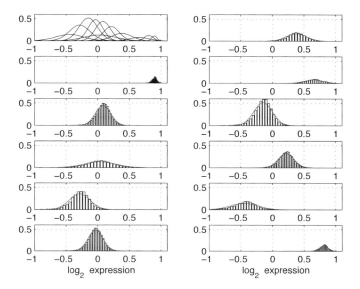

Fig. 11.6. Decomposition of probability density function into 11 Gaussian components, obtained by interpreting the data for 27 ALL and 11 AML patients in [102] as 38 independent repeats of measurements of gene expression and using the iterative formulas (11.23)–(11.25). The *upper left* plot shows the probability density functions of all components. All 11 componets are also presented separately together with histograms corresponding to genes qualified as belonging to those components

then the number of components in the mixture distribution is not of great importance and we can afford to apply mixtures with some excess of components. However, when we use mixture decomposition to study the structure of a process and we group genes into different clusters by their different patterns of behavior under different experimental conditions, the problem of deciding on the number of clusters becomes much more important. It can be demonstrated that the quality of clusters and their estimated confidences are rather sensitive to the choice of K [194]. If K is too small, different patterns become clustered together; if, on the other hand, K is too large, the model becomes overparametrized and some components are unnecessary and noninformative. Estimation of the number of components in a mixture distribution model can be done by penalizing overparametrization with the use of information criteria, such as Bayesian information criterion (BIC) [251, 36]. The BIC leads to a penalty for adding parameters (1/2)#parameters*log(#observations). So the BIC-corrected log likelihood functions (11.18) and (11.21) will take the form

$$L^{\text{BIC corrected}} = L - \frac{1}{2} \#\text{parameters} * \log(\#\text{observations}) \qquad (11.26)$$

where L on the right-hand side is given by either (11.18) or (11.21), and #parameters and #observations (or #measurements) are straightforward to determine. We can repeat the maximization of the likelihood (11.18) or (11.21) with different values of K (or, better, maximize the likelihood interactively while modifying K) and then find the value of K which maximizes (11.26). There is evidence [32] that the use of (11.26) leads to quite reliable estimates of K. A numerically efficient approach to estimating the number of components the use of the Metropolis–Hastings sampling algorithm, [41, 240] penetrates the space of the parameters of the mixture distributions and the space of possible values of K simultaneously. Another possibility is to use using infinite mixture models [193, 194], where the number of components K is neither limited nor penalized, but assumptions about the prior distributions of the parameters of the Gaussian components are made, analogous to those presented in the maximum likelihood approach in Sect. 11.3.2.

11.5 Class Prediction and Class Discovery

In this section we make some general remarks concerning class prediction (supervised classification) and class discovery (unsupervised classification). These remarks will be developed further and supported by computational examples in later sections.

Two tasks related to the classification of expression profile data are class prediction and class discovery [102]. Class prediction uses information about the expression profiles and the known classification of the data sets or experiments to construct classifiers applicable to future data. When the expression profiles belong to two known classes A and B, a very simple and rather efficient solution is often programmed into microarray scanner software: to pick out genes whose expression is strongly correlated with their class, i.e., underexpressed in A and overexpressed in B, or vice versa. Checking a large number of genes for correlation with a partition of the samples into two (or possibly more) classes touches on the statistical problem of multiple testing [2, 28, 97, 269]. Other approaches involve linear or nonlinear classification (some of these were presented in Chap. 4), [67], artificial neural networks, Boolean or Bayesian networks, fuzzy logic classifiers, etc.

Class discovery, i.e., unsupervised classification, concerns telling sets of data apart without using any prior information about the number of classes and/or on the categorization of data. Class discovery is not used for constructing classifiers for future data but, rather, (i) to confirm that the information implied in the design of the experiment, is also encoded in the gene expression profiles collected, and (ii) to explore the data from the angle of existence of unknown relations and mechanisms and to formulate hypotheses explaining these mechanisms. Successful class discovery that is consistent with the prior knowledge about the data and the experiment significantly increases the confidence that inference based on the measured gene expression profiles will prove reliable and robust. The numerical procedure most often applied for class discovery is a hierarchical clustering algorithm (discussed in Chap. 4) [67, 72, 196, 234, 234] which allows one to infer a tree for the arrays and/or genes based on distance matrix. Clusters are then obtained by cutting the branches of the tree at some level. Several variants of this method are possible, depending on the definition of the distance function and on the assumption about how the distance between clusters will follow from the distances between the members of the clusters [67, 235]. Other approaches to class discovery, such as K-means clustering, self organizing maps (SOMs), and Kohonen neural networks. are also used in analyses [16].

11.6 Dimensionality Reduction

A characteristic property of experiments with DNA microarrays is the very large number of genes, which can reach the order of tens of thousands, versus the relatively small numbers of samples (microarrays). This can be an obstacle to extraction of information from the experiments. Techniques of dimensionality reduction by selection of the genes that capture most of the variation in the data are therefore very often applied to DNA microarray data [5, 72, 266, 278]. These techniques include two main approaches: principal component analysis (PCA) and partial least squares analysis (PLS). The computational aspects of these methods involving singular value decomposition of the matrices of measured expressions and formulating the PLS analysis as a sequence of optimization problems were discussed in Chap. 4. See also [131, 103]. These methods are either combined with class discovery and prediction, discussed in the next section, or used as a primary source of information about which genes are correlated most with the process analyzed.

PCA aims at identifying major directions in the data space by singular value decomposition of the data matrix and by limiting the analysis to the genes most strongly correlated with the principal directions. PCA is an unsupervised approach; it explores the structure of the data without prior knowledge on experimental conditions, outcomes of experiments, and so forth. The PLS method has the same aim as PCA, to reduce dimensionality, but in contrast to PCA, it needs a measurement or output vector to introduce structure into the data space. In the PLS method, orthogonal directions in the data space are computed on the basis of maximization of the variance between the output vector and a gene expression vectors (or a combination of them). The use of the PLS method requires access to a continuous measurement related to each array. Since this is not always available, the use of the PLS method with microarray data is not reported in the literature on statistics for DNA microarrays as often as the use of PCA. Examples of the application of the PLS method to reduce the dimensionality of DNA microarray data can be found in [207, 208], where a survival-time variable was used as an output vector in the PLS algorithm.

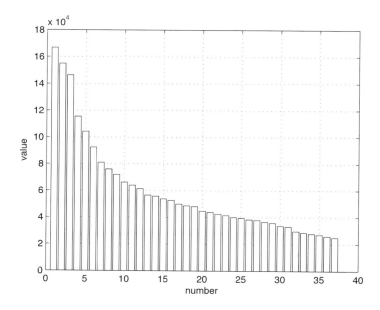

Fig. 11.7. Computed spectrum of singular values for the data set from [102]

11.6.1 Example of Application of PCA to Microarray Data

We shall illustrate dimensionality reduction for DNA microarray data by again using the dataset from [102]. As previously, we have included in the analysis only genes with strictly positive values of the expression level in all 38 scans, which confined the number of genes analyzed from the 6817 genes in Hu680 to 2568. Taking logarithms of the expression levels, we obtained a matrix X with $m = 2568$ rows and $n = 38$ columns. Each row in the matrix X corresponds to one gene, and each column corresponds to one experiment (a sample from a patient with either ALL or AML). We centered each of the rows (i.e., we subtracted from each element of the row the mean across the row) and applied the SVD (4.18) to the resulting matrix $X^{row\text{-}centered}$. We used the "economy" variant of SVD (4.44), which substantially speeds up computations. The number of nonzero singular values was $r = 37$, one singular value was equal to zero, which is related to the centering of the rows of the matrix X. The computed spectrum of singular values is shown in Fig. 11.7. The top ten singular values capture 65% of the total variance in the data.

11.7 Class Discovery

As already said, several different variants of class discovery procedures are applied as steps in data analysis; they are aimed at exploring the data structure and confirming the agreement of the results of unsupervised analysis with the existing knowledge about the data. For example, in [102], devoted to gene expression profiles of blood samples of AML and ALL patients, the two-cluster SOM technique was applied to all 38 expression profiles without assuming any knowledge about the data. The results of this unsupervised classification were in good consistency with the prior knowledge on the data. Namely, in the SOM classification only one AML case was misclassified as ALL, two ALLs were misclassified as ALL and for two samples, one ALL and one AML, the classification was not resolved (see Fig. 4A in [102]).

The basic interest concerning the data in [102] is whether there are differences of expression profiles between samples, and, more important, whether there are differences in expression profiles between two groups of ALL and AML patients. One must cope with the problems already outlined: the same gene can be marked as present in one patient's sample and absent (missing) in another; or can be measured as positive in one sample and erroneous in another. It may seem desirable to include as much of the data in the computational algorithm as possible, but practical experience shows that researchers must make a compromise between quantity and quality of data. Including genes that are weakly correlated or uncorrelated with the process studied, with many erroneous and noisy measurements, results in the addition of a source of disturbance, which, rather than providing new observations, obscures the relations the research program is after.

Therefore, studies based on gene expression data use steps aimed at excluding genes which add noise rather then contribute with useful information. Some typical approaches are:

(i) Exclude genes with multiple erroneous measurements, or labeled as absent in most (or all) of the experiments.
(ii) Estimate variances, by computing standard deviations, of genes across different measurements with different experimental conditions. Use the hypothesis that the genes with highest variances are the most informative and, in the further analysis, use only genes with high enough standard deviations.
(iii) Perform principal component analysis and include to further analysis only genes with a high enough affinity to the principal components.

11.7.1 Hierarchical Clustering

In the interpretation of results of experiments involving DNA microarrays, hierarchical clustering is probably the most often used algorithm for class discovery. As described in Chap. 4, there are several variants of hierarchical clustering, depending on the definition of the distance and the definition of linkage when clusters are formed. So there is a lot of flexibility in tuning the most convenient version of the algorithm to our needs. Also, with the help of appropriate software [310, 315, 343], the results of hierarchical clustering can be displayed graphically as trees, with a lot of information presented in a convenient and comprehensive manner.

Here we show an example of nonsupervised data exploration using data preprocessing and hierarchical clustering, again based on the study of ALL versus AML gene expression profiles [102]. We have reanalyzed the data from [102], applying the following steps in the analysis:

(i) We included into the analysis genes with strictly positive values of the expression level in all 38 scans, which reduced the number of genes analyzed from the 6817 genes on the Hu680 chip to 2568.
(ii) The data was log transformed.
(iii) We assumed the hypothesis that the useful information, concerning classification between ALL and AML, is carried by the genes with the largest variation between samples. We selected 110 genes with the highest values of standard deviations.
(iv) For the 110 genes selected in step (iii), we applied a hierarchical clustering algorithm.

The above procedure is based on the heuristic assumption that genes with more variation are more informative for data classification. The hierarchical clustering in step (iv) was performed with the options of Euclidean distance and complete linkage. The result of hierarchical clustering based on 110 genes from step (iv) is shown in Fig. 11.8. A classification into two groups can be obtained by cutting the tree at the highest level. One can see that the unsupervised classifier performs quite well, comparably to the class predictor described in [102]. There is one mistake of classification in Fig. 11.8: the AML sample number 29 is clustered together with ALL samples 1–27.

11.8 Class Prediction. Differentially Expressed Genes

Class prediction involves using information about the classification of the samples under study or about the different experimental conditions for different samples, in conjunction with the gene expression profiles, to classify samples obtained in further experiments. A very simple and rather efficient solution is often programmed into microarray scanner software: to pick out genes whose expression is strongly correlated with classes. Then use these genes as predictors for future classification experiments. The genes whose expressions differ most significantly between different samples or between different experimental conditions are said to be differentially expressed. Lists of these genes not only are of value as predictors or candidates for predictors for the construction of classifiers, but also provide insights into the processes involved in the samples studied, owing to existing knowledge about their functions and interrelations; see Sect. 11.10.

For example, many experiments involving the use of DNA microarrays compare gene expression profiles in cancer and normal cells. The comparisons

11.9 Multiple Testing, and Analysis of False Discovery Rate (FDR)   341

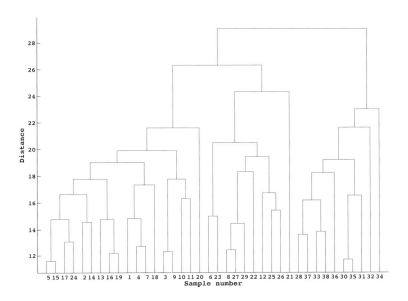

Fig. 11.8. The result of hierarchical clustering based on 110 genes in the example described in the text

lead to the publication of lists of cancer-versus-normal differentially expressed genes, often ordered with respect to their differentiating power determined by several approaches, for example, by the p-value of a statistical test. Genes can then be compared across different studies, as in [238] and the associated Website containing the collection of microarray datasets of human cancers [328].

Classifiers based on rather small numbers of the top differentially expressed genes selected in sample comparison studies are usually very effective in predicting taxonomies related to future measurements.

11.9 Multiple Testing, and Analysis of False Discovery Rate (FDR)

Creating a list of genes expressed differentially in a sample A versus a sample B most often involves multiple calls of procedures for testing statistical hypotheses. Since every statistical test accepts or rejects the associated hypothesis with some probability and the testing is repeated many times (typically, the number of tests equals the number of genes), an issue arises of controlling the rate of statistical errors of type I (false discoveries) among the results of the tests. Let us define the family-wide error rate (FWER) as the probability of making at least one (i.e., one or more) type I error (false discovery) in multiple testing.

The idea of control of false discoveries in multiple testing can now be expressed by introducing a constraint on the value of FWER, FWER $< \alpha$. Enforcing FWER $< \alpha$ must be done by introducing corrections concerning the significance levels of individual tests. The corrections, given by the Bonferroni or Sidak formulas or their variants [297], are known to be very conservative. A more flexible approach is related to introducing a concept of false discovery rate (FDR) [28] equal to the expected proportion of false positives (false discoveries) among the tests where the null hypothesis was rejected (the discoveries).

In [28], a multiple testing procedure based on FDR control was proven to be significantly less conservative than procedures based on the FWER. Further developments were proposed in [97, 269] and, in the context of microarray data, in [2].

Here we sketch the FDR approach published in [2]. This method uses a decomposition of the probability density function related to the distribution of p-values of statistical tests to estimate and control the FDR. Let us assume that a series of DNA microarray experiments have been conducted to compare samples A and B. For each of the genes on the microarray chip, a set of measurements is available and, on the basis of these measurements, a statistical test is performed, with the null hypothesis H_0 being "the distributions $p_A(x)$ and $p_B(x)$ of the expressions of gene X for samples A and B, are equal i.e., $p_A(x) = p_B(x)$". A computer procedure typically included in statistical software packages, returns a p-value of the test which leads either to rejecting H_0 at the significance level α if $p < \alpha$, or to "no premise to reject H_0" if $p > \alpha$ (typically, $\alpha = 0.05$).

Note that the p-value of a statistical test is a random variable with a distribution supported on the segment $[0, 1]$. Consider two situations:

(i) H_0 is true, or in other words $p_A(x) = p_B(x)$. In this situation the p-value of the test is obtained from inverting the cumulative distribution function of the test statistics, so the distribution of p-values is uniform over the segment $[0, 1]$. (Compare this situation Exercise 8.)

(ii) H_0 is not true, or in other words $p_A(x)$ and $p_B(x)$ are different distributions. If we do not assume a particular form of the distributions $p_A(x)$ and $p_B(x)$ we cannot derive any specific form for the distribution of p-values here. However, in the case when $p_A(x) \neq p_B(x)$, intuitively the p-values of the test should cluster close to zero, since H_0 should most often be rejected. So the distribution is no longer uniform.

In conclusion, the above probability density function, which we denote by $f(p)$, associated with the distribution of p-values of the statistical test applied in (i) and (ii), is a mixture of a uniform component $f^u(p)$ corresponding to (i) and a component related to (ii) which the authors of [2] propose to approximate by a mixture of a number of beta distributions, i.e.,

11.9 Multiple Testing, and Analysis of False Discovery Rate (FDR)

$$f(p) = w_u f^u(p) + \sum_{j=1}^{J} w_{\beta j} f^{\beta j}(p, a_j, b_j). \tag{11.27}$$

In the above equation $f^u(p)$ is the probability density function of a uniform distribution; $f^{\beta j}(p, a_j, b_j)$ denotes the probability density function of the jth beta distribution; w_u and $w_{\beta j}$, $j = 1, \ldots, J$ are weighting coefficients; $w_u + \sum_{j=1}^{J} w_{\beta j} = 1$, and a_j, b_j, $j = 1, \ldots, J$ are parameters of the beta distributions, (see Chap. 2). There is no direct probabilistic argument for using beta distributions as models for the p-values of statistical tests. However, beta distributions are supported on the unit interval and there is a lot of flexibility in fitting their shapes to data by changing the values of the parameters a and b, which makes them a good tool. The uniform distribution can be seen as a special case of the beta distribution with $a = b = 1$. In most cases it turns out that limiting the approximation to only one beta component, $j = 1$, is satisfactory for approximating $f(p)$ in (11.27).

In order to write down explicit expression for the false discovery rate, we summarize the possible situations in Table 11.1. The numbers $A(\alpha)$, $B(\alpha)$, $C(\alpha)$, and $D(\alpha)$ in Table 11.1 are random variables; they depend on the threshold chosen α, since the hypothesis H_0 is rejected if $p < \alpha$. Using $A(\alpha)$, $B(\alpha)$, $C(\alpha)$, and $D(\alpha)$ we can compute the false discovery rate as

$$FDR(\alpha) = E\left[\frac{A(\alpha)}{A(\alpha) + B(\alpha)}\right], \tag{11.28}$$

the ratio of the number cases where H_0 is rejected when it is true to the total number of cases where H_0 is rejected. We also define the discovery rate as

$$DR(\alpha) = E\left[\frac{B(\alpha)}{B(\alpha) + D(\alpha)}\right], \tag{11.29}$$

the ratio of the number of cases where a false H_0 is rejected to the total number of cases where H_0 is false. Using the decomposition (11.27), we can easily compute FDR and DR in (11.28) and (11.29) as

$$FDR(\alpha) = \frac{w_u F^u(\alpha)}{w_u F^u(\alpha) + \sum_{j=1}^{J} w_{\beta j} F^{\beta j}(\alpha, a_j, b_j)} \tag{11.30}$$

and

$$DR(\alpha) = \frac{\sum_{j=1}^{J} w_{\beta j} F^{\beta j}(\alpha, a_j, b_j)}{\sum_{j=1}^{J} w_{\beta j}} \tag{11.31}$$

(see Exercise 10). In the above equations, $F^u(\alpha)$ and $F^{\beta j}(\alpha)$ denote the cumulative distribution functions of the uniform and beta distributions.

Table 11.1. Numbers of cases $A(\alpha)$, $B(\alpha)$, $C(\alpha)$ and $D(\alpha)$ corresponding to possible situations associated to accepting or rejecting of the hypothesis H_0.

	H_0 rejected	H_0 not rejected
H_0 true	$A(\alpha)$	$C(\alpha)$
H_0 false	$B(\alpha)$	$D(\alpha)$

11.9.1 FDR analysis in ALL versus AML gene expression data

We now give an example of the use of the technique of FDR control again based on the study of ALL versus AML gene expression profiles [102]. As previously, we have included into the analysis only genes with strictly positive values of expression levels in all 38 scans, which reduced the number of genes analyzed to 2568. For all 2568 genes kept in the study we have performed Wilcoxon unpaired sign sum test between the 27 ALL and 11 AML samples [297]. A histogram of the p-values resulting from the 2568 tests is presented in the upper plot in Fig. 11.9. We then fitted the decomposition (11.27), with one beta component ($J = 1$). The weights obtained were $w_u = 0.54$ and $w_\beta = 0.46$, and a plot of the estimated probability density function is depicted by the solid line in the upper plot in Fig. 11.9. Using the decomposition (11.27), we computed expected false discovery rates and discovery rates $FDR(\alpha)$ and $DR(\alpha)$, given in (11.30) and (11.31). The resulting plots of $FDR(\alpha)$ and $DR(\alpha)$ are shown in Fig. 11.9 in the lower plot.

11.10 The Gene Ontology Database

The large numbers of genes with different products and functions require bioinformatic tools to make the information about them organized and available. Several databases mentioned in this book store sequences of nucleotides and related sequences of amino acids along with a wealth of background and support information concerning, for example, location, known functions of gene products, and homology between genes in different organisms. Biological studies of sequences and their functions are supported to a great extent by these depositories. For any sequence of nucleotides or amino acids, a researcher can, in just a few minutes, access a large amount of data concerning functions and products of all genes, proteins, and RNA sequences with some kind of similarity to it. Bioinformatic databases have a relational structure, with a dense network of links between items recorded in different depositories.

However, the processes of browsing through genomic and proteomic databases undertaken in many biological research projects have demonstrated the need for more consistent and more structured descriptions of gene products in the various databases. More precisely, many biological studies involving database searches for the purpose of comparison of experimental work performed in the laboratory with references in the literature and data in depositories have experienced inefficiencies and obtained misleading results, because

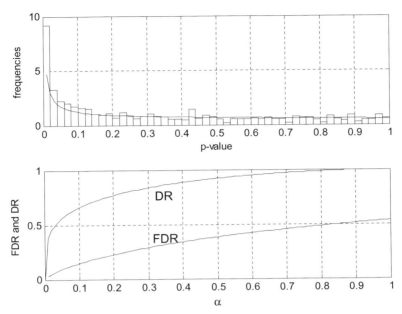

Fig. 11.9. An example of use of the technique of FDR control for the ALL versus AML data from [102]. *Upper plot*: Histogram of p-values resulting from 2568 Wilcoxon unpaired sign sum tests between the 27 ALL and 11 AML samples (bar plot) and estimated probability density function of the beta distribution (solid line). *Lower plot*: false discovery rate and discovery rate, $FDR(\alpha)$ and $DR(\alpha)$.

there was not enough structure and precision in the terminology. This situation has been responded to by recent initiatives to create ontology databases. In this section, we focus on the Gene Ontology (GO) database [312], which is probably the best known and most often referenced ontology database. By gene ontology we mean a standardized and structured vocabulary that describes genes and their products. Ontologies can also involve areas other than gene products.

It may seem that the development of efficient terminology and vocabulary should an inherent feature of progress in science, but in the presence of the massive amount research in the area, the Gene Ontology database maintains the homogeneity of the terminology and makes possible to compare and summarize the results of many studies.

11.10.1 Structure of GO

The terms in GO are organized as a tree structure with three main branches, or rather, GO consists of three ontologies named "molecular function", "biological process" and "cellular component". "Molecular function" describes the activities of gene products at the molecular level. For example, the terms in the category of "molecular function" include "nucleotide binding", "peptide receptor activity" or "lipid binding". The "biological function" ontology involves naming processes or series of events which can incorporate several molecular mechanisms. Some examples of GO terms classified as "biological process" are "apoptosis", "oxygen transport", "metabolism", or "response to stimulus". Finally, the "cellular component" GO terms include "cell nucleus", "membrane", "ribosome", and so forth.

The GO database can be explored or downloaded in many formats, including XML, and searched with the use of many different search protocols. There are also numerous programs and Web sites that perform GO searches along with standard queries, for example, BioMart [70], Babelomics [4, 7], and AmiGO [304].

As an example, we can do a very simple query involving the TEL1 yeast gene mentioned in Sect. 8.6. By typing in the term "TEL1" and choosing the option "gene symbol/name", we find under TEL1_YEAST the summary

(IDA) – protein kinase activity–molecular function,
(TAS) – phosphorylation–biological process,
(TAS) – response to DNA damage stimulus–biological process,
(TAS) – nucleus–cellular component,

describing GO terms related to the yeast TEL1 gene. The abbreviations indicate, to some extent, the certainty level of the classification, IDA stands for "Inferred from Direct Assay" and TAS means "Traceable Author Statement".

11.10.2 Other Vocabularies of Terms

The development of vocabulary terms in GO described above is being accompanied by similar progress in several related areas. Responding to this development in the organization of terminology, ontology browsers, such as those mentioned above, [7, 70, 304] are including vocabulary terms from new areas in their repertoire of analyses. The new vocabulary terms include genetic and metabolic pathways, protein domains and functional sites, transcription factors, and regulatory elements. The developments in ontology browsers involve both the addition of new vocabularies and the creation of mappings between these vocabularies.

A fast-growing dictionary of terms involves genetic pathways, signaling, metabolic, and regulatory. A large database containing data on the genetic pathways is KEGG, the Kyoto Encyclopedia of Genes and Genomes, [138, 323]. KEGG is a Website that organizes databases and associated software, integrating PATHWAY, BRITE, GENES, and LIGAND. Therefore it includes information both on genetic pathways and on related biochemical and biological processes. A common way of using KEGG terms is by analyzing lists of words given by names of pathways corresponding to a set of genes.

Another dictionary of terms concerns interPro motifs related to protein families, domains and functional sites. The related database is run by EBI, European Bioinformatic Institute, [314].

One more example is cisRED database of cis regulatory elements database [107], [308]. The cisRED database holds conserved sequence motifs identified by genome scale motif discovery, similarity, clustering, co-occurrence and co-expression calculations.

11.10.3 Supporting Results of DNA Microarray Analyses with GO and other Vocabulary Terms

The methodologies commonly applied in DNA microarray assays lead to comparisons of large numbers of genes and to grouping of genes into classes or clusters, using a criterion based on correlations or similarities of their patterns of expression. Although many meaningful insights have been obtained with this approach, owing to the large biological variation there is still a lot of uncertainty and ambiguity regarding the meaning of the measured gene expressions. Therefore including GO classes and terms in DNA-microarray-based studies allows automated confronting and combining of the experimental results obtained with the "background knowledge". Since the GO database is actually a tree of names, it can be very easily searched foe single terms as well as for lists of terms, trees, graphs, etc. This makes it a very convenient tool for creating or, rather, adding interpretations to summaries and comparisons of gene expression.

Some of the most obvious approaches to supporting DNA expression analysis by means of GO searches are the following:

(i) By using some criterion, obtain a list of genes differentially expressed in two experiments and use the GO terms of the genes in the list to infer possible mechanisms and processes.
(ii) Obtain two lists of genes, for example, a list A of genes upregulated in experiment A and a list B of genes upregulated in experiment B, and compare these two lists by use of their GO terms.
(iii) Obtain from GO a list of genes related to, for example, one biological process, apoptosis, and study the pattern of gene expression in some microarray measurements, limited to the list obtained from GO.

11.11 Exercises

1. This is a study of the noise-filtering method based on (11.16). Assume that we measure some signals S_i according to the model

$$S_i = X + \Delta_i$$

where it is known that the signal X is distributed exponentially and that Δ_i is a normally distributed, independent error. Develop a method for estimating $\hat{X}$ on the basis of a series of measurements $s_1, s_2, ..., s_K$.

2. Derive (11.15) for updating the estimate of the variance of distribution (11.9) under the hypothesis of the Wishart prior distribution (11.13).
3. Derive (11.23)–(11.25) for iterations of an EM algorithm to estimate the parameters of a Gaussian mixture for modeling repeated measurements of gene expression. Use the solution to Exercise 14 in Chap. 2.
4. Derive equations analogous to (11.23)–(11.25) or to (2.87)–(2.89) for iterations of EM algorithm for estimating parameters of a Gaussian mixture, under the assumption that successive measurements, numbered $1, 2, \ldots, R$, of gene expressions by use of DNA microarrays are conducted under changing experimental conditions. We impose the requirement that genes remain clustered to the same component over all experiments, but not that parameters of Gaussian components (means and variances) remain constant over all experiments.
5. Using the equations derived in Exercise 3, develop a computer program for estimating the parameters of Gaussian mixtures.
6. Download the gene expression dataset concerning the yeast cell cycle experiments presented in [51] (http://genomics.stanford.edu) and use programs developed in Exercise 5 to study the clusters of coexpressed genes.
7. Use the BIC correction of the likelihood (11.26) to estimate the number of components in the mixture distribution obtained in Exercise 6.
8. Assume that the probability density functions $p_A(x)$ and $p_B(x)$ are standard normal distributions. Perform 1000 random experiments involving (1) generating 20 realizations of random variable with pdf $p_A(x)$ and 20 realizations of random variable with pdf $p_B(x)$, and (2) calling a procedure of comparing $p_A(x)$ and $p_B(x)$ by the t-test. What is the distribution of p-values obtained in 1000 repetitions of steps (1) and (2)?
9. Develop a procedure and a computer program for decomposing an observed distribution of p-values of a statistical test as described in (11.27).
10. Derive (11.30) and (11.31). Use the program from Exercise 8 to draw curves of $E(FDR)$ and $E(DR)$ for given distribution of p-values corresponding to multiple statistical testing.
11. Download one of the datasets posted in the database [238]. Apply a normalization procedure with the use of the RMA method [335]. Score differentially expressed genes by the p-values of a statistical test (the unpaired Wilcoxon sum rank test can be used) and use programs developed in Exercises 8 and 9 to analyze the false discovery rate.
12. Perform a gene ontology study of the gene names in the clusters of coexpressed genes obtained in Exercise 6. Use the software developed in [4] to obtain summaries of the ontology terms.

12
Bioinformatic Databases and Bioinformatic Internet Resources

The origin of bioinformatics was associated with the need for maintenance of databases containing biological, biochemical, and clinical data. Historically, the process of collecting bioinformatic sequences started with amino acid sequences in proteins [61] and then proceeded by coordinating and standardizing submissions by the institutions engaged in maintaining the database. Currently there is fast growth in the volume of the data stored in bioinformatic databases. Bioinformatic databases are also growing in number. There are dense links between different databases, since they often contain information pertaining to several aspects of, for example, the same sequence of amino acids in a protein. Also, numerous Internet resources offer services related to searching and browsing bioinformatic databases and to various algorithms for data processing and inference based on bioinformatic data.

In this book, we referenced many databases containing bioinformatic data and we have shown samples of their resources. In this short chapter we present a view of the types and structures of bioinformatic databases and other bioinformatic resources and on the relations between them. We provide a list of bioinformatic Web sites and a classification of the types of bioinformatic databases and bioinformatic resources, based on their content and functions. Our list of databases and bioinformatic sites is not comprehensive; rather, it contains those most widely known and some samples of others. Also, our view on their classification may not be unquestionable, owing to the large variety of data and functions and the dense links between databases. Our aim when presenting the overview below was to provide a picture of the Internet resources related to bioinformatics.

In the references, we have provided Internet addresses of many databases and other bioinformatic sites. However, they should be treated with caution owing to the to constant evolution in the field.

12.1 Genomic Databases

The best known genomic database is GenBank, the database of genomic sequences maintained by the NCBI (National Center for Biotechnology Information) [326]. It contains all annotated nucleic acid and amino acid sequences. Its contents are mirrored by two other databases, the EMBL (European Molecular Biology Laboratory) database and DDBJ (DNA Data Bank of Japan). Apart from presenting and annotating sequences, these databases offer many functions related to searching and browsing sequences. They perform services concerning submitting new sequences to GenBank, and also contain links to various bioinformatic internet sites.

Many databases contain more specialized information on genomic sequences. Examples of such databases are the Single Nucleotide Polymorphisms (SNP) Consortium database for biomedical research [339], the databases of highly conserved DNA motifs [324], and the cisRED database [308] of gene promoter and regulatory sequences. The database [320] serves for standardizing the nomenclature for genes.

12.2 Proteomic Databases

Owing to the correspondences between amino acid and codon sequences, there are strong links between protein and nucleotide databases. As mentioned above, amino acid sequences of proteins are available at GenBank along with nucleotide sequences. Data on sequences of amino acids, on the taxonomy, functional aspects of proteins, protein families and domains, as well as data on known secondary and 3D structures of proteins, are stored in proteomic databases, [317, 329, 342, 345] databases, Swiss-Prot, Uni-Prot, and ExPASy (Expert Protein Analysis System) comprise information including the function, classification, amino acid sequences in proteins and structures of proteins. The PDB (Protein Data Bank) database [329] contains annotated data on the spatial structures of proteins and biological macromolecules. It also includes data on their sequences, functions, and related diseases.

There are also many databases specializing in particular aspects of proteins, protein functions, experiments involving proteins etc., such as databases of protein 2D gels [301], restriction enzymes [333], and secondary structures [319].

12.3 RNA Databases

Information on sequences of ribonucleotides in RNA, coding and noncoding RNA sequences, the functions of RNA molecules and their spatial structures, is available in the databases [318, 321, 329, 334]. Rfam database [334] stores noncoding RNA (ncRNA) families. Rfam also contains multiple sequence alignments and covariance models. The GtRNA database stores genomic tRNA ribonucleotide sequences and secondary structures. The Jena index of RNA structures [321] provides a lot of information about RNA, including indexes of the locations of molecular structures in the PDB database [329]. Data on ribonucleic acid sequences in RNA can also be found in GenBank [326].

12.4 Gene Expression Databases

Numerous databases contain data on expression levels measured in various experiments. Here we list sow some of the best known [325, 307, 326, 328]. They are aimed at making possible the sharing of data in the new field of microarray experiments. A part of the NCBI database, NCBI Gene Expression Omnibus (GEO), [326] is a database including links to microarray-based experiments measuring mRNA, genomic DNA, and protein abundances, as well as non array techniques such as serial analysis of gene expression (SAGE), and mass spectrometric proteomic data. The MGED database contains datasets from many experimental studies involving gene expression. It also contains links to gene expression data-processing procedures and ontologies. The databases CGED (Cancer Gene Expression Database) [307] and ONCOMINE [328] deliver published cancer gene expression data to the research community.

12.5 Ontology Databases

Probably the most extensively used ontology database is GO (Gene Ontology) [312], which provides controlled vocabularies for supporting analyses of gene expression measurements and other molecular-biology experiments. However, other ontologies are also developing at a fast rate. One database containing links to many Internet ontologies is OBO (Open Biomedical Ontologies) [327]. It provides Web addresses of many sites containing biomedical structured vocabularies, including GO, the Generic Model Organism Project (GMOD), Microarray Gene Expression Data (MGED), The National Center for Biomedical Ontology vocabulary.

12.6 Databases of Genetic and Proteomic Pathways

The area of genetic pathways is a very fast-growing field for bioinformatic data. One large database containing data on the genetic pathways is KEGG, the Kyoto Encyclopedia of Genes and Genomes [323], a Web site that organizes databases and associated software. The BioCarta database [305] supports proteomic studies by providing information on proteomic pathways, as well as on reagents, antibodies, proteins, cells and cell-based assays. Pathway databases offer a graphical presentation of their contents, which provides a useful support for qualitative understanding of signaling, regulatory, and other mechanisms.

12.7 Programs and Services

A large variety of programs for performing bioinformatic computations are available on the Internet. Here we list some areas and related Web sites.

The Internet bioinformatic services that are most important to biologists and most frequently used by them are probably those for aligning sequences against sequence databases. This task is performed by variants of the BLAST program, [140, 141, 250] or earlier program FASTA [68, 288]. Sequence databases contain links to these programs. Related to this area are also programs for performing multiple (block) alignments of sequences, e.g., CLUSTAL W [309, 314].

A well-known program and Internet site for inferring phylogenetic trees is PHYLIP [330]. The Web site also includes addresses of other Internet resources related to phylogenetics.

Examples of programs for annotating genome sequences are GENESCAN [313] or MEME [324]. Also, the NCBI Web site contains a simple service for searching of open reading frames.

Several servers are aimed at performing assembly of DNA sequences from reads, for example, Atlas Genome Assembly [117], Arachne [20], Celera Assembler [203], Jazz [62], Phusion [201], PCAP [125], and Euler [224], [316].

Internet servers for predicting the secondary and tertiary structures of proteins and RNA are, for example, [311, 322, 331, 336, 346].

There are many programs concerned with molecular geometry and visualization, for example, 3DNA [302] for nucleic acid structures or more general program for showing biomolecules Ras Mol [332]. Examples of visualizations of molecular structures obtained by using programs 3DNA and Ras Mol are presented in Figs. 12.1 and 12.2.

There are many different programs related to algorithms related to gene expression data, including image processing [337], normalization [306, 335], classification and clustering [306, 310, 343], and searches of ontologies [1, 304, 7, 306, 303]. An example of gene expression data from cDNA microarray, represented in the form of an image is presented in Fig. 12.3.

12.8 Clinical Databases

In this last section we mention repositories of clinical data. These databases are growing at a very fast rate owing to their importance in developing knowledge about disease etiologies, therapies, treatments, and so forth. They collect data describing clinical cases, diagnoses, therapy protocols, recovery, and

12.8 Clinical Databases   353

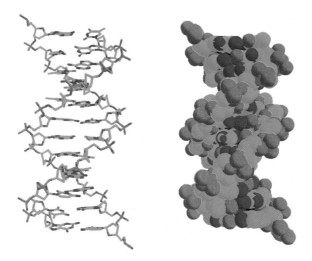

Fig. 12.1. Representation of the spatial structure of atoms and bonds in a DNA polymer obtained with the use of publicly available programs 3DNA [302] and Ras Mol [332]. The representation on the left, called "sticks", depicts atomic bonds, while the one on the right, called "balls" shows the approximate atomic radii. The colors correspond to atom types in the following way: red, oxygen; green, carbon; purple, phosphorus; blue, nitrogen

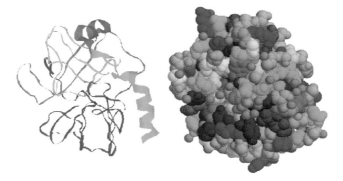

Fig. 12.2. Graphical presentation of the enzyme trypsin obtained with the use of spatial coordinates of atoms from Protein Data Bank (accession symbol 2ptn), and the molecular-graphics program Ras Mol. *Left*: A view resulting from choosing the Ras Mol option "ribbons" for enhancing secondary structures. *Right*: A view resulting from using "spacefill" option. Different colors represent different amino acids. The meaning of colors is as follows (see also Table 9.1): ASP, GLU, bright red; LYS, ARG, blue; PHE, TYR, mid blue; ALA, dark grey; HIS, pale blue; CYS, MET; yellow; SER, THR, orange; ASN, GLN, cyan; LEU, VAL, ILE, green; TRP, pink; PRO, flesh

354   12 Bioinformatic Databases and Bioinformatic Internet Resources

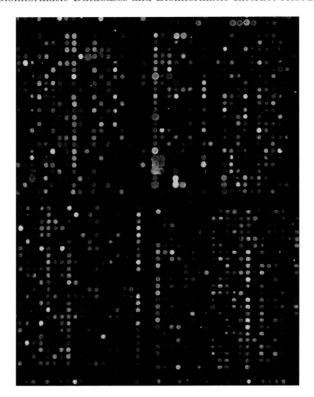

Fig. 12.3. An example of an image of a cDNA microarray, from the Website containing the data accompanying [3]

survival. Different types of data are brought together: patients records, diagnostic tests including medical images, treatment plans and follow up data on patient cohorts concerning recovery, complications and survival. There are many aspects of the development of clinical databases: quality, completeness and volume of data, availability and so forth [33]. An example of a clinical database is the European EURODIAB database related to type 1, childhood diabetes [74, 105], resulting from many years of collecting clinical cases of type 1 diabetes mellitus. Along with storing patient records, treatments, and survival data, clinical databases can also organize and help with access to tissue banks. An example is the GENEPI (GENEtic Pathways for the Prediction of the Effects of Ionising Radiation) database, which stores clinical data concerning therapeutic radiation and, at the same time, can help in arranging access to tissue samples [21].

References

1. Affymetrix (2002), *GeneChip Expression Analysis, Data Analysis Fundamentals*, Affymetrix Manual.
2. Allison D. B., Gadbury G., Heo M., Fernandez J., Lee, C. K., Prolla T. A., Weindruch R. (2002), A mixture model approach for the analysis of microarray gene expression data. Comput. Statist. Data Anal., vol. 39, pp. 1–20.
3. Alizadeh A. A., Eisen M. B., Davis R. E., Ma C., Lossos I. S., Rosenwald A., Boldrick J. C., Sabet H., Tran T., Yu X., et al. (2000), Distinct types of diffuse largeB-cell lymphoma identified by gene expression profiling, Nature, vol. 403, pp. 503–511.
4. Al-Shahrour F., Diaz-Uriarte R., Dopazo J. (2004), FatiGO: a web tool for finding significant associations of Gene Ontology terms with groups of genes. Bioinformatics, vol. 20, pp.578–580.
5. Alter O., Brown P.O., Botstein D., (2001), Processing and modeling genome-wide expression data using singular value decomposition. Proc. SPIE vol. 4266, pp. 171–186.
6. Andrade M. A. (ed.) (2003), *Bioinformatics and Genomes*, Springer.
7. Al-Shahrour F., Minguez P., Vaquerizas J. J. M., Conde L., Dopazo J. (2005), BABELOMICS: a suite of web tools for functional annotation and analysis of groups of genes in high-throughput experiments, Nucleic Acids Res., vol. 33 (Web server issue), pp. W460–W464.
8. Angluin D., Valiant L. G. (1979), Fast probabilistic algorithm for Hamiltonian circuits and matchings, J. Comput. System Sci., vol. 18, pp. 155–193.
9. Asai K., Hayamizu S., Handa K. (1992), Prediction of protein secondary structure by the hidden Markov model. Bioinformatics, vol. 9, pp. 141–146.
10. Ashburner, M., Ball, C. A., Blake, J. A., Botstein, D., Butler, H., Cherry, J. M., Davis, A. P., Dolinski, K., Dwight, S. S., Eppig, J. T., et al. (2000), Gene ontology: Tool for the unification of biology. The Gene Ontology Consortium. Nat. Genet., vol. 25., pp. 25–29.
11. Bahlo M., Griffiths R.C. (2000), Inference from gene trees in a subdivided population. Theor. Popul. Biol., vol. 57, pp. 79–95.
12. Bailey T.L., Gribskov M. (1998), Methods and statistics for combining motif match scores, J. Comput. Biol., vol. 5, pp. 211–221.
13. Bains W., Smith G. C. (1988), A novel method for DNA sequence determination, J. Theor. Biol., vol. 135, pp. 303–307.

14. Baker D., Sali A. (2001), Protein structure prediction and structural genomics. Science, vol. 294, pp. 93–96.
15. Baldi P., Brunak S. (1998), *Bioinformatics (Adaptive Computation and Machine Learning)*, MIT Press.
16. Baldi P., Hatfield G.W. (2002), *DNA Microarrays and Gene Expression, from Experiments to Data Analysis and Modeling*, Cambridge University Press.
17. Ballard D.H., (1981), Generalizing the Hough transform to detect arbitrary shapes. Patt. Recogn., vol. 13, pp. 111–122.
18. Barrett G. C., Elmore D. T. (1998), *Amino Acids and Peptides*, Cambridge University Press.
19. Batey R. T., Rambo R. P., Doudna J. A. (1999), Tertiary motifs in RNA structure and folding. Angew. Chem. (Int. Ed.), vol. 38, pp. 2326–2343.
20. Batzoglou S., Jaffe D. B., Stanley K., Butler J., Gnerre S., Mauceli E., Berger B., Mesirov J. P., Lander E. S. (2002), ARACHNE: A whole-genome shotgun assembler. Genome Res., vol. 12, pp. 177–189.
21. Baumann M., Holscher T., Begg A. C. (2003), Towards genetic prediction of radiation responses: ESTRO's GENEPI project. Radiother. Oncol., vol. 69, pp. 121–125.
22. Baxevanis A., Ouellette B. F. (eds.) (1998), *Bioinformatics, A Practical Guide the Analysis of Genes and Proteins*, Wiley.
23. Beer D. G., Kardia S. L. R., Huang C. C., Giordano T. J., Levin A. M., Misek D.E., Lin L., Chen G.A., Gharib T.G., et al., (2002), Gene expression profiles predict survival of patients with lung adenocarcinoma. Nat. Med., vol. 8, pp. 816–824.
24. Beerli P., Felsenstein J. (2001), Maximum likelihood estimation of a migration matrix and effective population sizes in n subpopulations using a coalescent approach, Proc. Natl. Acad. Sci. USA, vol. 98, no. 8, pp. 4563–4568.
25. Bellman R. E., (1970), *Introduction to Matrix Analysis*, McGraw-Hill.
26. Bellman R. E., Dreyfus S. E. (1962), *Applied Dynamic Programming*, Princeton University Press.
27. Bellman R., Kalaba R. E., Lockett J. A. (1966), *Numerical Inversion of the Laplace Transform*, Elsevier.
28. Benjamini Y., Hochberg Y. (1995), Controling the false discovery rate: a practical and powerful approach to multiple testing. J. R. Statist. Soc., Ser. B, vol. 57, pp. 125–133.
29. Berg J. M., Tymoczko J. L., Stryer L. (2002), *Biochemistry*, W. H. Freeman.
30. Berndt P., Hobohm U., Langen H. (1999), Reliable automatic protein identification from matrix-assisted laser desorption/ionization mass spectrometric peptide fingerprints. Electrophoresis, vol. 20, pp. 3521–3526.
31. Bhat S., Purisima E. O. (2006), Molecular surface generation using a variable-radius solvent probe. Proteins: Struct. Funct. Bioinformat., vol. 62, pp. 244–261.
32. Biernacki C., Govaert G., (1999), Choosing models in model-based clustering and discriminant analysis. J. Statist. Comput. Simul., vol. 64, pp. 49–71.
33. Black N., Payne M., DoCDat Development Group (2003), Directory of clinical databases: improving and promoting their use. Qual. Saf. Health Care vol. 12, pp. 348–352.
34. Bohm H. J. (1994), The development of a simple empirical scoring function to estimate the binding constant for a protein–ligand complex of known three-dimensional structure. J. Comput.-Aided Mol. Des., vol. 8, pp. 243-256.

35. Bolstad B. M., Irizarry R. A., Astrand M., Speed, T.P. (2003), A comparison of normalization methods for high density oligonucleotide array data based on bias and variance. Bioinformatics, vol. 19, no. 2, pp. 185–193.
36. Bernardo J. Smith A. (1994), *Bayesian Theory*, Wiley, New York.
37. Billingsley P. (1995), *Probability and Measure*, Wiley New York.
38. Bowie J. U., Luthy R., Eisenberg, D. (1991), A method to identify protein sequences that fold into a known three-dimensional structure. Science, vol. 253, pp. 164–170.
39. Boyer R. S., Moore J. S. (1977), A fast string searching algorithm, Commun. ACM, vol. 20, pp. 762–772.
40. Bredel M., Jacoby E. (2004), Chemogenomics: an emerging strategy for rapid target and drug discovery. Nature Rev., vol. 5, pp. 262–275.
41. Broet P., Richardson S., Radvanyi F. (2002), Bayesian hierarchical model for identifying changes in gene expression from microarray experiments. J. Comput. Biol., vol. 9., pp. 671-683.
42. Brooks B. R., Bruccoleri R. E., Olafson B. D., States D. J., Swaminathan S., Karplus M. (1983), CHARMM, a program for macromolecular energy, minimization, and dynamics calculations. J. Comput. Chem. vol. 4, pp. 187–217.
43. Burge C., Karlin S. (1998), Finding the genes in genomic DNA, Curr. Opin. Struct. Biol., vol. 8, pp. 346-354.
44. Burge C., Karlin S. (1997), Prediction of complete gene structures in human genomic DNA. J. Mol. Biol., vol. 268, pp. 78–94.
45. Burges, C. J. C. (1998), *A Tutorial on Support Vector Machines for Pattern Recognition*, Kluwer Academic, Boston.
46. Burges C., Vapnik V. (1995), A new method for constructing artificial neural networks. Technical Report, AT&T, Bell Laboratories.
47. Burrows M., Wheeler D. J. (1994), A block sorting lossless data compression algorithm. Technical Report 124, Digital Equipment Corporation, Palo Alto, CA.
48. Bystroff C., Baker D. (1998), Prediction of local structure in proteins using a library of sequence–structure motifs. J. Mol. Biol., vol. 281, pp. 565–577.
49. Cann R. L., Stoneking M., Wilson A. C. (1987), Mitochondrial DNA and human evolution. Nature, vol 325, pp. 31–36.
50. Chao K. M., Pearson W. R., Miller W. (1992), Aligning two sequences within a specified diagonal band. Comput. Appl. Biosci., vol. 8, pp. 481–487.
51. Cho R. J., Campbell M. J., Winzeler E. A., Steinmetz L., Conway A., Wodicka L., Wolfsberg T. G., Gabrielian A. E., Landsman D., Lockhart D. J., Davis R. W. (1998), A genome-wide transcriptional analysis of the mitotic cell cycle, Mol. Cell, vol. 2, pp. 65–73.
52. Choi V., Farach-Colton M. (2003), Barnacle: An assembly algorithm for clone-based sequences of whole genomes. Gene, vol. 320, pp. 165–176.
53. Cleveland W. S. (1979), Robust locally weighted regression and smoothing scatterplots. J. Amer. Statist. Assoc., vol. 74, pp. 829-836.
54. Cook W. J. (1998), *Combinatorial optimization*, Wiley, New York.
55. Corpet F., Michot B. (1994), RNAlign program: alignment of RNA sequences using both primary and secondary structures. Comput. Appl. Biosci., vol. 10, pp. 389–399.
56. Coutsias E. A., Seok C., Dill K. A. (2004), Using quaternions to calculate RMSD. J. Comput. Chem., vol. 25, pp. 1849-1857.

References

57. Cox D.R., Oakes, D. (1984), *Analysis of Survival Data*, Chapman and Hall, London.
58. Crick F. (1970), Central dogma of molecular biology, Nature, vol. 227, pp. 561–563.
59. Cuff J. A., Clamp M. E., Siddiqui A. S., Finlay M. Barton G. J. (1998), Jpred: a consensus secondary structure prediction server. Bioinformatics, vol. 14, pp. 892–893.
60. Dai H., Meyer M, Stepaniants S., Ziman M, Stoughton R. (2002), Use of hybridization kinetics for differentiating specific from non-specific binding to oligonucleotide microarrays. Nucleic Acids Res., vol. 30, no. 16, p. e86.
61. Dayhoff, M.O., Schwartz R.M., Orcutt B.C., (1978), A model for evolutionary change in proteins, in. *Atlas for protein sequence and structure*, vol. 5, suppl. 3, National Biomedical Research Foundation, Georgetown University, Washongton D.C., pp. 345–358.
62. Dehal P., Satou Y., Campbell R. K., Chapman J., Degnan B., De Tomaso A., Davidson B., Di Gregorio A., Gelpke M., Goodstein D.M., et al. (2002), The draft genome of *Ciona intestinalis*: Insights into chordate and vertebrate origins. Science, vol. 298, pp. 2157–2167.
63. Demptster A. P., Laird N. M., Rubin D. B. (1977), Maximum likelihood from incomplete data via the EM algorithm (with discussion). J. R. Statist. Soc., Ser. B, vol. 39, pp. 1–38.
64. Dijkstra E. W. (1959), A note of two problems in connection to graphs. Numer. Math., vol. 1, pp. 269-271.
65. Dreyfus S. E., Law A. M. (1977), *The Art and Theory of Dynamic Programming*, Academic Press.
66. Ditkin V. A., Prudnikov A. P. (1965), *Integral Transforms and Operational Calculus*, Pergamon Press.
67. Duda R. O., Hart P. E. (1973), *Pattern Classification and Scene Analysis*, Wiley. (Second edition, 2003).
68. Dumas J. P., Ninio J. (1981), Efficient algorithm for folding and comparing nucleic acid sequences, Nucleic Acids Res., vol. 10, no. 1., pp. 197–206.
69. Durbin R., Eddy S. R., Krogh A., Mitchison G. (1999), *Biological Sequence Analysis: Probabilistic Models of Proteins and Nucleic Acids*, Cambridge University Press.
70. Durinck S., Moreau Y., Kasprzyk A., Davis S., De Moor B., Brazma A., Huber W. (2005), BioMart and Bioconductor: a powerful link between biological databases and microarray data analysis. Bioinformatics, vol. 21, pp. 3439–3440.
71. Eidhammer I., Jonassen I., Taylor W. R. (2004), *Protein Bioinformatics: An Algorithmic Approach to Sequence and Structure Analysis*, Wiley.
72. Eisen M. B., Spellmen P. T., Brown P. O. (1998), Cluster analysis and display of genome-wide expression patterns, Proc. Natl. Acad. Sci. USA, vol. 95., pp. 14863–14868.
73. Eisen M. B., Brown P. O. (1999), DNA arrays for analysis of gene expression, Methods Enzymol., vol. 303, pp. 179–205.
74. EURODIAB ACE Study Group (2000), Variation and trends in. incidence of childhood diabetes in Europe, Lancet, vol. 355, pp. 873–876.
75. Ewens W. J., (1979), *Mathematical Population Genetics*, Springer.
76. Ewens W. J., Grant G. R., (2001), *Statistical Methods in Bioinformatics*, Springer.

77. Fasulo D., Halpern A., Dew I., Mobarry C. (2002), Efficiently detecting polymorphisms during the fragment assembly process. Bioinformatics vol. 18, (Suppl. 1) pp. S294–S302.
78. W. Feller (1968), *An Introduction to Probability Theory and Its Applications* vols 1 and 2, Wiley.
79. Felsenstein J. (1992), Estimating effective population size from samples of sequences, inefficiency of pairwise and segregating sites as compared to phylogenetic estimates. Genet. Res., vol. 59, pp. 139–147.
80. Felsenstein J. (1981), Evolutionary trees from DNA sequences: a maximum likelihood approach, J. Mol. Evol., vol. 17, pp. 368–376.
81. Felsenstein J. (1985), Confidence limits on phylogenies: An approach using the bootstrap. Evolution vol. 39, pp. 783–791.
82. Felsenstein J., (2004), *Inferring Phylogenies*, Sinauer Associates.
83. Felsenstein J., Churchill G. (1996), A hidden Markov model approach to variation among sites in rate of evolution. Mol. Biol. Evol., vol. 13, pp. 93–104.
84. Ferragina P., Manzini G. (2000), Opportunistic data structures with applications, 41st *Symposium on Foundations of Computer Science*, pp. 390–398.
85. Fickett J. W. (1995), ORFs and genes: how strong a connection?, J. Comput. Biol., vol. 2, pp. 117–123.
86. Fickett J. W., Torney D. C., Wolf D. R. (1992), Base compositional structure of genomes. Genomics, vol. 13, pp. 1056–1064.
87. Fisz M., (1963), *Probability Theory and Mathematical Statistics*, Wiley, New York.
88. Fitch W. (1971), Toward defining the course of evolution: minimum change for a specified tree topology. Syst. Zool., vol. 20. pp. 406–416.
89. Fletcher R. (1987), *Practical methods of optimization*, Wiley
90. Fodor S. P. A., Rava R. P., Huang X. C., Pease A. C., Holmes C. P., Adams C. L. (1993), Multiplexed biochemical assays with biological chips. Nature, vol. 364, pp. 555–556.
91. Folk M. J., Zoellick B., Riccardi G. (1998), *File Structures: an Object-Oriented Approach with C++*, Addison-Wesley.
92. Franklin R., Gosling R. G. (1953), Molecular configuration in sodium thymonucleate. Nature, vol. 171, pp. 740–741.
93. Fu Y. X., Li W.H. (1993), Maximum likelihood estimation of population parameters. Genetics, vol. 134, pp. 1261–1270.
94. Garey M. R., Johnson D. S. (1979), *Computers and intractability*, Freeman New York.
95. Gasteiger J., Marsilli M. (1978), A new model for calculating. atomic charges in molcules. Tetrahedron Lett., vol. 34, pp. 3181–3184.
96. Geller S. C, Gregg J. P., Hagerman P., Rocke D. (2003), Transformation and normalization of oligonucleotide microarray data. Bioinformatics, vol. 19, pp. 1817–1823.
97. Genovese C., Wasserman L. (2002), Operating characteristics and extensions of the false discovery rate procedure. J. R. Statist. Soc. Ser. B, vol. 64., pp. 499–517.
98. Gesteland R. F., Cech T. R., Atkins J. F. (eds.), (1993), *The RNA World : the Nature of Modern RNA Suggests a Prebiotic RNA*, Cold Spring Harbor Laboratory Press.
99. Ghosh, D., Chinnaiyan A. M. (2002), Mixture modelling of gene expression data from microarray experiments. Bioinformatics, vol. 18, pp. 275–286.

100. Ghosh, D. (2004), Mixture models for assessing differential expression in complex tissues using microarray data. Bioinformatics, vol. 20, pp. 1663–1669.
101. Ginalski K., Grishin N. V., Godzik A., Rychlewski R. (2005), Practical lessons from protein structure prediction, Nucleic Acids Res., vol. 33, pp. 1874–1891.
102. Golub T. R. Slonim D. K., Tamayo P., Huard C., Gaasenbeek M., Mesirov J. P., Coller H., Loh M. L., Downing J. R., Caligiuri M. A., Bloomfield C. D., Lander E. (1999), Molecular classification of cancer: class discovery and class prediction by gene expression monitoring, Science, vol. 286, pp. 531–537.
103. Golub G, van Loan C. F. (1996), *Matrix Computations*, 3rd edn. Johns Hopkins University Press.
104. Graham R. L., Knuth D. E., Patashnik O. (1989), *Concrete Mathematics*, Addison-Wesley.
105. Green A., Gale E. A., Patterson C. C. (1992), Incidence of childhood-onset insulin-dependent diabetes mellitus: the EURODIAB ACE study. Lancet vol. 339, pp. 905–909.
106. Griffiths-Jones S., Bateman A., Marshall M., Khanna A., Eddy S. R. (2003), Rfam: an RNA family database, Nucleic Acids Res., vol. 31, pp. 439–441.
107. Griffith O. L., Pleasance E. D., Fulton D. L., Oveisi M., Ester M., Siddiqui A. S., Jones S.J. (2005), Assessment and integration of publicly available SAGE, cDNA microarray, and oligonucleotide microarray expression data for global coexpression analyses. Genomics, vol. 4, pp. 476–488.
108. Griffiths R.C. (1989), Genealogical tree probabilities in the infinitely many sites model, J. Math. Biol., vol. 27, pp. 667–680.
109. Griffiths R.C., Tavare S., (1994), Sampling theory for neutral alleles in a varying environment, Phil. Trans. R. Soc. Lond. B., vol. 344, pp. 403–410.
110. Griffiths R. C., Tavare S. (1995), Unrooted genealogical tree probabilities in the infinitely many sites model. Math. Biosci., vol. 127, pp. 77–98.
111. Gusfield D. (1997), *Algorithms on strings, trees and sequences*, Cambridge University Press.
112. Hanson K. M., Wolf D. R. (1996), Estimators for the Cauchy distribution, in G. R. Heidbreder (ed.), *Maximum Entropy and Bayesian Methods*, Kluwer, pp. 255-263.
113. Hartemink A. J., Gifford D. K., Jaakkola T. S., Young R. A. (2001), Maximum likelihood estimation of optimal scaling factors for expression array normalization, *SPIE BiOS*, San Jose, California, January 2001, http://citeseer.ist.psu.edu/452719.html
114. Hartl D., Clark A. (1989), *Principles of Population Genetics*, Sinauer Associates.
115. Hasegava M., Kishino M., Yano T. (1985), Dating of the human-ape splitting by a molecular clock of mitochondrial DNA. J. Mol. Evol., vol. 22, pp. 160-174.
116. Hastings W. K., (1970), Monte Carlo sampling method using Markov chains and their applications. Biometrica, vol. 57, pp. 1317–1340.
117. Havlak P., Chen R., Durbin K. J., Egan A., Ren Y., Song X-Z. Weinstock M., Gibbs R. A. (2004), Tha Atlas genome assembly system, Genome Res., vol. 14, pp. 721–732.
118. Hawkins J. D., (1996), *Gene structure and Expression*, Cambridge University Press.
119. Healy J., Thomas E. E., Schwartz J. T., Wigler M. (2003), Annotating large genomes with exact word matches. Genome Res., vol. 13, pp. 2306–2315.

120. Hecker Y., Bolle R. (1994), On geometric hashing and the generalized Hough transform. IEEE Trans. Syst. Man Cybernet., vol. 24, pp. 1328–1338.
121. Heilig R., Eckenberg R., Petit J. Louis., Fonknechten N., Da Silva C., Cattolico L., Levy M., Barbe V., de Berardinis V., Ureta-Vidal A., Pelletier E., Vico V., Anthouard V., Rowen L., et al. (2003), The DNA sequence and analysis of human chromosome 14. Nature, vol. 421, pp. 601–607.
122. Henikoff S., Henikoff J. G. (1992), Amino acid substitution matrices from protein blocks. Proc. Natl. Acad. Sci. USA, vol. 89, pp. 10915–10919.
123. Hoare C. A. R. (1961), ACM Algorithm 64: Quicksort. Commun. ACM, vol. 4, no. 7, p. 321.
124. Horn B. K. P. (1987), Closed form solution of absolute orientation using unit quaternion. J. Opt. Soc. Am., A, vol. 4, pp. 629–642.
125. Huang X., Wang J., Aluru S., Yang S. P. Hillier L. (2003), PCAP: A whole-genome assembly program. Genome Res., vol. 13, pp. 2164–2170.
126. International Human Genome Sequencing Consortium (2001), Initial sequencing and analysis of the human genome. Nature, vol. 409, pp. 860-921.
127. Huson D. H., Reinert K., Kravitz S. A., Remington K. A., Delcher A. L., Dew I. M., Flanigan M., Halpern A. L., Lai Z., Mobarry C. M., Sutton G. G., Myers E. W. (2001), Design of a compartmentalized shotgun assembler for the human genome. Bioinformatics, vol. 17, pp. S132–S139.
128. Idury R. M., Waterman M. S. (1995), A new algorithm for DNA Sequence Assembly, J. Comput. Biol., vol. 2, pp. 291–306.
129. Iosifescu M. (1980), *Finite Markov Processes and Their Applications*, Wiley.
130. Irizarry R. A., Hobbs B., Collin F., Beazer-Barclay Y. D., Antonellis K. J., Scherf U., Speed T. P. (2003), Exploration, normalization, and summaries of high density oligonucleotide array probe level data. Biostatistics, vol. 4, no. 2. pp. 249–264.
131. Jackson J. E. (2003), *A User's Guide to Principal Components*, Wiley.
132. Jain A. (1989), *Fundamentals of Digital Image Processing*, Prentice-Hall.
133. Jarzab B., Wiench M., Fujarewicz K., Simek K., Jarzab M., Oczko-Wojciechowska M., Wloch J., Czarniecka A., Chmielik E., Lange D., Pawlaczek A., Szpak S., Gubala E., Swierniak A. (2005), Gene expression profile of papillary thyroid cancer: sources of variability and diagnostic implications. Cancer Res., vol. 65, pp. 1587–1597.
134. Johnson N. L., Kotz S., Balakrishnan N. (1994), *Continuous Univariate Distributions*, Wiley.
135. Jukes T. H., Cantor C. R. (1969), Evolution of protein molecules, in H. N. Munro (ed.) *Mammalian Protein Metabolism*, Academic Press, New York, pp. 21–132.
136. Kabsch W. (1978), A discussion of the solution for the best rotation to relate two sets of vectors. Acta Cryst., vol. A34, pp. 827–828.
137. Kabsch W., Sander C. (1983), Dictionary of protein secondary structure: Pattern recognition of hydrogen-bonded and gometrical features. Biopolymers, vol. 22, pp. 2577–2637.
138. Kanehisa M., Goto S. (2000), KEGG: Kyoto Encyclopedia of Genes and Genomes. Nucleic Acids Res., vol. 28, pp. 27–30.
139. Karas M., Bachmann D., Bahr U., Hillenkamp F. (1987), Matrix-assisted ultraviolet laser desorption of non-volatile compounds. Int. J. Mass Spectrom. Ion Phys., vol. 78, pp. 53–68.

140. Karlin S., Altschul S. F. (1990), Methods for assessing the statistical significance of molecular sequence features by using general scoring schemes. Proc. Natl. Acad. Sci. USA, vol. 87, pp. 2264–2268.
141. Karlin S., Altschul S. F. (1993), Applications and statistics for multiple high-scoring segments in molecular sequences. Proc. Natl. Acad. Sci. USA, vol. 90, pp. 5873–5877.
142. Kececioglu J.D., Myers E.W. (1995), Exact and approximate algorithms for the sequence reconstruction problem. Algorithmica, vol. 13, p. 7.
143. Kececioglu J.D., Myers E.W. (1995), Combinatorial algorithms for DNA sequence assembly, Algorithmica, vol. 13 p. 51.
144. Kendall M.G., Stuart A., Ord J.K., Arnold S.F., O'Hagan A., Forster J. (1991, 1999, 2004), *Kendall's Advanced Theory of Statistics*, vols. 1, 2A, 2B, Oxford University Press New York.
145. Kent W. J., Haussler D. (2001), Assembly of the working draft of the human genome with GigAssembler. Genome Res. vol. 11, pp. 1541–1548.
146. Kimmel M., Axelrod D. E. (2002), *Branching processes in biology*, Springer, New York.
147. Kimmel M., Chakraborty R., King J. P., Bamshad M., Wattkins W. S., Jorde L. B. (1998), Signatures of population expansion in microsatellite repeat data. Genetics, vol. 148, pp. 1921–1930.
148. Kimura M. (1980), A simple method for estimating evolutionary rate in a finite population due to mutational production of neutral and nearly neutral base substitution through comparative studies of nucleotide sequences. J. Molec. Biol., vol. 16, pp. 1501–1531.
149. King J. P., Kimmel M., Chakraborty R. (2000), A power analysis of microsatelite-based statistics for inferring past population growth. Mol. Biol. Evol., vol. 17, pp. 1859–1868.
150. Kingman J. F. C. (1982), The coalescent. Stoch. Proc. Appl., vol. 13, pp. 235–248.
151. Kingman J. F. C. (1993), *Poisson Processes*, Oxford University Press.
152. Kirkpatrick S., Gelatt C. D., Vecchi M. P. (1983), Optimization by simulated annealing. Science, vol. 220, pp. 671–680.
153. Klosterman P. S., Tamura M., Holbrook S. R., Brenner S. E. (2002), SCOR: a structural classification of RNA database. Nucleic Acids Res. vol. 30, pp. 392-394.
154. Knuth D. E. (1973), *The Art of Computer Programming*, Addison-Weseley.
155. Knuth D. E., Morris J. H., Pratt V. R. (1977), Fast pattern matching in strings. SIAM J. Comput., vol. 6, no. 2, pp. 323–350.
156. Korn T. M., Korn G. A. (1968), *Mathematical Handbook for Scientists and Engineers*, McGraw-Hill.
157. Koski T., Koskinen T. (2001), *Hidden Markov Models for Bioinformatics*, Kluwer Academic.
158. Kotz S., Balakrishnan N. Johnson N. L. (2000), *Continuous multivariable distributions*, vol. 1, *Models and applications*, Wiley.
159. Krovat E. M., Steindl T., Langer T. (2005), Recent Advances in Docking and Scoring. Curr. Comput.-Aided Drug Design, vol. 1, pp. 93–102.
160. Kuhner M. K., Yamato J., Felsenstein J. (1995), Estimating effective population size and mutation rate from sequence data using Metropolis–Hastings sampling. Genetics, vol. 140, pp. 1421–1430.

161. Kuhner M. K., Yamato J., Felsenstein J. (1998), Maximum likelihood estimation of population growth rates based on coalescent. Genetics, vol. 149, pp. 429–434.
162. Lacroix Z., Critchlow T. (2003), *Bioinformatics: Managing Scentific Data*, Morgan Kaufmann.
163. Lander E. S., Waterman M. S. (1988), Genomic mapping by fingerprinting random clones: A mathematical analysis. Genomics, vol. 2, pp. 231–239.
164. Lange K. (1998) *Numerical Analysis for Statisticians*, Springer, New York.
165. Lauffenberger D. A., Linderman J.J. (1996), *Receptors*, Oxford University Press, New York.
166. Leach A. R. (1996), *Molecular Modeling. Principles and Applications*, Longman.
167. Lee B., Richards F. M. (1971), The interpretation of protein structures: estimation of static accessibility. J. Mol. Biol., vol. 55, pp. 379–400.
168. Lesk A.M. (2002), *Introduction to Bioinformatics*, Oxford University Press.
169. Letsinger, R.L., Mahadevan, V. (1966), Stepwise synthesis of oligodeoxyribonucleotides on an insoluble polymer support. J. Am. Chem. Soc., vol. 88, pp. 5319–5324.
170. Lichtarge O. Bourne H. R., Cohen F. E. (1996), An evolutionary trace method defines binding surfaces common to protein families. J. Mol. Biol., vol. 257, pp. 342–358.
171. Lichtarge O., Bourne H. R., Cohen F. E. (1996), Evolutionarily conserved G binding surfaces support a model of the G protein-receptor complex. Proc. Natl. Acad. Sci. USA, vol. 93, pp. 7507–7511.
172. Li Y, Hunter R. L., McIver R. T (1994), High-resolution mass spectrometer for protein chemistry. Nature, vol. 370, pp. 393–395
173. Li W. (1999), Statistical properties of open reading frames in complete genome sequences. Comput. Chem., vol. 23, pp. 283–301.
174. Li W. H. (1997), *Molecular Evolution*, Sinauer Associates.
175. Li J, Zhang Z., Rosenzweig J., Wang Y. Y., Chan D. W. (2002), Proteomics and bioinformatics approaches for identification of serum biomarkers to detect breast cancer. Clin. Chem., vol. 48, pp. 1296–1304.
176. Li X., Pan X. M. (2001), New method for accurate prediction of solvent accessibility from protein sequence. Proteins, vol. 42, pp. 1–5.
177. Li X., Waterman M. S. (2003), Estimating the repeat structure and length of DNA sequences using l-mers. Genome Res. vol. 13, pp. 1916–1922.
178. Lins L., Thomas A., Brasseur R. (2003), Analysis of accessible surface of residues in proteins. Protein Sci., vol. 12, pp. 1406–1417.
179. Lipman D. J., Pearson W. R. (1985), Rapid and sensitive protein similarity searches. Science, vol. 227, pp. 1435–1441.
180. Lippert R. A. (2005), Space-efficient whole genome comparisons with Burrows–Wheeler transforms, J. Comput. Biol., vol. 12, no. 4, pp. 407–415.
181. Lowe T. M., Eddy S. R. (1997), tRNAscan-SE: A program for improved detection of transfer RNA genes in genomic sequence. Nucleic Acids Res. vol. 25, pp. 955–964.
182. Luenberger D. G. (2003), *Linear and Nonlinear Programming*, Kluwer.
183. Luenberger D. G. (1968), *Optimization by Vector Space Methods*, Wiley.
184. Madigan D., York J. (1995), Bayesian graphical models for discrete data. Internat. Statist. Rev., vol. 63, pp. 215–232.

185. Manber U., Myers G. (1993), A new method for on-line searches. SIAM J. Comput., vol. 22, pp. 935–948.
186. Mattick J. S. (2003), Challenging the dogma: the hidden layer of non-protein-coding RNAs in complex organisms. BioEssays, vol. 25, pp. 930–939.
187. Mattick J. S. (2001), Non-coding RNAs: the architects of eukaryotic complexity. EMBO rep., vol. 2, pp. 986–991.
188. McLachan A.D. (1982), Rapid comparisons of protein structures. Acta Cryst., vol. A38, pp. 871–873.
189. McLachan G. J., Bean R. W., Peel, D. (2002), A mixture model-based approach to the clustering of microarray expression data. Bioinformatics, vol. 18, pp. 413–422.
190. McLachan G. J., Krishnan T. (1997), *The EM Algorithm and Extensions*, Wiley.
191. McLachan G. J., Peel W. (2000), *Finite Mixture Distributions*, Wiley.
192. McMurry J. (1999), *Organic Chemistry*, 4th edn., Brooks/Cole.
193. Medvedovic M., Yeung K. Y., Bumgarner R. E. (2004), Bayesian mixtures for clustering replicated microarray data. Bioinformatics, vol. 20, pp. 1222–1232.
194. Medvedovic M., Sivaganesan S., (2002), Bayesian infinite mixture model based clustering of gene expression profiles. Bioinformatics, vol. 18, pp. 1194–1206.
195. Metropolis N., Rosenbluth A. W., Rosenbluth M. N., Teller A.H. (1953), Equations of state calculations by fast computing machines. J. Chem. Phys., vol. 21, pp. 1087–1092.
196. Michaels G. S., Carr D. B., Askenazi M., Fuhrman S., Wen X., Somogyi R. (1998), Cluster analysis and data visualization of large-scale gene expression data, *Pacific Symposium on Biocomputing*, vol. 3, pp. 42–53.
197. Michener C. D., Sokal, R. R. (1957), A quantitative approach to a problem in classification. Evolution, vol. 11, pp. 130–162.
198. Morey C., Avner P. (2004), Employment opportunities for non-coding RNAs. FEBS Lett., vol. 567, pp. 27–34.
199. Morrison D. R. (1965), PATRICIA–practical algorithm to retrieve information coded in alphanumeric, J. ACM, vol. 15, pp. 514–534.
200. Mount D.W. (2000), *Bioinformatics, Sequence and Genome Analysis*, Cold Spring Laboratory Press.
201. Mullikin J. C., Ning Z. (2003), The Phusion assembler, Genome Res., vol. 13, pp. 81–90.
202. Myers E. (1995), Toward simplyfying and accurately formulating fragment assembly, J. Comput. Biol., vol. 2, p. 275.
203. Myers E. W., Sutton G. G., Delcher A. L., Dew I. M., Fasulo D. P., Flanigan M. J., Kravitz S. A., Mobarry, C.M., Reinert, K.H., Remington K. A., et al. (2000), A whole-genome assembly of Drosophila. Science vol. 287, pp. 2196–2204.
204. Needleman S. B., Wunsch C. D. (1970), A general method applicable to the search for similarities in the amino acid sequence of two proteins. J. Mol. Biol., vol. 48, pp. 443–453.
205. Nelder J. A., Mead R. (1965), A simplex method for function minimization, Comput. J., vol. 7, pp. 308–313.
206. Neurath H., Wolters J., (1992), The alignment editor ALIGNOS-integrating tool for a cooperative databanks for structural RNAs., Bioinformatics vol. 1, pp. 22-25.

207. Nguyen D. V., Rocke D. (2002), Partial least squares proportional hazard regression for application to DNA microarray survival data. Bioinformatics, vol. 18, no. 12, pp. 1625–1632.
208. Nguyen D. V., Rocke D. (2002), Tumor classification by partial least squares using microarray gene expression data. Bioinformatics, vol. 18, no. 1, pp. 39–50.
209. Nirenberg M. W., Matthaei J. H. (1961), The dependence of cell-free protein synthesis in *E. coli* upon naturally occurring or synthetic polyribosomes. Proc. Nat. Acad. Sci. USA, vol. 47, pp. 1588–1602.
210. Nussinov R., Pieczenik G., Griggs J. R., Kleitman D. J. (1978), Algotihms for loop matchings. SIAM J. Appl. Math., vol. 35, pp. 68–82.
211. Nussonov R., Wolfson H. (1991), Efficient detection of three-dimensional structural motifs in biological macromolecules by computer vision techniques. Proc. Nat. Acad. Sci. USA, vol 88, pp. 10495–10499.
212. Olofsson P. (2005), *Probability, Statistics and Stochastic Processes*, Wiley.
213. Ouyang M., Welsh W. J., Georgopoulos P. (2004), Gaussian mixture clustering and imputation of microarray data. Bioinformatics, vol. 20, pp. 917–923.
214. Palzkill T. (2001), *Proteomics*, Kluwer.
215. Pao D. C. W., Li H. F., Jaykumar R. (1992), Shapes recognition using the straight line Hough transform: Theory and generalization. IEEE Trans. PAMI, vol. 14, pp. 1076–1089.
216. Paule M. R. (ed.) (1998), *Transcription of ribosomal RNA genes by eukaryotic RNA polymerase I*, Springer.
217. Pearlman D. A., Case D. A., Caldwell J. W., Ross W. R., Cheatham T.E., III, DeBolt S., Ferguson D., Seibel G., Kollman P. (1995), AMBER, a computer program for applying molecular mechanics, normal mode analysis, molecular dynamics and free energy calculations to elucidate the structures and energies of molecules. Comp. Phys. Commun., vol. 91, pp. 1–41.
218. Pechkova E., Nicolini C. (2003), *Proteomics and Nanocrystallography*, Kluwer.
219. Pe'er I., Shamir R. (2000), Spectrum alignment: efficient resequencing by hybridization, *Proceedings of the Eighth International Conference on Intelligent Systems in Molecular Biology (ISMB '00)*, pp. 260–268.
220. Percus J. K. (2001), *Mathematics of genome analysis*, Cambridge University Press.
221. Pevsner J. (2003), *Bioinformatics and Functional Genomics*, Wiley.
222. Pevzner P., (2000), *Computational Molecular Biology*, MIT Press.
223. Pevzner P., Tang H. (2001), Fragment assembly with double barreled data. Bioinformatics, vol. 17, suppl. 1, pp. S225–S233.
224. Pevzner P., Tang H., Waterman M. (2001), An Eulerian path approach to DNA fragment assembly. Proc. Nat. Acad. Sci. USA, vol. 98, pp. 9748–9753.
225. Pevzner P., Tang H., Waterman M. (2001), A new approach to fragment assembly in DNA sequencing, *Proceedings of the Fifth International Conference on Computational Biology (Recomb 2001)*, Montreal, pp. 256–265.
226. Petricoin E., III, Ardekani A., Hitt B., Levine P., Fusaro V., Steinberg S., Mills G., Simone C., Fishman D., Kohn E. (2002), Use of proteomic patterns in serum to identify ovarian cancer. Lancet, vol. 359, pp. 572–577.
227. Polanska J., Widlak P., Rzeszowska J., Kimmel M., Polanski A., (2006), Gaussian mixture decomposition of time-course DNA microarray data, in L. Brusch, G. de Vries, H. P. Herzel (eds.), *Cellular Biophysics, Regulatory Networks, and Data Analysis, Modeling and Simulation in Science, Engineering and Technology*, Birkhäuser.

228. Polanski A., Borek A. (1994), The decomposition of the parameter space for recognition of the polygonal shapes. Machine Graph. Vision, vol. 3, pp. 287–297.
229. Polanski A., Kimmel M., Chakraborty R. (1998), Application of a time-dependent coalescent process for inferring the history of population changes from DNA sequence data. Proc. Natl. Acad. Sci. USA, vol. 95, pp. 5456–5461.
230. Polanski A, Kimmel M. (2003), New explicit expressions for relative frequencies of SNPs with application to statistical inference on population growth. Genetics, vol. 165, pp. 399–409.
231. Pollastri G., Przybylski D., Rost B., Baldi P. (2001), Improving the prediction of protein secondary structure in three and eight classes using recurrent neural networks and profiles, Proteins: Structure, Function, and Genetics, vol. 47, pp. 228–235.
232. Primrose S.B. (1998), *Principles of Genome Analysis*, Blackwell Science.
233. Pybus O. G., Rambaut A., Harvey P. H. (2000), An integrated framework for the inference of viral population history from reconstructed genealogies, Genetics, vol. 155, pp. 1429–1437.
234. Quackenbush J. (2002), Microarray data normalization and transformation. Nature Genet., vol. 32, pp. 496–501.
235. Quackenbush J. (2001), Computational analysis of microarray data. Nature Rev. Genet., vol. 2, pp. 418–427.
236. Rabiner L. R. (1989), A tutorial on hidden Markov models and selected applications in speech recognition. Proc. IEEE, vol. 77, pp. 257–286.
237. Ramachandran G. N., Ramakrishnan C., Sasisekharan V. (1963), Stereochemistry of polypeptide chain configurations. J. Mol. Biol, vol. 7, pp. 95–99.
238. Rhodes D. R. et al., Rhodes D. R., Yu J., Shanker K., Deshpande N., Varambally R., Ghosh D., Barrette T., Pandey A., Chinnaiyan A.M. (2004), ONCOMINE, a cancer microarray database and integrated data mining platform. Neoplasia, vol. 6, no. 1, pp. 1–6.
239. Richards F. M. (1977), Areas, volumes, packing, and protein structures. Annu. Rev. Biophys. Bioeng., vol. 6, pp. 151–176.
240. Richardson S., Green P. J. (1997), On Bayesian analysis of mixtures with an unknown number of components. J. R. Statist. Soc., Ser. B., vol. 59, pp. 731–792.
241. Rivas E., Eddy S. R. (1999), A dynamic programming algorithm for RNA structure prediction including pseudoknots. J. Mol. Biol., vol. 285, pp. 2053-2068.
242. Roberts R. J., Vincze T., Posfai J., Macelis, D. (2005), REBASE-Restriction enzymes and DNA methyltransferases. Nucleic Acids Res., vol. 33, pp. D230–D232.
243. Rogers A. R., Harpending H. (1992), Population growth makes waves in the distribution of pairwise genetic differences. Mol. Biol. Evol., vol. 9., pp. 552–569.
244. Rosen K. (2002), *Discrete Mathematics and its Applications*, McGraw-Hill.
245. Rost B., Yachdav G., Liu J., (2003), The PredictProtein Server. Nucleic Acids Res., vol. 32, Web server issue, pp. W321–W326.
246. Saitou N, Nei M. (1987), The neighbor-joining method: a new method for reconstructing phylogenetic trees. Mol. Biol. Evol., vol. 4, pp. 406–225.

247. Salamov A. A., Solovyev V. V. (1995), Prediction of protein secondary structure by combining nearest-neighbor algorithms and multiple sequence alignments. J. Mol. Biol., vol. 247, pp. 11–15.
248. Sander C., Schneider R. (1991), Database of homology-derived protein structures and the structural meaning of sequence alignment. Proteins: Struct. Funct. Genet., vol. 9, pp. 56–68.
249. Sanger F., Nicklen S., Coulsen A. R., (1977), DNA sequencing with chain terminating inhibitors. Proc. Natl. Acad. Sci., vol. 74, pp. 5463–5467.
250. Schaffer A. A, Aravind L., Madden T. L., Shavirin S., Spouge J. L, Wolf Y. I., Koonin E. V., Altschul S. F. (2001), Improving the accuracy of PSI-BLAST protein database searches with composition-based statistics and other refinements. Nucleic Acids Res., vol. 29, pp. 2994–3005.
251. Schwarz G. (1978), Estimating the Dimension of a Model. Ann. Statist., vol. 6, pp. 461–464.
252. Schulz G. E., Schrimer R. H. (1990), *Principles of Protein Structure*, Springer.
253. Serra M. J., Turner D. H. (1995), Predicting the thermodynamic properties of RNA. Methods Enzymol., vol. 259, pp. 242–261.
254. Shrake A., Rupley J. A. (1973), Environment and exposure to solvent of protein atoms. Lysozyme and insulin. J. Mol. Biol., vol. 79, pp.351–371.
255. Simek K., Kimmel M. (2003), A note on estimation of dynamics of multiple gene expression based on singular value decomposition. Math. Biosci. vol. 182, pp. 183–199.
256. Simons K. T., Kooperberg C., Huang E., Baker D. (1997), Assembly of protein tertiary structures from fragments with similar local sequences using simulated annealing and bayesian scoring functions. J. Mol. Biol., vol. 268, pp. 209–225.
257. Simons K. T., Bonneau R., Ruczinski I., Baker D. (1999), Ab initio protein structure prediction of CASPIII targest using ROSETTA. Proteins: Struct. Funct. Genet., Suppl. 3, pp. 171–176.
258. Singh D., Febbo P. G., Ross K., Jackson D. G., Manola J., Ladd C., Tamayo T., Renshaw A. A., D'Amico A. V., Richie J. P., et al. (2002), Gene expression correlates of clinical prostate cancer behavior. Cancer Cell, vol. 1, pp. 203–209.
259. Skaletsky H., Kuroda-Kawaguchi T., Minx P. J., Cordum H. S., Hillier L., Brown L. G., Repping S., Pyntikova T., Ali J, Bieri T., (2003), The male-specific region of the human Y chromosome is a mosaic of discrete sequence classes. Nature vol. 423, pp. 825–837.
260. Skiena S. S. (1998), *The Algorithm Design Manual*, Springer.
261. Smith T. F., Waterman M. S. (1981), Identification of common molecular subsequences. J. Mol. Biol., vol. 197, pp. 147–195.
262. Smith H. O., Wilcox K. W. (1970), A restriction enzyme from Hemophilus influenzae. Purification and general properties. J. Mol. Biol., vol. 51, p. 379.
263. Smyth B., (2003), *Computing Patterns in Strings*, Addison Wesley.
264. Soderlund C., Longden I., Mott R. (1997), FPC: a system for building contigs from restriction fingerprinted clones. Comput. Appl. Biosci., vol. 13, pp. 523–535.
265. Southern E. M. (1975), Detection of specific sequences among DNA fragments separated by gel electrophoresis, J. Mol. Biol., vol. 98, pp. 503–517.
266. Spellman P. T., Sherlock G., Zhang M.Q., Iyer V. R., Anders K., Eisen M. B., Brown P. O., Botstein D., Futcher B. (1998), Comprehensive identification of cell cycleregulated genes of the yeast *Saccharomyces cerevisiae* by microarray hybridization. Mol. Biol. Cell, vol. 9, pp. 3273–3297.

267. Steckel D. (2003), *Microarray Bioinformatics*, Cambridge University Press.
268. Stephen G. A., (1994), *String Searching Algorithms*, World Scientific.
269. Storey J. D., Tibshirani R. (2003), Statistical significance for genomewide studies. Proc. Natl. Acad. Sci. USA, vol. 100, pp. 9440–9445.
270. Sulson J., Mallet F., Durbin R., Horsnell T. (1988), Software for genome mapping by fingerprinting techniques. Comput. Appl. Biosci., vol. 4, pp. 125–132.
271. Teng S.H., Yao F. F. (1997), Approximating shortest superstrings. SIAM J. Comput., vol. 26, pp. 410–417.
272. Thomason A. (1989), A simple linear expected time for finding a Hamiltonian Path, Discrete Math., vol. 75, pp. 373-379.
273. Thompson J. D., Higgins D. G., Gibson T. J. (1994), CLUSTAL W: improving the sensitivity of progressive multiple sequence alignment through sequence weighting, position-specific gap penalties and weight matrix choice. Nucleic Acids Res, vol. 22. pp. 4673–4680.
274. Turner J. (1989), Approximation algorithms for the shortest common superstring problem. Inf. Comput., vol. 83, p. 20.
275. Varadhan S. R. S. (2001), *Probability Theory*, American Mathematical Society.
276. Venter J. C., Adams M. D., Myers E. W., Li P. W., Mural R. J., Sutton G. G., Smith H. O., Yandell M., Evans C. A., Holt R. A., et al. (2001), The sequence of the human genome. Science, vol. 291, pp. 1304–1351.
277. Vlassov A. V., Kazakov S. A., Johnston B. H., Landweber L. F. (2005), The RNA world on ice: A new scenario for the emergence of RNA information. J. Mol. Evol., vol. 61, pp. 264–273.
278. Wall M. E., Dyck P. A., Brettin T. S. (2001), SVDMAN–singular value decomposition analysis of microarray data. Bioinformatics, vol. 17, pp. 566–568.
279. Wang R., Lu Y., Fang X., Wang S. (2004), An extensive test of 14 scoring functions using the PDBbind refined set of 800 protein–ligand complexes, J. Chem. Inf. Comput. Sci., vol. 44, pp. 2114–2125.
280. Wang Y., Lu J., Lee R., Gu Z., Clarke R. (2002), Iterative normalization of cDNA microarray data. IEEE Trans. Inf. Technol. Biomed., vol. 6. pp. 29–37.
281. Waterman M.S., (1995), Introduction to computational biology. Chapman and Hall/CRC, Boca Raton, FA.
282. Watson J., Crick F. (1953), A structure for deoxyribose nucleic acid, Nature, vol. 171, pp. 737–738.
283. Watterson G. A. (1975), On the number of segregating sites in genetical models without recombination, Theor. Popul. Biol., vol. 7, pp. 387–407.
284. Weber J. L., Myers E. W. (1997), Human whole-genome shotgun sequencing, Genome Res., vol. 7, pp. 401–409.
285. Wendl M. C. Yang S. P. (2004) Gap statistics for the whole genome shotgun DNA sequencing projects. Bioinformatics vol. 20, no. 10, pp. 1527–1534.
286. Westhof E., Auffinger P. (2000), RNA Tertiary Structure, in R. A. Meyers (ed.), *Encyclopedia of Analytical Chemistry*, pp. 5222–5232.
287. White R.J. (2001), *Gene Transcription. Mechanisms and Control*, Blackwell Science.
288. Wilbur W. J., Lipman D. J. (1983), Rapid similarity searches of nucleic acid and protein data banks. Proc. Natl. Acad. Sci. USA, vol. 80, pp. 726–730.
289. Wilf H. S., (1990), *Generatingfunctionology*, Academic Press.
290. Wilf H. S. (2002), *Algorithms and Complexity*, A. K. Peters, http://www.cis.upenn.edu/~wilf/

291. Wilkins M. H. F., Stokes A. R., Wilson H. R, (1953), Molecular structure of deoxypentose nucleic acids, Nature, vol. 171, pp. 738–740.
292. Wolfson H., Rigoutsos I. (1997), Geometric hashing: an overview, IEEE Comput. Sci. Eng., vol. 4, pp. 10–21.
293. Wulfkuhle J. D., Liotta L. A., Petricoin E. (2003), Proteomic applications for the early detection of cancer. Nature Rev., vol. 3, pp. 267–275.
294. Yang, Z. (2000) *Phylogenetic Analysis by Maximum Likelihood (PAML) 3.0*, University College London, London.
295. Yeung K. Y., Fraley C., Murua A., Raftery A., Ruzzo L. (2001), Model-based clustering and data transformations for gene expression data. Bioinformatics, vol. 17, pp. 977–987.
296. Zangwill W. I. (1969), *Nonlinear Programming; a Unified Approach*, Prentice-Hall.
297. Zar J. H. (1999), *Biostatistical analysis*, Prentice Hall.
298. Zuker M., Sankoff D. (1984), RNA secondary structures and their prediction, Bull. Math. Biol., vol. 46, pp. 591–621.
299. Zuker M., Stiegler P. (1981), Optimal computer folding of large RNA sequeces using thermodynamics and auxiliary infomation. Nucleic Acids Res., vol. 9., pp. 133–148.
300. Zvelebil M. J., Barton G. J., Taylor W. R., Sternberg, M. J. (1987), Prediction of protein secondary structure and active sites using the alignment of homologous sequences, J. Mol. Biol., vol. 195, pp. 957–961.
301. 2D Protein database, *Haemophilus*, University of Aberdeen, http://www-.abdn.ac.uk
302. 3DNA. A software package for the analysis, rebuilding, and visualization of three-dimensional nucleic acid structures, http://rutchem.rutgers-.edu/~xiangjun/3DNA/
303. Affymetrix, http://www.affymetrix.com/index.affx
304. AmiGo Web page for gene ontology searches, http://www.godatabase.org/cgi-bin/amigo/go.cgi
305. BioCarta, http://www.biocarta.com/index.asp
306. Bioconductor, http://www.bioconductor.org/
307. CGED, cancer gene expression database, http://cged.hgc.jp/cgi-bin/input.cgi
308. cisRED, regulatory-elements database, http://www.cisred.org/
309. CLUSTAL W, EBI server for multiple sequence alignment on the web, http://www.ebi.ac.uk/clustalw
310. Cluster, hierarchical clustering algorithm, http://bonsai.ims.u-tokyo.ac.jp
311. DSSP, program for protein secondary structure assignment, http://swift.cmbi-.ru.nl/gv/dssp/
312. Gene Ontology Consortium, http://www.geneontology.org/
313. GENSCAN, Identification of complete gene structures in genomic DNA, http://genes.mit.edu/GENSCAN.html
314. EBI, European Bioinformatics Institue, http://www.ebi.ac.uk/
315. M. Eisen laboratory, Publications and software, http://rana.lbl.gov/
316. EULER Portal, http://nbcr.sdsc.edu/euler/about.htm
317. ExPASy, Expert protein analysis system, proteomics server, http://us-.expasy.org/
318. GtRNA, genomic tRNA database, http://lowelab.ucsc.edu/GtRNAdb/
319. HSSP, database of homology-derived secondary structure of proteins, http://swift.cmbi.kun.nl/swift/hssp/

320. HUGO, Human Gene Nomenclature Committee, http://www.gene.ucl.ac.uk/nomenclature/index.html
321. Index of RNA structures, http://www.imb-jena.de/IMAGE_RNA.html
322. Jpred, a consensus secondary structure prediction server, http://www.compbio.dundee.ac.uk/~www-jpred/
323. KEGG, Kyoto encyclopedia of genes and genomes, http://www.genome.jp/kegg/
324. MEME, multiple EM for motif elicitation, http://meme.sdsc.edu/meme/intro.html
325. MGED, microarray gene expression data society, http://www.mged.org/
326. National Center for Biotechnology Information, http://www.ncbi.nih.gov/index.html
327. OBO, open biomedical ontologies, http://obo.sourceforge.net/
328. ONCOMINE, cancer profiling database, http://www.oncomine.org/main/index.jsp
329. PDB, protein data bank, http://www.rcsb.org/pdb/
330. PHYLIP, a package of programs for inferring phylogenies, http://evolution.genetics.washington.edu/phylip.html
331. Predictprotein server, http://cubic.bioc.columbia.edu/predictprotein/
332. Ras Mol, molecular visualization freeware, http://www.umass.edu/microbio/rasmol/
333. ReBase, restriction enzymes database, http://rebase.neb.com/rebase/rebase.html
334. RNA families database of alignments and CMs, http://www.sanger.ac.uk/Software/Rfam/
335. RMA express, http://stat-www.berkeley.edu/users/bolstad/RMAExpress/RMAExpress.html
336. Rosetta (Robetta), full chain protein structure prediction server, http://robetta.bakerlab.org/
337. Scan Analyze, http://rana.lbl.gov/EisenSoftware.htm
338. SCOR, a structural classification of RNA database, http://scor.lbl.gov/
339. Single nucleotide polymorphisms consortium database, http://snp.cshl.org/
340. Stony Brook algorithm repository, http://www.cs.sunysb.edu/~algorith/
341. Suffix trees, Internet address for data compression and suffix trees publications and software, http://datacompression.info/SuffixTrees.shtml
342. Swiss-Prot, protein knowledgebase, http://us.expasy.org/sprot/sprot-top.html
343. TreeView, visualization of trees, http://genome-www.stanford.edu/~alok/TreeView
344. Turner laboratory, free energy and enthalpy tables for RNA folding, http://www.bioinfo.rpi.edu/zukerm/rna/energy/
345. UniProt, the universal protein resource, http://www.psc.edu/general/software/packages/uniprot/
346. Vienna RNA package, RNA secondary structure prediction and comparison, http://www.tbi.univie.ac.at/RNA/

Index

3′ position and direction, 214
5′ position and direction, 214

acceptance–rejection rule, 59
acceptor (in gene), 222
accumulator array, 94, 118
adenine, 214
agarose gel, 224
algorithm, 67
aligning sequence against database, 95, 183
alignment, 9, 155, 201, 229, 292–294, 303, 351, 352
alpha helix, 266
alternative splicing, 261
alternative to null hypothesis, 46
AMBER, 277
amine group, 265
amino acid, 2, 155, 173, 175, 176, 178, 188, 213, 221, 223, 253, 262, 266, 269, 293, 350
amino acid R group, 262
amino acid side chain, 262
amino acid substitution matrices, 173
anchored contig, 244
antisense strand, 218
artificial neural network, 101
artificial neuron, 100
automatic DNA sequencer, 228
average-linkage clustering, 107

back propagation, 101
backbone of protein, 265

backward algorithm for hidden Markov model, 61
Baum–Welch algorithm, 63
Bayes' rules, 15
Bernoulli trial, 22
beta distribution, 27
beta sheet, 267
binomial distribution, 22
binomial test, 46
BioCarta, 351
bioinformatic databases, 349
BLAST, 352
BLOSUM substitution matrices, 176
bond lengths, 278
Boolean satisfiability problem, 148
Boyer–Moore algorithm, 72
bulge, 302
Burrows–Wheeler transform, 85, 258

carboxyl group, 265
Cauchy distribution, 33
central dogma of molecular biology, 214
chain termination DNA sequencing, 226
Chapman–Kolmogorov equation, 56
characteristic function, 21
CHARMM, 277
Chinese postman problem, 239
chromosomes, 221
class discovery, 336
class NP-complete, 149, 233, 240
class NP-hard, 149, 240
class P, 149
class prediction, 336

class-NP, 149, 233, 240
classification, 98, 331, 336
clinical databases, 352
CLUSTAL W, 184, 352
clustering, 103
coalescence, 189, 202
codon, 223
collision resolution, 92
collisions in hashing, 92
combinatorial optimization, 147
comparative modeling, 293, 311
comparative sequence analysis, 303
complementary DNA arrays, 315
complete-linkage clustering, 107
computational prediction of protein structure, 290
computational prediction of RNA structure, 303
computer science, 67
concavity, 126
conditional probability, 14
constrained optimization, 128
contig, 229, 244
convexity, 126
correction of ratio-intensity plots for cDNA, 328
Cramer–Rao theorem, 35
cumulative probability distribution function, 16
cytosine, 214

DDBJ, 350
De Bruijn graph, 238
demographic scenarios, 205
deoxyribonucleic acid, see DNA
deoxyribose, 214
detailed balance condition, 56, 170
dideoxy nucleotides, 226
dihedral angle, 279
dimensionality reduction, 98, 107, 337
dipeptide, 265
displacement of rigid body, 282
distance-based tree, 190
distance methods, 189
distribution of random variable, 15
DNA, 1, 3, 77, 94, 107, 155, 156, 162, 163, 182, 202, 213, 214, 216, 217, 223, 232, 300, 316
DNA alignment, 123

DNA assembly, 95, 230
DNA backbone, 217
DNA cloning, 226
DNA microarray, 313
DNA molecular structure, 214
DNA polymorphisms, 252
DNA primer, 226
DNA replication, 202, 216
DNA sequence, 4, 82, 156, 202
donor (in gene), 222
donor DNA, 226
dot matrix for aligning DNA, 159
dynamic programming, 140, 307, 310

Edman degradation, 274
electrophoresis, 224, 272
electrostatic interactions, 279
EM algorithm, 13, 41, 198, 250, 257
EMBL, 350
ergodic Markov chain, 54
Euler angles, 283
Euler path, 239
Euler superpath problem, 239
Eulerian graph, 240
EURODIAB, 354
evolutionary trace method, 201, 295
exon, 222
ExPASy, 350
expectation maximization method see also EM algorithm, 37
expectation of random variable, 19
exponential distribution, 26

false discovery rate, 342
family-wide error rate, 341
fast search, 72
FASTA, 352
feature extraction, 97
Felsenstein–Churchill algorithm, 201
Felsenstein nucleotide substitution model, 164
Felsenstein trees, 194
Fisher information, 35
Fisher–Wright process, 203
fluorescent dyes, 313
forward algorithm for HMM, 61
functional sites in proteins, 201, 294
FWER, 341

gamma distribution, 27
gap in DNA coverage, 244
Gauss–Newton iteration, 140
GenBank, 253, 350, 351
gene, 221
gene expression, 1, 220, 313
gene expression databases, 351
gene ontology, 345
GENEPI, 354
generalized Hough transform, 118
genetic and proteomic pathways, databases of, 351
genetic code, 222
genetic drift, 202
genome annotation, 252
genome assembly, 230
genome coverage, 243
genome sequencing, 223
genome structure, 220
genomic databases, 213, 350
genomics, 213
geometric distribution, 21, 23, 203
geometric hashing, 119
GO, 345
gradient algorithm, 139
guanine, 214

hairpin loop, 302
Hamiltonian path problem, 148, 234
hash tables, 94
hashing, 91
hidden Markov model, 13, 60, 255, 257
hierarchical clustering, 106, 315
HKY model, 165
HMM, 60, 255, 257
Hough transform, 117
Human Genome Consortium Project, 229
hybridization, 235, 313, 315
hydrogen bond, 214, 217, 266, 281, 302
hypergeometric distribution, 25

immunoblot, 273
independence, 14
inequality constraints, 131
infinite-alleles mutation model, 204
infinite-sites mutation model, 204
intensity matrix of time-continuous Markov chain, 57

internal loop, 302
intron, 222
invariant distribution, 54
isoelectric focusing, 272
isoelectric point, 272

Jensen's inequality, 38
Jukes–Cantor model, 164

K-means algorithm, 104, 315
KEGG, 351
Kekulé representation, 268
Kuhn–Tucker theorem, 131

l-mers, 238
l-tuples, 238
lagging strand, 218
Lagrange multiplier theorem, 130
law of total probability, 15
leading strand, 218
least-squares method, 134, 289
Lennard–Jones potential, 279
ligand, 295
linear classifiers, 98
linear form, 134
linear programming, 99, 136
linear regression, 134
local balance condition, 56, 170

MALDI, 273
marginal distribution, 17
Markov chain, 49
Markov chain Monte Carlo method, 13, 36, 57, 198, 208, 295
Markov process, 49
Markov property, 50
mass spectrometry, 273
matrix-assisted laser desorption ionization, 273
maximum likelihood method, 28, 189
maximum likelihood trees, 194
maximum-parsimony method, 189
maximum-parsimony trees, 198
MC3 method, 60
MCMC method, 36, 57, 198, 208, 295
messenger RNA, 218, 221, 261, 300
Metropolis–Hastings algorithm, 58, 198, 208
minimization of RMSD, 283

minimum-variance parameter estimation, 35
mixed normal distribution, 45, 332
mixed Poisson distribution, 44, 244
mixture distribution, 43, 244, 314, 331
mixture of beta distributions, 342
molecular clock trees, 191
molecular dynamics, 281
molecular field, 276
molecular modeling, 276
molecular phylogenetics, 187
molecular surface, 290
moments of random variable, 20
moments, method of, 31
most recent common ancestor, 203
mRNA, 218, 221, 261, 300
multibranched loop, 302
multinomial distribution, 25
multiple alignment, 183, 201, 294
multiple testing, 341
mutation, 202

NCBI, 94, 253, 350, 351
Needleman–Wunsch algorithm, 178
negative binomial distribution, 23
neighbor-joining tree, 193
Nelder–Mead algorithm, 137
neural network, 101
neuron, 100
Newton–Raphson iteration, 140
nitrogen bases, 214
NMR, 275
noncoding RNA, 301
nonparametric bootstrap method, 200
nonparametric tests, 48
normal distribution, 26
normalization procedures for DNA microarrays, 321
Northern blot, 302
NP, 149
NP-complete, 149, 233, 240
NP-hard, 149, 240
nuclear magnetic resonance, 275
nucleotide substitution models, 163
nucleotides, 215
null hypothesis, 46
Nussinov algorithm, 304

oligonucleotide array, 315

ontology databases, 351
open reading frame, 255, 352
operational taxonomic unit, 188
optimization, 123
ORF, 255, 352
OTU, 188
overlap graph, 232

PAGE, 272
pairwise differences, 209
palindrome, 83, 259
PAM substitution matrix, 174
parametric tests, 47
parametric transform, 116
partial atomic charges, 277
partial-least-squares method, 115, 337
Patricia trie, 76
pattern analysis, 97
PCA, 107, 337
PCR, 225
PDB, 262, 275, 350
peptide bond, 265
periodic states of Markov chain, 54
persistent states of Markov chain, 52
phosphate groups, 214
phosphodiester bonds, 217
photolithography technology, 316
PHYLIP, 187, 352
phylogenetic tree, 187
PLS method, 115, 337
Poisson distribution, 24, 210
Poisson process, 24, 26, 209, 244
polyA sites, 222
polyacrylamide gel electrophoresis, 272
polyadenylated sites, 222
polyadenylation, 300
polymerase chain reaction, 225
polypeptide chain, 266
primer, 226
principal component analysis, 107, 337
principal directions, 108
probability, 13
probability density function, 16
probability-generating function, 21
protein, 1, 8, 155, 174, 201, 213, 220, 253, 254, 261, 263, 268, 275, 294, 300, 313, 344
protein 2D gel, 272
protein annotation, 292

protein backbone, 265
protein immunoblot, 273
protein isoelectric point, 272
protein primary structure, 266
protein quaternary structure, 271
protein secondary structure, 266
protein tertiary structure, 268
protein, active-sites prediction, 294
protein–ligand binding analysis, 295
proteomic databases, 350
pseudoknot, 310
pulley principle, 197

quadratic form, 134
quadratic programming, 103, 137
quantile normalization, 327
quaternion, 282
quicksort algorithm, 69

Ramachandran plot, 269
random variable, 15
Rao–Blackwell theorem, 37
recurrent mutation model, 204
recursive optimization, 137
repetitive structure of DNA, 232
replication, 202, 216
restriction enzymes, 224
restriction enzymes, fingerpring with, 231
reversible Markov chain, 55
ribonucleic acid, see RNA
ribonucleotide, 221, 299
ribosome, 220, 222, 261, 299, 346
ribosome RNA, 301
RMA, 326
RMSD, 282
RNA, 155, 162, 182, 213, 219, 254, 299
RNA databases, 350
RNA primary structure, 302
RNA secondary structure, 302
RNA stem, 302
RNA tertiary structure, 302
RNA world hypothesis, 300
robust multiarray analysis, 326
root mean square deviation, 282
rooted tree, 188
rotation matrix, 283
rRNA, 301

sample space, 13
sampling l-mers, 247
SDS gel electrophoresis, 272
sense strand, 218
separating hyperplane, 99
sequence alignment, 9, 155, 201, 229, 292–294, 303, 351, 352
sequence alignment by dynamic programming, 178
sequence assembly, 95, 230
sequence overlap, 95
sequence overlap detection, 230
sequence-tagged site, 229, 232
sequencing by hybridization, 235
shortest path, 145
shortest-superstring problem, 148, 233
shotgun sequencing, 228
significance level, 46
simulated annealing, 60, 295
single-linkage clustering, 107
single-nucleotide polymorphisms, 211, 252, 350
singular values, 108
singular-value decomposition, 108, 282, 337
Smith–Waterman algorithm, 182
SNP, 211, 252, 350
solid-phase chemical DNA synthesis, 316
solvent-accessible surface, 290
sorting, 68
Southern blot, 224
spectral alignment problem, 241
spiked-in control RNA, 323
splicing, 221, 300
standard deviation, 20
start codon, 222
stationary distribution, 54, 164
statistical hypotheses, 45
statistical tests, 45
statistics, 13
stepwise mutation model, 204
stop codons, 222
string search, 70
STS, 229, 232
suboptimal algorithms of combinatorial optimization, 150
sufficient statistic, 36
suffix array, 80

suffix tree, 77
suffix trie, 77
support vector machine, 102
SVD, 108, 282, 337
SVM, 102
Swiss-Prot, 350

t-test, 48
taxonomic unit, 188
thymine, 214
time complexity, 148
time-continuous Markov chains, 56
time of flight, 273
TOF, 273
topology of tree, 188
torsion angles, 279
transcription, 219, 313
transfer RNA, 221, 261, 301
transient states of Markov chain, 52
transition intensity matrix, 195
transition probabilities, 50
transition probability matrix, 50
translation, 219, 313
translation vector, 283
traveling salesman problem, 148
tree reconstruction, 187
trie, 75
tRNA, 221, 261, 301
TU, 188

Turing machine, 68
type I and II statistical errors, 49

ultrametric distance, 191
uniform distribution, 31
unrooted tree, 188
untranslated region of gene, 222
unweighted pair group method with
    arithmetic mean, 193, 212
UPGMA, 193, 212
UPGMA algorithm, 193, 212
uracil, 219, 299
UTR, 222

valence angles, 278
van der Waals interactions, 279
variance, 20
vector DNA, 226
vector random variables, 16
Viterbi algorithm, 62

Watson–Crick DNA model, 214
weighted RMSD, 283
Western blot, 273
WGS method, 229
whole-genome shotgun, 229
Wilcoxon test, 48
WRMSD, 283

X-ray diffraction, 254, 275

Printing: Krips bv, Meppel
Binding: Stürtz, Würzburg